Elementary Lie Group Analysis and Ordinary Differential Equations

WILEY SERIES IN MATHEMATICAL METHODS IN PRACTICE

ADVISORY EDITORS

Bruno Brosowski
Johann Wolfgang Goethe Universität, Frankfurt am Main, Germany

Gary F. Roach
University of Strathclyde, Glasgow, Scotland, UK

Quarternionic and Clifford Calculus for Physicists and Engineers
K. Gürlebeck and W, Sprössig

Linear Semi-infinite Optimization
M.A. Goberna and M.A. López

Applied Nonlinear Semigroups
A. Belleni-Morante and A.C. McBride

Elementary Lie Group Analysis and Ordinary Differential Equations
Nail H. Ibragimov

Elementary Lie Group Analysis and Ordinary Differential Equations

Nail H. Ibragimov
University of North-West
Mmabatho, South Africa

JOHN WILEY & SONS
Chichester · New York · Weinheim · Brisbane · Singapore · Toronto

Copyright © 1999 by John Wiley & Sons Ltd.
Baffins Lane, Chichester
West Sussex PO19 1UD, England

National   Chichester (01243) 779777
International   (+44) 1243 779777

e-mail (for orders and customer service enquiries): cs-books@wiley.co.uk

Visit our Home Page on http://www.wiley.co.uk
or
http://www.wiley.com

All rights reserved. No part of this publication may be reproduced, stored in a
retrieval system, or transmitted, in any form or by any means, electronic, mechanical,
photocopying, recording, scanning or otherwise, except under the terms of the Copyright,
Designs and Patents Act 1988 or under the terms of a licence issued by the Copyright
Licensing Agency, 90 Tottenham Court Road, London, W1P 9HE, UK, without the
permission in writing of the publisher.

Other Wiley Editorial Offices

John Wiley & Sons, Inc., 605 Third Avenue,
New York, NY 10158-0012, USA

Wiley-VCH Verlag GmbH, Pappelallee 3,
D-69469 Weinheim, Germany

Brisbane • Singapore • Toronto

Library of Congress Cataloging-in-Publication Data

Ibragimov, N. Kh. (Naĭl Khaĭrullovitch)
  Elementary Lie group analysis and ordinary differential equations
Nail H. Ibragimov.
      p.   cm. — (Wiley series in mathematical methods in practice ;
v. 4)
  Includes index.
  ISBN 0-471-97430-7
  1. Differential equations — Numerical solutions.   2. Lie groups.
I. Title.   II. Series: Mathematical methods in practice ; v. 4.
QA372.I36   1999
515'.352 — dc21   98-33416
                                                                    CIP

British Library Cataloguing in Publication Data

A catalogue record for this book is available from the British Library
ISBN 0 471 97430 7

Produced from camera-ready copy supplied by the author.

Contents

Prologue . . . xiii

Series Preface . . . xv

Preface . . . xvii

Part I. Introduction to differential equations . . . 1

1 **Model differential equations** . . . 3
   1.1 Introduction . . . 3
   1.2 Problems from calculus and geometry . . . 5
       1.2.1 Integration by quadrature . . . 5
       1.2.2 Differential equations for families of curves . . . 6
       1.2.3 Minimum surface of revolution . . . 8
   1.3 Natural phenomena . . . 9
       1.3.1 Free fall of a body near the earth . . . 9
       1.3.2 Meteoroid . . . 10
       1.3.3 A model of rainfall . . . 11
       1.3.4 Motion of planets . . . 12
   1.4 Elementary physics in everyday life . . . 13
       1.4.1 Educated farmer . . . 13
       1.4.2 Outflow from a funnel . . . 14
       1.4.3 Heating and air conditioning . . . 16
       1.4.4 Electrical instruments. Equation of van der Pol . . . 19
       1.4.5 Mechanical vibrations . . . 19
       1.4.6 Collapse of driving shafts . . . 20
   1.5 Ecology . . . 22
       1.5.1 Radioactive decay . . . 22
       1.5.2 Population growth . . . 22
       1.5.3 Predator and prey . . . 23
       1.5.4 Competing species . . . 23
   Problems . . . 24

2 Elementary methods of integration — 25
2.1 First-order equations — 25
- 2.1.1 Linear equations. Superposition of solutions — 25
- 2.1.2 Separation of variables. A nonlinear superposition — 28
- 2.1.3 Notation of exterior differential calculus — 30
- 2.1.4 Exact equations — 31
- 2.1.5 Integrating factor — 32
- 2.1.6 Homogeneous equations — 33
- 2.1.7 Equations of the type $y' = f(\frac{ax+by+c}{kx+ly+m})$ — 33
- 2.1.8 The Riccati equation — 34
- 2.1.9 The Bernoulli equation — 35

2.2 Higher-order equations — 35
- 2.2.1 Integration of the equation $y^{(n)} = f(x)$ — 35
- 2.2.2 Equations $F(y, y', \ldots, y^{(n)}) = 0$. Reduction of order — 36
- 2.2.3 General and intermediate integrals — 37
- 2.2.4 $F(x, y, y', \ldots, y^{(n)}) = 0$ with F a total derivative — 37
- 2.2.5 The equations $y^{(n)} = f(y^{(n-1)})$ and $y^{(n)} = f(y^{(n-2)})$ — 38
- 2.2.6 Equations $F(x, y^{(k)}, y^{(k+1)}, \ldots, y^{(n)}) = 0$ — 39
- 2.2.7 Linear equations. Method of variation of parameters — 39
- 2.2.8 Linear equations with constant coefficients — 40

2.3 Systems of first-order linear equations — 42
Problems — 44

3 General properties of solutions — 45
3.1 Existence and uniqueness theorems — 45
- 3.1.1 Classical solutions. Cauchy's problem — 45
- 3.1.2 The method of successive approximations — 46
- 3.1.3 Systems of first-order equations — 49
- 3.1.4 Higher order equations — 50
- 3.1.5 Analytic solutions. The method of majorants — 50
- 3.1.6 Systems of analytic equations of the first order — 52
- 3.1.7 Example: The exponential map — 53

3.2 Equations of the first order not solved for y' — 55
- 3.2.1 Equations algebraic in y and y' — 55
- 3.2.2 Equations $x = f(y')$ and $y = f(y')$. Parametric solutions — 56
- 3.1.3 Lagrange's equation — 58
- 3.1.4 Clairaut's equation. The envelope of the general integral as a singular solution — 58

3.2 Global behavior of solutions — 60
- 3.2.1 Singularities of first-order equations — 60
- 3.2.2 Equations in the complex domain. Movable and fixed singular points — 62

Problems — 64

4 Partial differential equations of the first order — 65
4.1 Notation — 65
4.1.1 Equations of the first order and classical solutions — 65
4.1.2 Linear, quasi-linear and nonlinear equations — 66
4.2 Integration of linear equations — 67
4.2.1 First integrals of systems of ordinary differential equations — 67
4.2.2 Homogeneous linear partial differential equations — 69
4.2.3 Non-homogeneous equations — 71
4.3 Quasi-linear equations — 73
4.3.1 Laplace's method — 73
4.3.2 Extension of Laplace's method to many variables — 74
4.3.3 Reduction to a homogeneous linear equation — 75
4.3.4 Integral surfaces as loci of characteristic curves — 78
4.4 Nonlinear equations — 80
4.4.1 Complete, general and singular integrals — 80
4.4.2 Completely integrable systems. The Lagrange-Charpit method — 81
4.4.3 Solution of Cauchy's problem via complete integrals — 84
4.4.4 Monge's theory of characteristics — 84
4.4.5 Example: Light rays as visible characteristics — 86
4.4.6 Cauchy's method — 87
4.4.7 Hamilton's equations – characteristics of the Hamilton-Jacobi equation — 88
4.5 Systems of homogeneous linear equations — 88
4.5.1 Basic notions — 88
4.5.2 Complete systems — 91
4.5.3 Integration of complete systems — 93
Problems — 95

Part II. Fundamentals of Lie group analysis — 97

5 Gateway to modern group analysis: Historical survey — 99
5.1 Appearance of groups in the theory of equations — 99
5.2 Sophus Lie and symmetry analysis of differential equations — 100
5.2.1 His life story — 100
5.2.2 Symmetry groups, Lie algebras and integration of ordinary differential equations — 102
5.2.3 Group classification of differential equations — 104
5.2.4 Linearization of second-order equations — 104
5.2.5 Nonlinear superposition — 105
5.3 Tangent transformations — 105
5.3.1 Contact transformations and their applications — 105
5.3.2 Infinitesimal contact transformations — 107

		5.3.3	Higher order tangent transformations: Lie and Bäcklund's discussion 107
		5.3.4	Modern theory of Lie-Bäcklund transformation groups . . . 109
	5.4	Applied group analysis 110	
		5.4.1	Symmetry and conservation laws 110
		5.4.2	Restoration of group analysis in the 1960s 111
		5.4.3	Invariant and partially invariant solutions 111
		5.4.4	Symmetry of fluids 113
	5.5	New trends 113	
		5.5.1	Invariance principle in boundary-value problems 113
		5.5.2	"Hidden" symmetries, in particular non-local symmetries . 114
		5.5.3	Approximate symmetries 115
		5.5.4	Group theoretic modelling 116
		5.5.5	Miscellany 118
	Problems 118		

6 Preliminaries on transformations and groups 119

	6.1	Transformations in elementary mathematics 119	
		6.1.1	Algebra: Solution of equations 119
		6.1.2	Geometry: Similarity and calculation of areas 121
	6.2	Transformations in $\mathbb{R}^n$ 122	
		6.2.1	Changes of coordinates and point transformations 122
		6.2.2	Definition of a group 123
		6.2.3	Subgroups 124
	6.3	A display of transformation groups 124	
		6.3.1	Transformations of the straight line 124
		6.3.2	Groups in the plane 126
		6.3.3	Motions, conformal mappings and other groups in $\mathbb{R}^n$ . . . 129
		6.3.4	Different types of groups: Continuous, discontinuous and mixed. Path curves. Global and local groups 132
	Problems 134		

7 Infinitesimal transformations and local groups 135

	7.1	One-parameter groups 135	
		7.1.1	Notation and assumptions 135
		7.1.2	Definition of a local group 136
		7.1.3	Representation of local groups in a canonical parameter . . 137
		7.1.4	Infinitesimal transformations 138
		7.1.5	The Lie equations. The exponential map 138
		7.1.6	Determination of a canonical parameter 140
		7.1.7	Invariants 141
		7.1.8	Canonical and semi-canonical variables 143
		7.1.9	Three methods of construction of one-parameter groups with known infinitesimal generators 145

CONTENTS   ix

       7.1.10 Table of usual one-parameter groups in the plane,
their generators, invariants and canonical variables 149
  7.2 Invariant equations 150
       7.2.1 The infinitesimal test 150
       7.2.2 Invariant representation of invariant manifolds 151
       7.2.3 Examples on Theorem 7.2.2 153
  7.3 Introduction to Lie algebras 155
       7.3.1 Definition of Lie algebras of operators 155
       7.3.2 Notation. Table of commutators 156
       7.3.3 Subalgebra and ideal 157
       7.3.4 Quotient algebra 158
       7.3.5 Derived algebras 158
       7.3.6 Solvable Lie algebras 159
       7.3.7 Isomorphic and similar Lie algebras 159
       7.3.8 Non-isomorphic structures and non-similar realizations
in the plane of 1, 2, and 3-dimensional Lie algebras 162
  7.4 Multi-parameter groups: Outline of basic notions 167
       7.4.1 Definition 167
       7.4.2 Composition via one-parameter groups 168
       7.4.3 Basis of invariants 169
       7.4.4 Regular and singular invariant manifolds 169
  7.5 Approximate transformation groups 170
       7.5.1 Motivation 170
       7.5.2 Notation 172
       7.5.3 One-parameter approximate transformation groups 173
       7.5.4 Approximate group generator 174
       7.5.5 A preparatory lemma 174
       7.5.6 Lie equations in the first order of precision 175
       7.5.7 The approximate exponential map 177
       7.5.8 Examples on the exponential map 179
  Problems 181

8 Calculus of differential algebra   **183**
  8.1 Main variables and total derivatives 183
  8.2 The universal space $\mathcal{A}$ of modern group analysis 184
       8.2.1 Differential functions and the space $\mathcal{A}$ 184
       8.2.2 Successive derivatives of differential functions 185
       8.2.3 One independent and one differential variable.
Faà de Bruno's formula 185
  8.3 Extended point transformation groups. Contact transformations . 187
       8.3.1 Transformations of the plane 187
       8.3.2 One independent and several differential variables 190
       8.3.3 Point transformations involving many variables 190
       8.3.4 Properties of extended generators 192

	8.3.5	Differential invariants. Invariant differentiation 192
	8.3.6	Change of derivatives under differential substitutions 195
	8.3.7	Infinitesimal contact transformations 196
	8.3.8	Irreducible contact transformation groups in the plane . . . 197
8.4	Operators and identities in $\mathcal{A}$ 198	
	8.4.1	The Euler-Lagrange operator. Test for a total derivative . . 198
	8.4.2	Lie-Bäcklund operators 201
	8.4.3	Operators N^i associated with Lie-Bäcklund operators 203
	8.4.4	The fundamental identity 203
8.5	The frame of differential equations 204	
Problems 206		

9 Symmetry of differential equations   207

9.1	Notation and assumptions 207	
9.2	Determination of infinitesimal symmetries 208	
	9.2.1	Two definitions of a symmetry group 208
	9.2.2	Determining equations 210
	9.2.3	Intrinsic symmetries of mathematical models 211
9.3	Samples for solution of determining equations 213	
	9.3.1	Ordinary differential equations: Theorems on the maximum number of symmetries and examples 213
	9.3.2	Extended generators and calculation of symmetries of partial differential equations with two independent variables . 216
	9.3.3	Solution of the determining equations for a system 219
	9.3.4	Determination of equations admitting a given group 220
	9.3.5	Contact symmetries 223
9.4	Invariant solutions 224	
	9.4.1	Lie's theory of invariant solutions 224
	9.4.2	Inherited symmetries of equations for invariant solutions . . 229
9.5	Approximate symmetries of equations with a small parameter . . . 231	
	9.5.1	Definition of an approximate symmetry group 231
	9.5.2	Determining equations 231
	9.5.3	Calculation of infinitesimal approximate symmetries 232
9.6	On discontinuous and non-local symmetries 235	
9.7	Symmetry and conservation laws 236	
	9.7.1	Principle of least action. Euler-Lagrange equations 236
	9.7.2	Differential algebraic proof of Noether's theorem 238
	9.7.3	Examples from classical mechanics 240
	9.7.4	Derivation of Kepler's laws from symmetries 241
	9.7.5	Conservation laws of relativistic mechanics 243
Problems 247		

10 Invariants of algebraic and differential equations   249
    10.1 Invariants of algebraic equations 250
        10.1.1 Preliminaries 250
        10.1.2 Tschirnhausen's transformation. An approach to the Galois group 251
        10.1.3 Infinitesimal method 255
    10.2 Linear ordinary differential equations 257
        10.2.1 Preliminaries 257
        10.2.2 Infinitesimal method 258
    10.3 Nonlinear ordinary differential equations 261
    10.4 Linear partial differential equations 261
        10.4.1 The Laplace invariants 261
        10.4.2 The Ovsyannikov invariants 262
    10.5 The Maxwell equations 262
    Problems 263

Part III. Basic integration methods   265

11 First-order equations and systems   267
    11.1 Integration by using an infinitesimal symmetry 267
        11.1.1 Lie's integrating factor 267
        11.1.2 Method of canonical variables 269
        11.1.3 Group interpretation of variation of parameters 270
    11.2 Systems admitting nonlinear superposition 271
        11.2.1 Superposition formulas for nonlinear systems 271
        11.2.2 Main theorem. The Vessiot-Guldberg-Lie algebra 272
        11.2.3 Examples 272
        11.2.4 Projective interpretation of the Riccati equation 274
        11.2.5 Linearizable Riccati equations 275
        11.2.6 Decoupling and integration of systems using Vessiot-Guldberg-Lie algebras 279
        11.2.7 Application: An invariant solution of nonlinear equations modelling laser systems 279
    Problems 282

12 Second-order equations   283
    12.1 Reduction of order 283
        12.1.1 Canonical variables 283
        12.1.2 Invariant differentiation 284
    12.2 Integration by means of two symmetries 285
        12.2.1 Consecutive integration 285
        12.2.2 Canonical forms of two-dimensional Lie algebras 286
        12.2.3 Lie's integration algorithm 289

12.2.4 A sample for implementation of the algorithm	291
12.2.5 Application to the non-homogeneous linear equation	292
12.3 Lie's linearization test	295
12.3.1 Main theorem	295
12.3.2 Examples	297
12.4 Integration using approximate symmetries	299
Problems	300

13 Higher-order equations — 301

13.1 Third-order equations	301
13.1.1 Equations admitting a solvable Lie algebra	301
13.1.2 Equations admitting a non-solvable Lie algebra	303
13.2 Group theoretical background of Euler's method for linear equations with constant coefficients	304
13.2.1 Symmetries and invariant solutions	304
13.2.2 The case of multiple roots	305
Problems	306

Answers — 307

Notes — 325

Index — 343

Prologue

It was mid-August of 1870. The day was one of those restless days of the first month of the Franco-Prussian war when people were distrustful. The French police arrested a suspicious-looking young man who was wandering in lonely places in the forest of Fontainebleau (near Paris) stopping now and then to make notes and drawings in his notebook. *"He was of tall stature and had the classic Nordic appearance. A full blond beard framed his face and his grey-blue eyes sparkled behind his eyeglasses. He gave the impression of unusual physical strength"* (E. Cartan). The police searched him and found a map, letters in German and papers full of mysterious formulas, complexes, diagrams and names. He was suspected of being a German spy and imprisoned.

The arrested man was Sophus Lie, a young Norwegian mathematician, visiting the mathematical capitals of Europe. He had received a travel grant from the University of Christiania for his first mathematical paper (1869).

Sophus Lie had to stay in prison in Fontainebleau for 4 weeks before his French colleague Gaston Darboux learned about the incident and arrived on behalf of the Academy of Sciences with a release order signed by the Minister of Home Affairs.

Lie himself had taken things truly philosophically and made good use of his time in prison. For, as he recounted later, in these forced leisure days he had plenty of peace and quiet to concentrate on his problems and advance them essentially. In a letter to his Norwegian friend Ernst Motzfeldt written directly after the release, Sophus Lie remarked: *"I think that a mathematician is well suited to be in prison."*

Series Preface

Mathematics is a fundamental and enabling discipline in most applied sciences. More than ever, both students and research workers in many fields are finding that in order to meet the modern day demands of their discipline they need a sound and practical appreciation of the various mathematical methods which are available for their use.

By tradition, many topics in mathematics are first taught in pure mathematics courses, where they are developed theoretically, and then again in applied mathematics courses where practical applications are the main concern. This can often cause confusion to the applied scientist. The present series is a response to the demand to reconcile these two approaches and, in addition, is intended to meet the perceived and increasing demand for mathematics as applied to specific disciplines and industries.

The books in this series will each present a well motivated and persuasive account, involving many practical examples, of the mathematical techniques required within each title's special topic area. They aim to provide mathematical support for workers in a wide range of research environments.

Bruno Brosowski
Universität Frankfurt/Main, Germany

Gary Roach
University of Strathclyde, UK

Preface

The present book provides students and teachers with an easy to follow and comprehensive introduction to differential equations and Lie group analysis with an emphasis on useful tools rather than on general theory. In fact, this is the first self contained university text on ordinary differential equations (after Sophus Lie's classical *Lectures* [5.14]), where all the basic integration methods are derived from group-invariance principles. The text is designed to meet the needs of beginners and provides material for several undergraduate and postgraduate courses.

About Lie group analysis

The mathematical discipline known as *Lie Group Analysis* was originated in the 1870s by an outstanding mathematician of the nineteenth century, Sophus Lie.

One of Lie's striking achievements was the discovery that the majority of *ad hoc* methods of integration of ordinary differential equations could be explained and deduced simply by means of his theory. Moreover, Lie gave a classification of ordinary differential equations in terms of their symmetry groups, thereby identifying the full set of equations which could be integrated or reduced to lower-order equations by the new method. In particular, Lie's classification shows that the second-order equations integrable by his method can be reduced to merely four distinct canonical forms by changes of variables. Subjecting these four canonical equations to changes of variables alone, one obtains all known equations integrated by old methods as well as infinitely many unknown integrable equations.

Further development furnished ample evidence that the new theory provides a universal tool for tackling considerable numbers of differential equations when other means of integration fail. In fact, group analysis is the only universal and effective method for solving nonlinear differential equations analytically. Furthermore it augments intuition in understanding and using symmetry for formulation of mathematical models, and often discloses possible approaches to solving complex problems. For these reasons group analysis should be part of curricula in applied mathematics, the sciences and engineering.

However, the philosophy of Lie groups has not enjoyed widespread acceptance in the past and the subject has been neglected, to a great extent, in university programs. Consequently, students, teachers and engineers still use *ad hoc* methods presented in traditional texts and voluminous catalogues of special types of equations, instead of dealing with Lie's few canonical equations. *"Often the less there is to justify a traditional custom, the harder it is to get rid of it"* (M. Twain).

About this book

The purpose of this book is to provide students, teachers and applied scientists with an easy to follow but comprehensive text and to prepare a background for research in modern group analysis. The material is presented in a way that will enable every diligent reader to succeed in his study of the subject. Examples and worked exercises incorporated into the text are tailored to develop differential algebraic and group analytic skills. The problems formulated at the end of each chapter are regarded as an essential ingredient in the book. The problems where a challenge requiring special skills is involved are set off by asterisks *, while ** designate open research problems.

The major difficulty in writing a university text on *Lie group analysis and ordinary differential equations* is that a previous acquaintance with elements of the theory of differential equations is a prerequisite for Lie group theory. This state of affairs is also a considerable obstacle for beginners who intend to use Lie's *Lectures* [5.14]. I have striven to overcome the difficulty by presenting a concise *Introduction to differential equations*. It contains all the basic classical devices for integration of special types of equations, and provides the necessary background in the theory of ordinary and first-order partial differential equations.

The book is based on various courses given over the past twenty-five years to undergraduate and postgraduate students as well as professional audiences. In particular, I have used the notes of my lectures delivered at Novosibirsk University (1971-80), Collège de France (Paris, 1979, 1984, 1992), Moscow University and Moscow Institute of Physics and Technology (1987-93), University of the Witwatersrand (Johannesburg, 1994-97) and Stanford University (1996).

The modern group analysis of differential equations is made up of a great number of classical and recent results. A text on this subject must therefore, unless it is to become tedious, be extremely selective. I drew upon many sources for this book and carefully selected the material. The Notes provide a brief account of relevant historical circumstances as well as biographical and bibliographical notes for each chapter. The reader indifferent to the historical background can simply ignore references to the Notes at the end of the book, whereas those interested in a complete account of results and proper references are referred to the recent *Handbook* [5.35] and the references therein.

References to Definitions, Theorems, Remarks, etc. bear the section number, e.g. Theorem 7.2.1 refers to the Theorem in Section 7.2.1. References to the Notes are numbered by chapter and are given in square brackets.

I acknowledge financial support from the Foundation for Research Development of South Africa. My thanks are due to Dr. Fazal Mahomed who found time in the midst of many academic engagements to read and re-read the text. I am deeply grateful to my wife Raisa and my daughters Sania and Alia for their support and understanding of this venture. Finally, I thank the staff of John Wiley for their skill and cooperation.

Nail H. Ibragimov

Part I

INTRODUCTION TO DIFFERENTIAL EQUATIONS

> *The theory of differential equations is the most important discipline in the whole modern mathematics.*
> **S. Lie, 1895**

This preparatory part is designed to meet the needs of beginners and provides the necessary background in *elementary ordinary and first-order partial differential equations* which is assumed in the succeeding parts of the book.

Chapter 1 contains various examples of formulation of problems in terms of differential equations whose discussion discloses a great variety of types of ordinary differential equations occurring in mathematical modelling. Furthermore, it introduces the reader to the traditional vocabulary.

The basic practical devices commonly used for the integration of special types of equations are presented in a compact form in Chapter 2. The underlying group invariance principles will be explained in Part III.

Chapters 3 and 4 provide a review of indispensable definitions and theorems used throughout the book.

Chapter 1

Model differential equations

1.1 Introduction

Differential equations, in a proper sense, have appeared in mathematics in the 1680s in the works of the creators of the differential and integral calculus, I. Newton [1.1] and G.W. Leibnitz [1.2]. Newton's *Principles* [1.3] contains numerous differential equations formulated and integrated in the framework of elementary geometry. The term *differential equation* was mentioned by Leibnitz for the first time in his letter to Newton (1676) and then used in his publications after 1684.

Since then, the formulation of fundamental natural laws and of technological problems in the form of rigorous mathematical models is given frequently, even prevalently, in terms of differential equations. These equations relate the behavior of certain unknown functions (termed *dependent variables*) at a given point (time, position, etc. called *independent variables*) to their behavior at neighboring points. In other words, mathematical models formulated in terms of differential equations involve independent variables and dependent variables together with their derivatives. A differential equation is said to be of the nth order if it involves derivatives of this order but not higher.

If the dependent variables are functions of a single independent variable (such as time), the equations are termed *ordinary differential equations*. The renowned Newton's second law, e.g. for a particle in an external force field $\boldsymbol{F}$,

$$\frac{\mathrm{d}(m\boldsymbol{v})}{\mathrm{d}t} = \boldsymbol{F}, \qquad (1.1)$$

falls precisely into this category. Here time t is the independent variable, m and $\boldsymbol{v} = (v^1, v^2, v^3)$ denote the particle mass (in general, m is not constant) and its vector velocity, respectively. Newton's equation is a first-order equation (specifically, a system of three equations of the first order) if attention is focused on the velocity components v^i regarded as the dependent variables, provided that the force $\boldsymbol{F}$ depends upon t and $\boldsymbol{v}$ alone. However, for motions in the force field of the general form $\boldsymbol{F} = \boldsymbol{F}(t, \boldsymbol{x}, \boldsymbol{v})$, (1.1) is a second-order equation for the position vector $\boldsymbol{x} = (x^1, x^2, x^3)$ of the particle. Indeed consider, for simplicity sake, the case of constant mass m and substitute $\boldsymbol{v} = \boldsymbol{x}'$ where the prime denotes differentiation with respect to t. Then equation (1.1) is rewritten as the following system of three second-order equations:

$$m\frac{d^2\boldsymbol{x}}{dt^2} = \boldsymbol{F}(t, \boldsymbol{x}, \boldsymbol{x}'). \tag{1.2}$$

If, on the other hand, unknown functions depend on several independent variables, so that the equations in question relate the independent variables, the dependent variables and their partial derivatives, then one deals with *partial differential equations*. A celebrated representative of this category is d'Alembert's equation [1.4] for small transversal vibrations of strings:

$$\frac{\partial^2 u}{\partial t^2} - k^2 \frac{\partial^2 u}{\partial x^2} = 0, \tag{1.3}$$

where k^2 is a positive constant. This is a second-order partial differential equation with two independent variables, time t and the coordinate x along the string. It is also known as the one-dimensional wave equation. Similar partial differential equations with many spatial variables occur frequently in mathematical physics, e.g. the propagation of light disturbances is governed by the three-dimensional wave equation:

$$\frac{\partial^2 u}{\partial t^2} - k^2 \left(\frac{\partial^2 u}{\partial x^2} + \frac{\partial^2 u}{\partial y^2} + \frac{\partial^2 u}{\partial z^2} \right) = 0.$$

A model of small transversal vibrations of uniform slender rods provides a partial differential equation of the fourth order:

$$\frac{\partial^2 u}{\partial t^2} + \mu \frac{\partial^4 u}{\partial x^4} = f, \tag{1.4}$$

where f is a total force acting on the rod and μ is a positive constant.

The mathematical model of thermal diffusion (due to J.B.J. Fourier, 1811) provides a partial differential equation known as the heat conduction equation. It is written, e.g. in the one-dimensional case (one spatial variable), in the form

$$\frac{\partial u}{\partial t} - k^2 \frac{\partial^2 u}{\partial x^2} = 0.$$

The above classical examples illustrate so-called linear partial differential equations with constant coefficients. Simple linear equations with variable coefficients occurring in practical affairs are provided, e.g. by recent mathematical models in finance. For example, the Black-Scholes model (1973) in stock option pricing is described by the equation

$$u_t + \frac{1}{2} A^2 x^2 u_{xx} + B x u_x - C u = 0, \tag{1.5}$$

where A, B, and C are the parameters of the model.

1.2. PROBLEMS FROM CALCULUS AND GEOMETRY

The aim of this chapter is to introduce the student to the field of ordinary differential equations by providing typical examples. They arise from problems in mathematics or enjoy widespread acceptance as mathematical models of natural phenomena, physics and engineering, nature conservation, population and community analysis, etc.

1.2 Problems from calculus and geometry

1.2.1 Integration by quadrature

Let us begin with the fundamental differential equation of the integral calculus solved in the pioneering works of Newton and Leibnitz.

Suppose we are given a continuous function $f(x)$ and are asked to find a function $y = F(x)$ whose derivative y' is $f(x)$. That is, $y' \equiv \mathrm{d}F(x)/\mathrm{d}x = f(x)$. The function $F(x)$ is called an *antiderivative* of $f(x)$ or an *integral of $f(x)$ with respect to x*. Thus, the problem is to solve the first-order *ordinary differential equation*:

$$\frac{\mathrm{d}y}{\mathrm{d}x} = f(x). \tag{1.6}$$

Its *general solution* is written

$$y = \int f(x)\mathrm{d}x + C, \tag{1.7}$$

where $\int f(x)\mathrm{d}x = F(x)$ is any integral of $f(x)$ with respect to x, and C is an arbitrary constant known as the *constant of integration*. By specifying the constant of integration, one obtains a *particular solution*. Thus, the differential equation (1.6) has an infinite number of solutions given by the one-parameter family of integrals (1.7).

Notation. In calculus, the symbol $\int f(x)\mathrm{d}x$ of the *indefinite integral* is a standard notation for all solutions to the equation (1.6), i.e. $\int f(x)\mathrm{d}x = F(x)+C$ with $F(x)$ denoting a particular integral of $f(x)$. In the theory of differential equations, however, a different interpretation is commonly used. Namely, $\int f(x)\mathrm{d}x$ is identified with any particular integral $F(x)$ of the function $f(x)$. This is the interpretation which is adopted in the present book.

In the notation (1.7), the general solution, e.g. to a second-order differential equation,

$$\frac{\mathrm{d}^2 y}{\mathrm{d}x^2} = f(x) \tag{1.8}$$

is written

$$y = \int \mathrm{d}x \int f(x)\mathrm{d}x + C_1 x + C_2 \tag{1.9}$$

with two arbitrary constants C_1 and C_2.

Remark. In the classical literature, the integral formula (1.7) is termed the quadrature. Consequently, the differential equation (1.6) is said to be *integrable by quadrature*. The same terminology applies to the equation

$$\frac{dy}{dx} = h(y)$$

since its solution is given by the integral formula similar to (1.7),

$$x = \int \frac{dy}{h(y)} + C \equiv H(y) + C,$$

whence $y = H^{-1}(x - C)$, where H^{-1} denotes the inverse to H.

1.2.2 Differential equations for families of curves

Let x and y be the independent and dependent variables, respectively. Let the primes denote differentiation with respect to x, so that y' and $y^{(n)}$ denote the first- and the nth-order derivatives, respectively.

Consider, in the (x, y) plane, a family of curves,

$$y = f(x, C_1, \ldots, C_n) \tag{1.10}$$

given, in general, implicitly by

$$\Phi(x, y, C_1, \ldots, C_n) = 0. \tag{1.11}$$

By the definition of implicit functions, equation (1.11) with y replaced by the function f from (1.10) is satisfied identically in x (from some interval) for all admissible values of the parameters C_k, $k = 1, \ldots, n$. Consequently, one can differentiate this identity with respect to x. Iterating the procedure n times yields:

$$\frac{\partial \Phi}{\partial x} + \frac{\partial \Phi}{\partial y} y' = 0,$$

$$\frac{\partial^2 \Phi}{\partial x^2} + 2 \frac{\partial^2 \Phi}{\partial x \partial y} y' + \frac{\partial^2 \Phi}{\partial y^2} y'^2 + \frac{\partial \Phi}{\partial y} y'' = 0,$$

$$\cdots\cdots\cdots\cdots\cdots\cdots\cdots\cdots$$

$$\frac{\partial^n \Phi}{\partial x^n} + \cdots\cdots\cdots + \frac{\partial \Phi}{\partial y} y^{(n)} = 0. \tag{1.12}$$

Elimination of the parameters C_k from equations (1.11) and (1.12) yields an nth-order ordinary differential equation

$$F(x, y, y', \ldots, y^{(n)}) = 0. \tag{1.13}$$

By construction, the function (1.10) provides a solution, depending on n arbitrary constants C_i, to the differential equation (1.13). Accordingly, (1.13) is termed the differential equation of the family of curves (1.10) or (1.11).

1.2. PROBLEMS FROM CALCULUS AND GEOMETRY

The above procedure can be simplified by using the following *differential algebraic* notation (see Chapter 8).

Notation. The *total differentiation* D_x of functions depending on a finite number of variables $x, y, y', y'', \ldots$ is defined by

$$D_x = \frac{\partial}{\partial x} + y'\frac{\partial}{\partial y} + y''\frac{\partial}{\partial y'} + \cdots + y^{(s+1)}\frac{\partial}{\partial y^{(s)}} + \cdots. \tag{1.14}$$

In this notation, equations (1.11) are written in a compact form:

$$D_x\Phi = 0, \quad D_x^2\Phi = 0, \quad \ldots, \quad D_x^n\Phi = 0. \tag{1.15}$$

Moreover, in this way one eludes the necessity of invoking equation (1.10) n times while deriving equations (1.11).

Exercise 1. Find the differential equation of the family of straight lines, $y = ax + b$ containing two parameters, a and b.

Solution. By setting $\Phi = y - ax - b$ equations (1.15) are written:

$$D_x\Phi \equiv y' - a = 0, \quad D_x^2\Phi \equiv y'' = 0.$$

Here the last equation does not contain the parameters. Hence, the differential equation (1.13) of straight lines is the simplest *linear* equation of the second order:

$$y'' = 0. \tag{1.16}$$

Exercise 2. Find the differential equation of the family of parabolas given in the form $\Phi \equiv y - ax^2 - bx - c = 0$ and depending on three parameters, a, b, and c.

Solution. Equations (1.15) are written:

$$D_x\Phi \equiv y' - 2ax - b = 0, \quad D_x^2\Phi \equiv y'' - 2a = 0, \quad D_x^3\Phi \equiv y''' = 0.$$

Thus the differential equation of the parabolas is the simplest linear equation of the third order:

$$y''' = 0. \tag{1.17}$$

Exercise 3. Find the differential equation of the family of circles given in the form $\Phi \equiv (x-a)^2 + (y-b)^2 - c^2 = 0$.

Solution. Equations (1.15) yield:

$$x - a + (y-b)y' = 0, \quad 1 + y'^2 + (y-b)y'' = 0, \quad 3y'y'' + (y-b)y''' = 0.$$

Here the third equation does not contain the parameters a and c, and it suffices to substitute there the expression $y - b = -(1 + y'^2)/y''$ found from the second equation. Hence, the family of circles is described by the *nonlinear* equation:

$$y''' - 3\frac{y'y''^2}{1 + y'^2} = 0. \tag{1.18}$$

Exercise 4. Find the differential equation of the family of hyperbolas given in the form $\Phi \equiv (y-a)(b-cx) - 1 = 0$.

Solution. Equations (1.15) yield:

$$(b-cx)y' - c(y-a) = 0, \quad (b-cx)y'' - 2cy' = 0, \quad (b-cx)y''' - 3cy'' = 0.$$

We find from the second equation that $b - cx = 2cy'/y''$ and substitute it into the third one to obtain the following differential equation of the hyperbolas:

$$y''' - \frac{3}{2}\frac{y''^2}{y'} = 0. \tag{1.19}$$

The differential expression in the left-hand side of equation (1.19) is known in the literature as the *Schwarzian derivative* or, simply, the *Schwarzian* [1.5].

1.2.3 Minimum surface of revolution

Consider curves $y = y(x)$ in the (x,y) plane connecting two fixed points, $P_1 = (x_1, y_1)$ and $P_2 = (x_2, y_2)$. One revolves the curves about the y axis to obtain surfaces. The problem is to find that curve for which the surface of revolution has a minimum area.

Since the area of a narrow strip of the surface is $2\pi x ds \equiv 2\pi x\sqrt{1 + y'^2}dx$, the total area of the surface of revolution is given by the integral

$$S = 2\pi \int_{P_1}^{P_2} x\sqrt{1+y'^2}dx.$$

Hence, one arrives at the following variational formulation of the problem: find the curve for which the variational integral $\int L(x, y, y')dx$ with the Lagrangian

$$L = x\sqrt{1+y'^2}$$

has a stationary value. The condition for a stationary value is equivalent to the *Euler-Lagrange equation* (see Section 9.7.1):

$$\frac{\partial L}{\partial y} - D_x\left(\frac{\partial L}{\partial y'}\right) = 0. \tag{1.20}$$

In our example

$$\frac{\partial L}{\partial y} = 0, \quad \frac{\partial L}{\partial y'} = \frac{xy'}{\sqrt{1+y'^2}},$$

and hence equation (1.20) is written

$$D_x\left(\frac{xy'}{\sqrt{1+y'^2}}\right) = 0.$$

Upon differentiation, one obtains the following nonlinear differential equation of the second order:

$$y'' + \frac{1}{x}(y' + y'^3) = 0. \tag{1.21}$$

The desired curve is the solution $y = y(x)$ of (1.21) (see Problem 1.3) satisfying the boundary conditions: $y(x_1) = y_1$, $y(x_2) = y_2$.

Remark. The formulation of the original problem by the differential equation (1.21) assumes that the function $y(x)$ is continuous, as are also its first and second derivatives. Accordingly, one may lose, e.g. curves composed of straight line segments. One should be aware of this kind of restriction which is in the nature of modelling by differential equations.

1.3 Natural phenomena

1.3.1 Free fall of a body near the earth

Consider the free fall of a body toward the earth under the assumption that, near the earth, gravity is constant and that it is the only force acting on the object. Let $m = $ const. be the mass of the object, h its height above the ground, t time, and $g \approx 981$ cm/s^2 the gravitational constant. In this notation, the force of gravity is $F = -mg$. Then Newton's equation (1.2) is written:

$$\frac{d^2 h}{dt^2} = -g.$$

This is an equation of the form (1.8) with constant $f = -g$. Consequently, the integration (1.9) yields:

$$h = -\frac{g}{2}t^2 + C_1 t + C_2.$$

By letting $t = 0$ in this solution and in the velocity $v \equiv h' = -gt + C_1$, one obtains the physical meaning of the integration constants, namely $C_2 = h_0$ is the initial position of the body and $C_1 = v_0$ is its initial velocity. Thus, the solution for the free fall near the earth can be written:

$$h = -\frac{g}{2}t^2 + v_0 t + h_0. \tag{1.22}$$

Exercise. A body at rest ($v_0 = 0$) falls from the height h_0. Find its terminal velocity v_*, i.e. the velocity when the body reaches the ground.

Solution. By formula (1.22), $h = h_0 - gt^2/2$, $v = -gt$. Denoting t_* the instant when the body reaches the ground ($h = 0$) and v_* its velocity at that instant, one obtains $v_* = -gt_*$, $h_0 = gt_*^2/2$, whence eliminating t_*:

$$v_* = -\sqrt{2gh_0}. \tag{1.23}$$

Here the minus sign appears owing to the fact that the h axis is directed upwards from the surface of the earth whereas the body falls toward the earth.

1.3.2 Meteoroid

The fall of a distant body (meteoroid) before entering the earth's atmosphere is defined by Newton's second law of dynamics (1.1) together with his *law of inverse squares* according to which a meteoroid and the earth attract each other by the force $F = GmM/r^2$, where G is the universal constant of gravitation, m and M denote the masses of a meteoroid and the earth, and r the distance between their centers. Let R be the radius of the earth. Then the value of the force of attraction on the surface of the earth is $F = GmM/R^2$. On the other hand, the gravitation force near the earth (i.e. the *weight* of a body with mass m) is written mg. Whence the equation $mg = GmM/R^2$, or $GmM = mgR^2$. Hence the object is attracted to the earth by the force $F = mgR^2/r^2$.

The mass m of a meteoroid is constant before entering the earth's atmosphere. Let us ignore the air resistance and assume that the mass does not change along the whole trajectory of a falling meteoroid. Then equation (1.2) is written:

$$\frac{d^2 r}{dt^2} = -\frac{gR^2}{r^2}. \tag{1.24}$$

The minus sign appears since r is directed from the earth to the meteoroid and hence it is opposite to the direction of the force of the gravitational attraction.

Exercise 1. Reduce the order of the equation (1.24).

Solution. By letting $dr/dt = v(r)$ and noting that

$$\frac{d^2 r}{dt^2} = \frac{dv}{dr}\frac{dr}{dt} = v\frac{dv}{dr} \equiv \frac{1}{2}\frac{d(v^2)}{dr},$$

equation (1.24) is rewritten

$$\frac{d(v^2)}{dr} = -\frac{2gR^2}{r^2}.$$

Whence upon integration by taking into account that, in our notation, the velocity is negative:

$$v = -\sqrt{\frac{2gR^2}{r} + C}, \quad C = \text{const.} \tag{1.25}$$

Exercise 2. Find the terminal velocity v_* (i.e. the velocity on the surface of the earth) of a meteoroid falling from a point at infinity where it was in rest.

Solution. Let us first specify the constant of integration in (1.25) by assuming that initially ($t = 0$) the meteoroid rested ($v_0 = 0$) at a distance r_0 from the center of the earth. Letting $t = 0$ and hence $v = 0$ in (1.25) yields that $C = -2gR^2/r_0$, and formula (1.25) becomes:

$$v = -R\sqrt{2g}\sqrt{\frac{1}{r} - \frac{1}{r_0}}.$$

Letting $r_0 = \infty$ and $r = R$, one obtains the terminal velocity:

$$v_* = -\sqrt{2gR}. \tag{1.26}$$

Hence the meteoroid reachs the ground with the same velocity as a body falling from the height h_0 equal to the radius R of the earth (compare (1.23) and (1.26)).

1.3. NATURAL PHENOMENA

1.3.3 A model of rainfall

As an approximation to this natural phenomenon, let us simulate two successive stages of rainfall, the first stage being the *development of raindrops in clouds* and the second one being the *fall of raindrops through the air*.

1. DEVELOPING DROPS. The onset of raindrops in clouds is imitated here by the free fall toward the earth of a spherical mass of water in saturated atmosphere under the force of gravity. The typical thickness of clouds producing precipitation is from 100 m to 4 km, but very thick clouds (cumulonimbus) may reach 20 km.

The mass m of a drop increases owing to condensation, the increment being proportional to time and to the surface area of the drop, viz. $dm = k4\pi r^2 dt$, where r is the radius of the drop and k is an empirical constant. On the other hand, the mass of a spherical drop of water (with density $\rho = 1$) is $m = 4\pi r^3/3$, whence $dm = 4\pi r^2 dr$. Hence $dr = k dt$, and Newton's second law (1.1) with $F = -mg$ is written:

$$k\frac{d(r^3 v)}{dr} = -gr^3. \tag{1.27}$$

The solution of this differential equation (see Problem 1.5) satisfying the initial condition, $v = v_0$ when $r = r_0$, has the form:

$$v = -\frac{gr}{4k}\left(1 - \frac{r_0^4}{r^4}\right) + \frac{r_0^3}{r^3}v_0. \tag{1.28}$$

Typical cloud droplets have the radius $r_0 \approx 10~\mu m$, while raindrops reach the earth with radii about 1 mm. Let us assume, in our simplified model, that the initial radius r_0 of a drop is infinitely small. Then we let $r_0 = 0$ in the solution (1.28) to obtain $v = -gr/(4k)$, or invoking the equation $r = kt$,

$$v = -\frac{1}{4}gt. \tag{1.29}$$

Hence the magnitude $|v|$ of the velocity of raindrops, at the stage of their developing in clouds, increases as a linear function of time.

2. FALLING RAIN. This stage is imitated by the fall of raindrops through the air toward the earth. It is assumed that gravity and air resistance are the only forces acting on the object, e.g. the evaporation of falling drops is ignored.

Let air resistance be a function, $f(v)$, of the velocity v of drops only. Let us denote by m the mass of a raindrop at the instant when the drop leaves the clouds and assume that it remains unaltered during the fall, $m = $ const. Then the velocity of the raindrop is determined, according to Newton's second law, by a differential equation of the first order:

$$m\frac{dv}{dt} = -mg + f(v), \tag{1.30}$$

together with the initial condition

$$v\big|_{t=t_*} = v_*, \tag{1.31}$$

where the notation $|_{t=t_*}$ means evaluated at $t = t_*$. Here t_* is the instant when the raindrop leaves the clouds and v_* is its terminal velocity in that instant. Provided that t_* and v_* are found from the first stage, one obtains the velocity of raindrops by solving the *initial value problem* (1.30)–(1.31).

Commonly, it is assumed that air resistance is proportional to the square of the velocity of a falling object provided that the object is not "very small" and that its velocity is less than that of sound but not infinitely small. However, under certain conditions air resistance can be approximated by a linear function of velocity as well. Thus, one can consider, as a reasonable model of rainfall, the following simple form of equation (1.30):

$$m\frac{\mathrm{d}v}{\mathrm{d}t} = -mg - \alpha v + \beta v^2, \tag{1.32}$$

where $\alpha \geq 0$ and $\beta \geq 0$ are empirical constants. The choice of the signs is in accordance with the fact that the air resistance opposes the force of gravity and that v is negative in our coordinate system which is directed upwards.

1.3.4 Motion of planets

The apparent motions of the planets appear to be irregular and complicated. However, it was obvious in the remote past that the heavens ought to exemplify mathematical beauty. This would only be the case if the planets moved in circles. Indeed, in Greek science one can find a hypothesis that all the planets, including the earth, go round the sun in circles [1.6]. J. Kepler [1.7] discovered, however, that planets move in ellipses, not in circles, with the sun at a focus, not at the center. He formulated in 1609 two of the cardinal principles of modern astronomy: *Keler's first law* (Fig. 1.1) and *Kepler's second law* (Fig. 1.2).

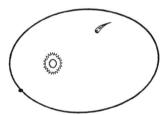
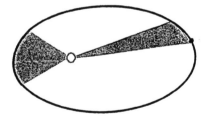

Figure 1.1 Kepler's first law: The orbit of a planet is an ellipse with the sun at one focus.

Figure 1.2 Kepler's second law: The areas swept out in equal times by the line joining the sun to a planet are equal.

Kepler's third law published in 1619 asserts that the ratio T^2/R^3 of the square of the period T and the cube of the mean distance R from the sun is the same for all planets.

Kepler's laws reduce the motion of planets to geometry and reveal, at a new level, a mathematical harmony in nature. From a practical point of view, it was important that Kepler gave an answer, based on empirical astronomy, to the question of *how* the planets move.

The geometry of the heavens provided by Kepler's laws challenged scientists to answer the question of *why* the planets obey these laws. The question required an investigation of the dynamics of the Solar system. The necessary dynamics had been initiated by Galileo [1.8] and developed into modern rational mechanics by Newton.

Meditating on the philosophical and mathematical essence of Kepler's laws, Newton formulated in *Principles* [1.3] his gravitation law. Accordingly, the force of attraction between the sun and a planet has the form

$$F = \frac{GmM}{r^3} x, \tag{1.33}$$

where G is the constant of gravitation, m and M are the masses of a planet and the sun, $x = (x^1, x^2, x^3)$ is the position vector of the planet considered as a particle, and $r = |x|$ is the distance of a planet from the sun. Hence, ignoring the motion of the sun under a planet's attraction, Newton's second law (1.2) yields:

$$\frac{d^2 x}{dt^2} = -\frac{GM}{r^3} x.$$

This is a vector form of the following system of three coupled scalar equations:

$$\frac{d^2 x^i}{dt^2} = -GM \frac{x^i}{((x^1)^2 + (x^2)^2 + (x^3)^2)^{3/2}}, \quad i = 1, 2, 3, \tag{1.34}$$

for three coordinates $x^i (i = 1, 2, 3)$ of the position vector x.

Newton derived Kepler's laws by solving the differential equations (1.33). It can be shown however that the Kepler's laws are direct consequences of specific symmetries of Newton's gravitation force [5.29] and can be derived without integrating the nonlinear equations (1.34). See Sections 5.5.2 and 9.7.4.

1.4 Elementary physics in everyday life

1.4.1 Educated farmer

The phenomenon of cooling (heating) by a surrounding medium is commonly used in everyday life. One immerses a body, for cooling (heating) it, in a medium of lower (higher) temperature than that of the body. The medium may be the surrounding air, a very large cold bath, a preheated oven, etc., while the body in question may be a thermometer, a hot metal plate to be cooled, blood plasma stored at low temperature to be warmed before using, milk and other liquids. It is assumed that the temperature T of the surrounding bath is unaffected by the immersed body and hence $T = $ const. It is assumed further that the temperature τ of the immersed body is the same in all its parts at each instant so that $\tau = \tau(t)$. Then what is known as *Newton's law of cooling* states simply that the rate of change of τ is proportional to the temperature difference $T - \tau$. Thus, Newton's law of cooling is written as the following ordinary differential equation of the first order:

$$\frac{d\tau}{dt} = k(T - \tau), \tag{1.35}$$

where k is a positive constant depending on the substance of the immersed body and that of the surrounding medium.

Example (Pasteurization). Recall that pasteurization is the partial sterilization of milk without boiling it and is based on Louis Pasteur's discovery that germs in milk temporarily stop functioning if every particle of the milk is heated to 64° C (147° F) and then the milk is quickly cooled.

Let us imagine an educated farmer who decided to pasteurize milk for the first time but, unfortunately, found that his thermometer was broken. Since our farmer is an educated one, we can fancy that he would solve the problem of warming the milk precisely to 64° having at his disposal only an oven and his watch, by using equation (1.35) instead of the broken thermometer as follows.

The farmer has firstly to determine the coefficient k. To that end, he places the milk stored at room temperature $\tau_0 = 25°$ C in the oven set, e.g. at $T = 250°$ C and waits until the milk boils. Suppose it took 15 min for the milk to boil. Now, letting the boiling temperature of milk be 90° C, the farmer uses the solution of equation (1.35) (see Problem 1.10),

$$\tau = T - Be^{-kt}. \tag{1.36}$$

At the initial moment $(t = 0)$ this equation is written $25° = 250° - B$ and specifies the constant of integration, $B = 225°$. Then the solution at $t = 15$ min yields:

$$90 = 250 - 225e^{-15k},$$

where we consider only the numerical values. Hence, $15k = -\ln(160/225)$, or $k \approx 34/1500$. Thus, equation (1.35) yields the following formula for the temperature of the milk placed in the oven at 250°:

$$\tau = 250 - 225e^{-34t/1500}.$$

For $\tau = 64$ it follows $-34t/1500 = \ln(186/225) \approx -0.19$, whence $t \approx 8.4$ min. Thus, the farmer should warm the milk for 8 min 24 s in the oven set at 250° C.

1.4.2 Outflow from a funnel

One can use a differential equation to describe the behavior of fluids channelled by a funnel. Consider, e.g., a conical funnel of angle θ at its apex with the hole of

1.4. ELEMENTARY PHYSICS IN EVERYDAY LIFE

radius r_0. According to the generally accepted law of hydraulics, the velocity of outflow of a fluid from a hole under hydrostatic pressure is given by the formula

$$v = \eta\sqrt{2gh} \text{ cm/s},$$

similar to (1.23), with an empirical coefficients η (e.g. for water, $\eta = 3/5$). Here h is the height of the fluid over the hole and $g \approx 981$ cm/s is the gravitational constant. To describe the outflow from the funnel, it suffices to determine the height $h = h(t)$ of the fluid in the funnel. To determine the unknown function $h(t)$, let us write down the balance of the fluid. The volume of the fluid that flows away from the hole in dt seconds is equal to $dV = \pi r_0^2 v dt \equiv \pi \eta r_0^2 \sqrt{2gh}\, dt$ cm^2. On the other hand, the decrease of the fluid volume, due to a negative increment $(-dh)$ of the height of the fluid in the funnel, is given by $dV = -\pi r^2(t)dh$, where $r(t)$ is the radius of the funnel at the height $h = h(t)$ and hence $r(t) = h\tan(\theta/2)$. It follows that the balance equation is

$$\pi \tan^2(\theta/2) h^2 dh + \pi \eta r_0^2 \sqrt{2gh}\, dt = 0.$$

Thus, we arrive at the differential equation

$$h^{3/2}\frac{dh}{dt} + \lambda = 0, \qquad (1.37)$$

where $\lambda = \eta\sqrt{2g}\, r_0^2/\tan^2(\theta/2)$ is a constant.

Exercise. The educated farmer (see the preceding section) made a drinking place for his cattle having at his disposal only a conical container of angle 60° at the apex with a hole suitable for appending a tube of any diameter. The farmer could fill the container by water from an artesian well. The maximum height of water in the container was 4 m. Using equation (1.37), the farmer determined the diameter d of the tube so that the water from the full container would flow to the drinking place in precisely 20 h. Carry out the calculations and find the diameter d.

Solution. We firstly substitute the numerical values $g = 981$, $\tan^2 30° = 1/3$ and $\eta = 3/5$ in definition of λ. The reckoning shows that $\lambda \approx 80 r_0^2$, hence integrating equation (1.37):

$$\frac{2}{5}h^{5/2} = C - 80 r_0^2 t.$$

The initial condition, $h\big|_{t=0} = 400$ cm yields $C = 2^{11} \times 5^4$, and hence

$$\frac{2}{5}h^{5/2} = 2^{11} \times 5^4 - 80 r_0^2 t.$$

Since the full container should empty in 20 h, i.e. $h = 0$ when $t = 20 \times 3600$ s, it follows from the above equation:

$$2^{11} \times 5^4 - 80 r_0^2 \times 20 \times 3600 = 0.$$

We rewrite this equation in the form

$$2^{11} \times 5^4 = 2^{10} \times 5^4 \times 9 r_0^2,$$

whence $r_0 = \sqrt{2}/3$. Thus, the diameter of the tube is $d = 2 r_0 \approx 1$ cm.

1.4.3 Heating and air conditioning

Newton's cooling law, appropriately adapted to real situations, provides a good approximation to modelling the temperature dynamics inside a building [1.9].

Let the inside temperature τ be an unknown function of time t. Let $T = T(t)$ be the outside temperature considered as a given function. We firstly note that Newton's law (1.35), where k is a *positive* constant, is in agreement with the natural expectation that the inside temperature τ increases ($d\tau/dt > 0$) when $T > \tau$, and decreases ($d\tau/dt > 0$) when $T < \tau$. The constant k has the dimension of t^{-1} and depends on the quality of the building, in particular on its thermal insulation. In common situations, $0 < k < 1$, and it is infinitely small in ideally insulated buildings.

Suppose that the building is supplied by a heater and by an air conditioner. Denote by $H(t)$ the rate of increase in temperature inside the building caused by the heater, and by $A(t)$ the rate of change (increase or decrease) in temperature caused by the air conditioner. Assuming that these are the only factors affecting the temperature in the building, we have the differential equation

$$\frac{d\tau}{dt} = k[T(t) - \tau] + H(t) + A(t). \tag{1.38}$$

As an example of an equation (1.38) with $A(t) \neq 0$, let us assume that a furnace supplies heating at a given rate $H(t) \geq 0$ and that the building is provided with a thermostat to keep the inside temperature around a desired (*critical*) temperature τ_c. If the actual temperature $\tau(t)$ is above τ_c, the air conditioner supplies cooling, otherwise it is off. Then $A(t) = l[\tau_c - \tau]$, where l is a positive empirical parameter, and the equation (1.38) is written in the form

$$\frac{d\tau}{dt} = k[T(t) - \tau] + H(t) + l[\tau_c - \tau] \tag{1.39}$$

with given functions $T(t)$ and $H(t)$ and the constants k, τ_d, l.

Exercise 1. In a still cold winter evening when the outside temperature stayed constant at $T_0 = -10°$ C, the electricity was shut-down in your house at 6 pm. This caused cessation of operation of your heater and air conditioner for the whole night. Suppose that the inside temperature at $t_0 = 6$ pm was $\tau_0 = 25°$ C. Unfortunately, however, your building was not well insulated (windows, etc.). Therefore, it took only an hour for the temperature to drop to $\tau \approx 19.5°$ C. What temperature do you expect in your bedroom at 6 am, provided that the outside temperature stays at $T_0 = -10°$ C the whole night?

Hint. Use the mathematical model (1.38) to find the temperature variation in your house during the night, in particular the temperature at $t_1 = 6$ am.

Solution. Since $H(t) = A(t) = 0$, equation (1.38) has the form (1.35) and hence its solution is given by (1.36), $\tau(t) = T_0 - Be^{-kt}$. At the initial moment $t = t_0$, it follows that $25 = -10 - Be^{-kt_0}$, or $B = -35e^{kt_0}$. Hence

$$\tau(t) = -10 + 35e^{-k(t-t_0)}.$$

1.4. ELEMENTARY PHYSICS IN EVERYDAY LIFE

Since $\tau|_{t=t_0+1} = 19.5°$, it follows that $19.5 = -10 + 35e^{-k}$, whence $k = \ln(1.18) \approx 1/6$. Thus, the variation of the inside temperature during the night is given by (in °C)

$$\tau(t) = -10 + 35e^{-(t-t_0)/6}, \quad t_0 \le t \le t_1.$$

In particular, the temperature at $t = t_1 = 6$ am is $-5.25°$ C. See Fig. 1.3.

Exercise 2. Suppose now that the previous accident happened when the outside temperature was $+8°$ C at $t_0 = 6$ pm decreasing uniformly to $-10°$ C at $t_1 = 6$ am. Find the temperature variation in your house during the night.

Solution. Here again $H(t) = A(t) = 0$ but, unlike the previous case, the outside temperature is unsteady and is given by $T(t) = 8 - \frac{3}{2}(t - t_0)$, $t_0 \le t \le t_1$. The differential equation (1.38) is written

$$\frac{d\tau}{dt} = k[8 - \frac{3}{2}(t - t_0) - \tau] \tag{1.40}$$

and has the general solution (see Section 2.1.1)

$$\tau(t) = 8 + \frac{3}{2k} - \frac{3}{2}(t - t_0) - Be^{-kt}, \quad B = \text{const.} \tag{1.41}$$

The initial condition, $\tau|_{t=t_0} = 25$ yields $B = (\frac{3}{2k} - 17)e^{kt_0}$. We know from Exercise 1 that, for the building in question, $k = 1/6$. Hence, $B = -8e^{kt_0}$. Thus, the variation of the inside temperature during the night is given by (in °C)

$$\tau(t) = 17 - \frac{3}{2}(t - t_0) + 8e^{-(t-t_0)/6}, \quad t_0 \le t \le t_1.$$

In particular, the temperature at $t_1 = 6$ am is around $0°$ C. See Fig. 1.4.

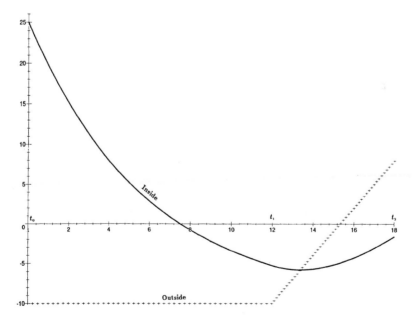

Figure 1.3 Temperature variation given in Exercise 1 and Problem 1.12 ($k = 1/6$).

18   1. MODEL DIFFERENTIAL EQUATIONS

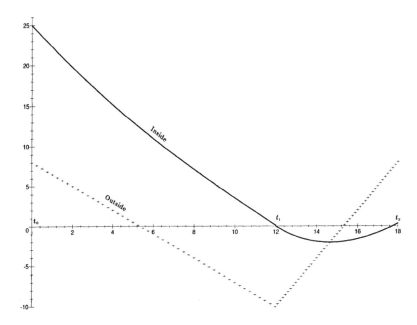

Figure 1.4 Temperature variation in Exercise 2 and Problem 1.13 ($k = 1/6$).

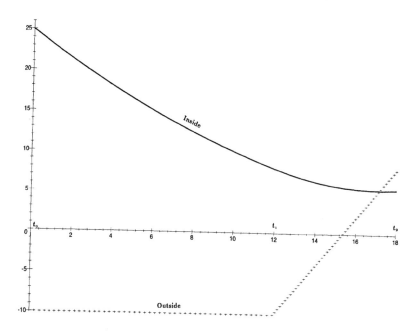

Figure 1.5 Temperature variation given in Problem 1.14 ($k = 1/18$).

1.4. ELEMENTARY PHYSICS IN EVERYDAY LIFE

1.4.4 Electrical instruments. Equation of van der Pol

A common illustrative example is the discharge of an electrical condenser through an inductive coil of wire. According to elementary laws of electricity [1.10], the phenomenon is described by the equations

$$C\frac{dV}{dt} = -I, \quad V - L\frac{dI}{dt} = RI, \tag{1.42}$$

where I is the *current* of the discharge, V the *voltage* (the potential difference between the terminals of the condenser), R the *resistance*, C the condenser's *capacity*, and L the coil's *inductance* . Here I and V are functions of time t, whereas R, C, and L are regarded as given constants. Hence, (1.42) is a system of first-order ordinary differential equations with two dependent variables, I and V, considered as unknown functions of t.

Let us denote the dependent variable V by y and its first and second derivatives with respect to t by y' and y''. In this notation, the second equation (1.42), upon substituting I from the first one, becomes a linear second-order ordinary differential equation:

$$ay'' + by' + cy = 0 \tag{1.43}$$

with constant coefficients $a = LC$, $b = RC$, and $c = 1$.

Replacing in the second equation (1.42) the classical Ohm's law by the so-called *generalized Ohm's law*, one obtains a nonlinear system,

$$C\frac{dV}{dt} = -I, \quad L\frac{dI}{dt} = V - h(I), \tag{1.44}$$

or an equivalent single nonlinear equation of the second order:

$$ay'' + y = -f(y'), \quad a = \text{const.} \tag{1.45}$$

By letting $f(y') = \varepsilon(y'^3 - y')$, one arrives at the equation of van der Pol used in the theory of triodes [1.11]:

$$ay'' + y = \varepsilon(y' - y'^3), \quad \varepsilon = \text{const.} \tag{1.46}$$

1.4.5 Mechanical vibrations

In everyday life, one encounters many types of vibrations. These are, e.g., the rustle of leaves and twigs of trees caused by wind, the bouncing motion of a car due to cracks in the road, water waves, the oscillation of a ship on waves, etc.

A common example of a differential equation describing small mechanical vibrations is provided by the physical problem of a heavy particle suspended from a coiled spring and oscillating in the vertical direction y about its position of equilibrium $y = 0$. The particle will be subject to a restoring force F_1 that is, according to Hooke's law, proportional to its displacement y from the position $y = 0$ and is opposite to the displacement, i.e. $F_1 = -ky$ with a positive constant coefficient k. In reality, when a body moves in a medium, there is also a damping (or friction) force F_2 which tends to retard the motion. Commonly, it is assumed that the force of friction is proportional to the magnitude of the velocity of the particle, but opposite in direction. Hence, $F_2 = -l\,dy/dt$, where the independent variable x is time and l is a positive coefficient of proportionality called the *damping constant*. Denote by $f(t)$ the total external force (due to wind, cracks in the road, etc.) regarded as a given function of time.

Thus, applying Newton's second law (1.2) with $F = F_1 + F_2 + f(t)$, the equation for small oscillations of a particle with mass m is written

$$m\frac{d^2y}{dt^2} + l\frac{dy}{dt} + ky = f(t), \qquad (1.47)$$

or

$$Ly = f(t), \quad \text{where } L = m\frac{d^2}{dt^2} + l\frac{d}{dt} + k.$$

The mechanical system is said to be *damped* if $l \neq 0$, and *undamped* otherwise. The motion is said to be *free* if $f(t) \equiv 0$, and *forced* otherwise.

Damped oscillations of a mechanical system with n degrees of freedom are described by a system of ordinary differential equations

$$\sum_{i=1}^{n}\left(m_{ij}\frac{d^2y^j}{dt^2} + l_{ij}\frac{dy^j}{dt} + k_{ij}y^j\right) = f^i(t), \quad i = 1,\ldots,n,$$

with constant coefficients m_{ij}, l_{ij}, and k_{ij}. It is written in the form (1.47) after introducing the vector notation $\boldsymbol{y} = (y^1,\ldots,y^n), \boldsymbol{f} = (f^1,\ldots,f^n)$ and the matrix differential operator L as follows:

$$L\boldsymbol{y} = \boldsymbol{f}(t), \qquad L = M\frac{d^2}{dt^2} + A\frac{d}{dt} + B, \qquad (1.48)$$

where M, A, and B are matrices with the entries m_{ij}, l_{ij}, and k_{ij}, respectively [1].

The above linear models are legitimate for small vibrations (called *harmonic oscillations*) when the amplitude of the vibrations is sufficiently small. Higher approximations, known as *anharmonic oscillations*, are described by nonlinear differential equations. For example, one-dimensional (free and undamped) anharmonic oscillations in the third order of precision are described by the equation

$$m\frac{d^2y}{dt^2} + ky + \alpha y^2 + \beta y^3 = 0. \qquad (1.49)$$

1.4.6 Collapse of driving shafts

At the beginning of this century, constructors of motor ships came across the troublesome phenomenon of a seemingly accidental "beating" and the possible

---

[1]Throughout the book, vectors $\boldsymbol{y}$ etc. are regarded as *column vectors* though they are written as rows. Accordingly, in the matrix $M = \|m_{ij}\|$, i and j are rows and columns, respectively.

1.4. ELEMENTARY PHYSICS IN EVERYDAY LIFE

collapse of shafts in power transmission systems. The strange phenomenon was explained by A.N. Krylov [1.12] using the theory of differential equations. The mathematical background of the problem is as follows.

According to the model of the vibrations of rods (1.4), the positions of equilibrium of a uniformly rotating cylindrical shaft are determined by the ordinary differential equation:

$$\mu \frac{d^4 u}{dx^4} = f,$$

where u is the shaft's displacement from its "trivial" position of equilibrium, $u = 0$, and f is the density of the centrifugal force acting on the shaft. To find f, consider a small element of the shaft dx and denote by p the weight of the shaft per unit length. Then the mass of the element dx is $dm = (p/g)dx$, where g is the gravitational constant. The centrifugal force df acting on dx due to the rotation by a constant angular velocity ω has the form $df = \omega^2 u dm = \omega^2 (p/g) u dx$. Hence $f = (p/g)\omega^2 u$, and the differential equation in question is written

$$\mu \frac{d^4 u}{dx^4} = \frac{p\omega^2}{g} u, \qquad (1.50)$$

where the positive constant μ depends on the material of the shaft.

Let the shaft revolve at two bearings located at $x = 0$ and $x = l$. Then $u|_{x=0} = u|_{x=l} = 0$. Furthermore, it can be shown that bearings are points of inflection of the function $u = u(x)$. Thus, one arrives at the problem of investigating the solutions of the differential equation (1.50) satisfying four *boundary conditions*:

$$u|_{x=0} = 0, \quad u|_{x=l} = 0, \quad \frac{d^2 u}{dx^2}\bigg|_{x=0} = 0, \quad \frac{d^2 u}{dx^2}\bigg|_{x=l} = 0. \qquad (1.51)$$

The phenomenon of "beating" occurs when the boundary value problem defined by equations (1.50)–(1.51) has a "nontrivial solution", i.e. a solution $u = u(x)$ that does not vanish identically in the interval $0 \le x \le l$.

Integration of equation (1.50) rewritten in the form

$$\frac{d^4 u}{dx^4} = \alpha^4 u, \quad \text{where} \quad \alpha^4 = \frac{p\omega^2}{g\mu} = \text{const.}, \qquad (1.52)$$

provides its general solution (see Section 2.2.8)

$$u = C_1 e^{\alpha x} + C_2 e^{-\alpha x} + C_3 \cos(\alpha x) + C_4 \sin(\alpha x), \quad C_i = \text{const.} \qquad (1.53)$$

The boundary conditions (1.51) yield

$$C_1 + C_2 + C_3 = 0, \quad C_1 e^{\alpha l} + C_2 e^{-\alpha l} + C_3 \cos(\alpha l) + C_4 \sin(\alpha l) = 0,$$

$$C_1 + C_2 - C_3 = 0, \quad C_1 e^{\alpha l} + C_2 e^{-\alpha l} - C_3 \cos(\alpha l) - C_4 \sin(\alpha l) = 0.$$

The reckoning shows that $C_1 = C_2 = C_3 = 0$ and $C_4 \sin(\alpha l) = 0$. If $C_4 = 0$, one arrives at the trivial solution, $u = 0$, so that the shaft is straight. On the other hand, by letting $\sin(\alpha l) = 0$, that is $\alpha = n\pi/l$, one obtains the cases when beating occurs. Then (1.53) yields $u = C_4 \sin(n\pi x/l)$. According to the definition of α, the equation $\alpha = n\pi/l$ yields that a collapse of the shaft is possible whenever its angular velocity approaches any one of the following *critical values*:

$$\omega_n = \frac{n^2\pi^2}{l^2}\sqrt{\frac{g\mu}{p}}, \quad n = 1, 2, \ldots. \tag{1.54}$$

1.5 Ecology

1.5.1 Radioactive decay

Radioactivity is a consequence of the breaking up of elements with high atomic weights such as uranium minerals [1.13]. The discovery of radioactivity provided new means, e.g. of determining geological time, etc. Artificial radioactivity is widely used in practical affairs – chemistry, medicine, nuclear energetics, etc. However, an industrial use of nuclear energy requires an inexorable vigilance by the population because of the danger of pollution by radioactive waste products.

A mathematical model of radioactive decay assumes that the rate of decay is proportional to the amount of a radioactive substance:

$$\frac{dU}{dt} = -kU. \tag{1.55}$$

Here U is the amount of a radioactive substance present at time t and $k > 0$ is a positive constant to be determined from an experiment.

The general solution to the differential equation (1.55) has the form

$$U(t) = U_0 e^{-k(t-t_0)}, \tag{1.56}$$

where the constant of integration U_0 is an initial $(t = t_0)$ amount of the substance. The empirical constant k depends on the radioactive matter in question. Usually, it is determined in terms of the so-called *half-life* defined as the interval of time $\Delta t = t - t_0$ after which the substance will have diminished to half of its original amount.

Example. It is known that the half-life of *radium* is $\Delta t = 1\,600$ years. Therefore, according to the formula (1.56), $U_0/2 = U_0 e^{-1\,600k}$, whence $k = (\ln 2)/1\,600 \approx 0.00043$. Thus, the radioactive disintegration of radium in t years is given by

$$U(t) = U_0 e^{-\frac{\ln 2}{1\,600}t}.$$

1.5.2 Population growth

The celebrated Malthusian principle of population [1.14] is mathematically rather simple and is based on the natural assumption that the rate of population growth is proportional to the population considered. Accordingly, it is formulated by

1.5. ECOLOGY

the differential equation (1.55) with a negative constant $k = -\alpha$. Hence the Malthusian principle provides *unlimited growth* of a population according to the exponential law:

$$P(t) = P_0 e^{\alpha(t-t_0)}, \qquad (1.57)$$

where P_0 and $P(t)$ denote the population in millions at the initial time $t = t_0$ and at an arbitrary time t, respectively. The main consequence of the *Essay* was that the realization of a happy society will always be hindered by the universal tendency of the population to outrun the means of subsistence.

It was soon seen, however, that Malthus' model is unrealistic and requires a modification. Subsequently, several modified population models have been considered in an attempt to find more realistic laws of population. One of them, known as the *logistic law,* is described by the following nonlinear equation:

$$\frac{dP}{dt} = \alpha P - \beta P^2, \qquad (1.58)$$

where the nonlinear term βP^2 can be interpreted as a kind of *social friction*. An analysis of this law shows that it is adequate to describe certain insect populations. However, its value as a general "law" of population growth is extremely limited as far as human population is concerned.

1.5.3 Predator and prey

A model of predator and prey suggested by Volterra and Lotka is formulated by a system of nonlinear ordinary differential equations of the first order,

$$\frac{du}{dt} = (a - bv)u, \quad \frac{dv}{dt} = (ku - l)v, \qquad (1.59)$$

where $a, b, k,$ and l are positive constants. Here v denotes a predator species and u its prey. It is assumed that the prey population provides the total food supply for the predators.

A qualitative analysis of solutions of the above system shows, e.g., that any biological system described by the Volterra-Lotka equations (1.59) ultimately approaches either a constant or periodic population [1.15].

1.5.4 Competing species

A community of two species, u and v, competing for a common food supply is governed by a system of first-order ordinary differential equations of the form

$$\frac{du}{dt} = f(u,v)u, \quad \frac{dv}{dt} = g(u,v)v, \qquad (1.60)$$

where f and g are given functions or, more generally, they satisfy certain natural conditions required by particular growth rate assumptions.

Problems

1.1. Check by substitution that
  (i) $y = ax + b$ is a solution of equation (1.16), $y'' = 0$;
  (ii) $y = ax^2 + bx + c$ is a solution of equation (1.17), $y''' = 0$;
  (iii) $y = a + (b - cx)^{-1}$ is a solution of equation (1.19), $2y'y''' = 3y''^2$.

1.2. Integrate the ordinary differential equations (1.16)–(1.19):
$$y'' = 0, \quad y''' = 0, \quad y''' = 3\frac{y'y''^2}{1+y'^2}, \quad y''' = \frac{3}{2}\frac{y''^2}{y'}.$$

1.3. Integrate the differential equation (1.21):
$$y'' + \frac{1}{x}(y' + y'^3) = 0.$$

1.4. Calculate the terminal velocity v_* of a meteoroid by formula (1.26), $v_* = -\sqrt{2gR}$, substituting the numerical values of the gravitational constant $g \approx 9,81$ m/s² and of the radius of the earth $R \approx 6\,360$ km. Give the answer in km/s.

1.5. Integrate the differential equation (1.27),
$$k\frac{\mathrm{d}(r^3 v)}{\mathrm{d}r} = -gr^3, \quad k, g = \mathrm{const}.$$

1.6. Use Problem 1.5 to find the solution (1.28) of equation (1.27) satisfying the initial condition, $v\big|_{r=r_0} = v_0$.

1.7. Integrate the differential equation (1.32) with $\beta = 0$ and find the solution satisfying the initial condition (1.31).

1.8. Integrate the differential equation (1.32) with $\alpha = 0$.

1.9. Integrate the differential equation (1.32) with arbitrary parameters α and β.

1.10. Integrate the differential equation (1.35).

11. Calculate the time of pasteurization if the farmer uses
  (i) milk stored at $\tau_0 = 25°$ C and sets the oven temperature at $T = 120°$ C;
  (ii) refrigerated milk at $\tau_0 = 8°$ C and sets the oven temperature at $T = 250°$ C;
  (iii) refrigerated milk at $\tau_0 = 8°$ C and sets the oven temperature at $T = 180°$ C.

1.12. Let, in the conditions of Exercise 1 in Section 1.4.3, the outside temperature increase uniformly from $-10°$ C at 6 am to $+8°$ C at noon. Find the variation of the inside temperature during this time.

1.13. Solve the previous problem in the conditions of Exercise 2 in Section 1.4.3.

14. Consider, in Exercise 1 in Section 1.4.3 and Problem 1.12, a building with $k = 1/18$. Find the variation of the inside temperature from $t_0 = 6$ pm to $t_2 = 12$ am.

1.15. Integrate the differential equation (1.55).

Chapter 2

Elementary methods of integration

This chapter is designed as a short account of basic *ad hoc* methods invented mainly in the seventeenth and eighteenth centuries. These classical devices are simple and efficient and therefore commonly used in the practice of integration of special types of ordinary differential equations [2.1].

2.1 First-order equations

Here we will consider ordinary differential equations of the first order,

$$y' = f(x, y), \qquad (2.1)$$

solved with respect to the derivative $y' = dy/dx$. Equivalently, these equations can be written in a differential form,

$$\omega \equiv M(x, y)dx + N(x, y)dy = 0. \qquad (2.2)$$

2.1.1 Linear equations. Superposition of solutions

The standard form of the general linear equation of the first order is

$$y' + P(x)y = Q(x). \qquad (2.3)$$

It is called *homogeneous* if $Q(x) = 0$, and *non-homogeneous* otherwise[1].

Consider the homogeneous equation

$$y' + Py = 0. \qquad (2.4)$$

---
[1] Throughout the book all considerations are local. This means, e.g., in the definition of homogeneous equations, that $Q(x) = 0$ identically in some neighborhood of a generic point x. It is also assumed that all functions are continuous, together with their derivatives involved in further discussions, unless otherwise stipulated.

After rewriting it in the differential form (2.2) and dividing by y,

$$\frac{dy}{y} + P dx = 0$$

the integration yields $\ln y + \int P(x) dx = \text{const}$. Hence, the solution is:

$$y = C e^{-\int P dx}, \quad C = \text{const}. \tag{2.5}$$

The non-homogeneous equation (2.3) can be solved by the method of *variation of parameters* [2.2]. In this method, one looks for the solution in the form (2.5) where the constant C is replaced by an unknown function $u(x)$:

$$y = u(x) e^{-\int P dx}.$$

Substitution into (2.3) yields:

$$\frac{du}{dx} e^{-\int P dx} = Q(x),$$

whence

$$u(x) = \int Q e^{\int P dx} dx + C, \quad C = \text{const}.$$

Thus, the solution of the general linear equation (2.3) is given by two quadratures:

$$y = \left(C + \int Q e^{\int P dx} dx \right) e^{-\int P dx}. \tag{2.6}$$

Let us take a particular solution of the homogeneous equation (2.4), e.g.,

$$y_0(x) = e^{-\int P dx}$$

and two particular solutions of the non-homogeneous equation (2.3), e.g.,

$$y_1(x) = e^{-\int P dx} \int Q e^{\int P dx} dx, \quad y_2(x) = e^{-\int P dx} + e^{-\int P dx} \int Q e^{\int P dx} dx.$$

Then the general solutions (2.5) and (2.6) are written, respectively, as follows:

$$y = C y_0(x) \tag{2.7}$$

and

$$y = (1 - C) y_1(x) + C y_2(x). \tag{2.8}$$

The formula (2.8) (or (2.7) for the homogeneous equation) represents the general solution as a *linear superposition* of particular solutions, $y_1(x)$ and $y_2(x)$. Accordingly, the functions $y_1(x)$, $y_2(x)$ are said to be a *fundamental set of solutions*, or a *fundamental system*. The following theorem states that one can replace $y_1(x)$, $y_2(x)$ by any two particular solutions $\bar{y}_1(x)$, $\bar{y}_2(x)$ such that $\bar{y}_1 \neq \bar{y}_2$.

2.1. FIRST ORDER EQUATIONS

Theorem. The superposition formula (2.8) does not depend upon the choice of particular solutions $y_1(x)$ and $y_2(x)$ provided that $y_1 \neq y_2$.

Proof. Let us write the general solution (2.6) in the form
$$y = Cg(x) + h(x),$$
where $g(x) = y_0(x)$ and $h(x) = y_1(x)$. Let
$$\bar{y}_1(x) = C_1 g(x) + h(x), \quad \bar{y}_2(x) = C_2 g(x) + h(x)$$
be two particular solutions corresponding to different values $C_1 \neq C_2$ of the parameter C. It follows that
$$g = \frac{\bar{y}_2 - \bar{y}_1}{C_2 - C_1}, \quad h = \frac{C_2 \bar{y}_1 - C_1 \bar{y}_2}{C_2 - C_1}.$$
Hence
$$y = C \frac{\bar{y}_2 - \bar{y}_1}{C_2 - C_1} + \frac{C_2 \bar{y}_1 - C_1 \bar{y}_2}{C_2 - C_1},$$
whence by setting $K = (C - C_1)/(C_2 - C_1)$ and rearranging,
$$y = (1 - K)\bar{y}_1(x) + K\bar{y}_2(x).$$
This formula coincides with (2.8) since C is arbitrary, and hence so is K.

Remark. Invoking the equation $y_2 - y_1 = y_0$, formula (2.8) can be written
$$y = y_1(x) + Cy_0(x),$$
where $y_1(x)$ and $y_0(x)$ solve equations (2.3) and (2.4), respectively. In words: the general solution of the non-homogeneous equation (2.3) is obtained from any of its particular solution by adding the general solution of the homogeneous equation (2.4).

Example. For the equation
$$y' - \frac{y}{x} = x,$$
a fundamental set of solutions is provided by
$$y_1(x) = x^2, \quad y_2(x) = x^2 + x,$$
and the superposition principle (2.8) yields the general solution:
$$y = (1 - C)x^2 + C(x^2 + x) = x^2 + Cx.$$
One obtains the same result by taking another fundamental set of solutions, e.g.
$$y_1(x) = x^2 - x, \quad y_2(x) = x^2 + x.$$
Then formula (2.8) yields:
$$y = (1 - C)(x^2 - x) + C(x^2 + x) = x^2 + (2C - 1)x,$$
whence by setting $C_1 = 2C - 1$, one arrives at the previous formula for the general solution, $y = x^2 + C_1 x$ with an arbitrary constant C_1. Thus, any change of a fundamental set of solutions affects only the constant of integration C.

2.1.2 Separation of variables. A nonlinear superposition

The technique of *separation of variables* [2.3] is a natural generalization of the integration procedure discussed in Section 1.2.1, and is applicable to equations of the type

$$y' = p(x)q(y). \tag{2.9}$$

Equation (2.9) rewritten in the differential form,

$$\frac{dy}{q(y)} = p(x)dx,$$

is readily integrable:

$$\int \frac{dy}{q(y)} = \int p(x)dx + C,$$

where C is an arbitrary constant of integration. Let

$$H(y) = \int \frac{dy}{q(y)}, \quad \phi(x) = \int p(x)dx.$$

Then the solution to equation (2.9) is given implicitly by $H(y) = \phi(x) + C$, whence

$$y = H^{-1}(\phi(x) + C), \tag{2.10}$$

where H^{-1} denotes the inverse to the function $H(y)$. The formula (2.10) provides the general solution to equation (2.9) by two quadratures.

Let us take a particular solution y_0, e.g. by letting $C = 0$ in (2.10):

$$y_0(x) = H^{-1}(\phi(x)).$$

Then the general solution (2.10) is written

$$y = H^{-1}(H(y_0) + C). \tag{2.11}$$

The presentation (2.11) of the general solution will be termed a *nonlinear superposition* for equation (2.9).

Theorem. The nonlinear superposition formula (2.11) does not depend upon a choice of a particular solution $y_0(x)$.

Proof. Since (2.10) is the general solution, any particular solution corresponds to a particular value C_1 of the arbitrary constant C. Let

$$y_1(x) = H^{-1}(\phi(x) + C_1)$$

be any particular solution. Then $\phi(x) = H(y_1) - C_1$, and the substitution into (2.10) yields:

$$y = H^{-1}(H(y_1) + K), \quad K = C - C_1.$$

This is precisely the formula (2.11) with y_0 replaced by y_1 and with C replaced by a new arbitrary constant K.

2.1. FIRST ORDER EQUATIONS

Example 1. Consider the equation

$$y' = p(x)y^2 \tag{2.12}$$

with arbitrary $p(x)$. Here

$$H(y) = \int \frac{dy}{y^2} = -\frac{1}{y}.$$

Since $z = H(y) = -1/y$ implies $y = -1/z$, it follows that

$$H^{-1}(z) = -\frac{1}{z}.$$

Therefore, formula (2.10) is written

$$y = \frac{1}{C - \phi(x)}.$$

Since $H(y_0) = -1/y_0$ and hence $H(y_0) + C = (Cy_0 - 1)/y_0$, the nonlinear superposition (2.11) for equation (2.12) has the form:

$$y = \frac{y_0(x)}{1 - Cy_0(x)}. \tag{2.13}$$

For equations of the type

$$p(x)q(y)dx + r(x)s(y)dy = 0,$$

the separation of variables is written

$$\int \frac{p(x)dx}{r(x)} + \int \frac{s(y)dy}{q(y)} = C.$$

Example 2. For the equation

$$x(y^2 - l^2)dx + y(x^2 - k^2)dy = 0, \quad k, l = \text{const.},$$

the separation of variables yields:

$$\int \frac{xdx}{x^2 - k^2} + \int \frac{ydy}{y^2 - l^2} = \ln|C|,$$

where the constant of integration is taken in the form $\ln|C|$ for simplicity in the final formula. Upon integration, $\ln|x^2 - k^2| + \ln|y^2 - l^2| = \ln|C|$ whence

$$(x^2 - k^2)(y^2 - l^2) = C.$$

In this example the nonlinear superposition (2.11) is conveniently written in an implicit form:

$$y^2 = Cy_0^2(x) + (1 - C)l^2. \tag{2.14}$$

2.1.3 Notation of exterior differential calculus

(This section furnishes the reader with a general view on the classical notion of exact equations. It is not indispensable, however, for understanding the succeeding material and can be omitted at the first reading.)

Recall that a differential form (specifically, a p-form) is [2.4]

$$\omega = \sum_{i_1 < \ldots < i_p} a_{i_1 \ldots i_p}(x) \mathrm{d}x^{i_1} \wedge \cdots \wedge \mathrm{d}x^{i_p}, \tag{2.15}$$

where $x = (x^1, \ldots, x^n) \in \mathbb{R}^n$, $\mathrm{d}x = (\mathrm{d}x^1, \ldots, \mathrm{d}x^n) \in \mathbb{R}^n$, and $a_{i_1 \ldots i_p}(x)$ are continuously differentiable functions. The summation is extended over all values $i_1, \ldots, i_p = 1, \ldots, n$ such that $i_1 < \cdots < i_p$.

The *exterior differential calculus* is defined by the law of *exterior multiplication* (wedge product):

$$\mathrm{d}x^i \wedge \mathrm{d}x^j = -\mathrm{d}x^j \wedge \mathrm{d}x^i \quad (\text{in particular,} \quad \mathrm{d}x^i \wedge \mathrm{d}x^i = 0), \tag{2.16}$$

and by the *exterior differentiation* of forms:

$$\mathrm{d}\omega = \sum_{i_1 < \ldots < i_p} \sum_{j=1}^{n} \frac{\partial a_{i_1 \ldots i_p}}{\partial x^j} \mathrm{d}x^j \wedge \mathrm{d}x^{i_1} \wedge \cdots \wedge \mathrm{d}x^{i_p}. \tag{2.17}$$

Accordingly, the differential $\mathrm{d}\omega$ is a $(p+1)$-form. The exterior differentiation and multiplication of forms obey the following rules:

$$\mathrm{d}^2 \omega \equiv \mathrm{d}(\mathrm{d}\omega) = 0, \tag{2.18}$$

$$\omega \wedge \eta = (-1)^{pq} \eta \wedge \omega, \tag{2.19}$$

$$\mathrm{d}(\omega \wedge \eta) = \mathrm{d}\omega \wedge \eta + (-1)^p \omega \wedge \mathrm{d}\eta, \tag{2.20}$$

where ω is a p-form and η a q-form. If $p = n$, then any n-form is written $\omega = a(x) \mathrm{d}x^1 \wedge \cdots \wedge \mathrm{d}x^n$, and its integral is defined by

$$\int \omega = \int a(x) dx^1 \cdots dx^n.$$

Definition. A form ω is said to be *closed* if $\mathrm{d}\omega = 0$, and *exact* if $\omega = \mathrm{d}\eta$ for some $(p-1)$-form η.

It follows from (2.18) that any exact form is closed. Conversely, any closed form is *locally exact*, i.e. it is exact in a neighborhood of the generic point $x \in \mathbb{R}^n$. This fact, known as the Poincaré lemma, will be used here in the following form.

Theorem. A differential form ω is locally exact, i.e.

$$\omega = \mathrm{d}\eta \tag{2.21}$$

in a neighborhood of the generic point, if and only if it is closed:

$$\mathrm{d}\omega = 0. \tag{2.22}$$

2.1.4 Exact equations

Let us apply the previous notation to the differential form ω in (2.2):

$$\omega \equiv M(x,y)dx + N(x,y)dy.$$

It is a 1-form with $n = 2$ and $x^1 = x, x^2 = y$. Invoking (2.17) and (2.16), the differential of ω is written

$$d\omega = \left(\frac{\partial N}{\partial x} - \frac{\partial M}{\partial y}\right) dx \wedge dy. \qquad (2.23)$$

On the other hand, the form ω is exact, i.e. $\omega = d\Phi$ or

$$M dx + N dy = \frac{\partial \Phi}{\partial x} dx + \frac{\partial \Phi}{\partial y} dy \qquad (2.24)$$

for a function $\Phi(x,y)$ of two variables, if and only if it is closed, i.e. (see (2.23))

$$\frac{\partial N}{\partial x} = \frac{\partial M}{\partial y}. \qquad (2.25)$$

The function Φ is found from equation (2.24) rewritten as a system of differential equations for the unknown function Φ:

$$\frac{\partial \Phi}{\partial x} = M(x,y), \quad \frac{\partial \Phi}{\partial y} = N(x,y). \qquad (2.26)$$

The system (2.26) is *overdetermined* but it is integrable provided that the condition (2.25) is satisfied [2.5]. Its solution is known from differential calculus and is given by

$$\Phi(x,y) = \int_{x_0}^{x} M(z,y) dz + \int_{y_0}^{y} N(x_0, z) dz. \qquad (2.27)$$

Accordingly, the first-order differential equation (2.2),

$$M(x,y)dx + N(x,y)dy = 0,$$

is said to be *exact* if its coefficients satisfy (2.25). The solution of the exact equation is given implicitly by

$$\Phi(x,y) = C, \qquad (2.28)$$

where Φ is defined by (2.27) and C is an arbitrary constant.

2.1.5 Integrating factor

Let the differential equation (2.2) be not exact. Then there exists a function $\mu(x, y)$ such that the equivalent equation,

$$\mu\omega \equiv \mu(M\,\mathrm{d}x + N\,\mathrm{d}y) = 0,$$

is exact. The function $\mu(x, y)$ is known as an *integrating factor* [2.6]. By the definition of exact equations (see (2.25)), the integrating factor satisfies the equation

$$\frac{\partial(\mu N)}{\partial x} = \frac{\partial(\mu M)}{\partial y}.$$

The solution of this equation with respect to $\mu(x,y)$ is not usually simpler than the integration of the original differential equation (2.2). However, the notion of an integrating factor may be useful in particular cases. For example, Lie group analysis provides a general formula for an integrating factor in the case of first-order equations with known infinitesimal symmetries (see Part III).

Example. Consider the equation

$$2xy\,\mathrm{d}x + (y^2 - 3x^2)\,\mathrm{d}y = 0.$$

Its coefficients, $M = 2xy$ and $N = (y^2 - 3x^2)$, do not satisfy (2.25), and hence it is not exact. Here an integrating factor is $\mu = 1/y^4$. Indeed,

$$\frac{\partial(\mu N)}{\partial x} \equiv \frac{\partial}{\partial x}\left(\frac{1}{y^2} - \frac{3x^2}{y^4}\right) = -6\frac{x}{y^4},$$

$$\frac{\partial(\mu M)}{\partial y} \equiv \frac{\partial}{\partial x}\left(\frac{2x}{y^3}\right) = -6\frac{x}{y^4} = \frac{\partial(\mu N)}{\partial x}.$$

Hence, the equation

$$\frac{2x}{y^3}\,\mathrm{d}x + \left(\frac{1}{y^2} - \frac{3x^2}{y^4}\right)\mathrm{d}y = 0$$

is exact. Formula (2.27) yields:

$$F(x, y) = \int_{x_0}^{x} \frac{2z}{y^3}\,\mathrm{d}z + \int_{y_0}^{y}\left(\frac{1}{z^2} - \frac{3x_0^2}{z^4}\right)\mathrm{d}z.$$

Whence, upon integration,

$$F(x, y) = \frac{z^2}{y^3}\bigg|_{x_0}^{x} + \left(-\frac{1}{z} + \frac{x_0^2}{z^3}\right)\bigg|_{y_0}^{y} = \frac{x^2}{y^3} - \frac{1}{y} + K(x_0, y_0),$$

where K is a constant depending on the "initial" values x_0 and y_0. Hence, the solution of the differential equation in question is given implicitly by

$$\frac{x^2}{y^3} - \frac{1}{y} = C, \quad \text{or} \quad x^2 - y^2 = Cy^3,$$

with an arbitrary constant C. To check the solution, we differentiate the solution formula to get $2x - 2yy' = 3Cy^2y'$, whence upon substitution of $C = (x^2 - y^2)/y^3$ we get the original differential equation: $2xy\,\mathrm{d}x + (y^2 - 3x^2)\,\mathrm{d}y = 0$.

2.1. FIRST ORDER EQUATIONS

2.1.6 Homogeneous equations

An equation (2.1) is said to be *homogeneous* if it is invariant under the dilation $\bar{x} = ax$, $\bar{y} = ay$ with the real parameter $a \neq 0$, in other words, if $f(ax, ay) = f(x, y)$ identically in x and y. The standard form of homogeneous equations is

$$y' = f\left(\frac{y}{x}\right). \tag{2.29}$$

Equation (2.29) is solved by the substitution

$$\frac{y}{x} = u, \quad \text{or} \quad y = xu(x).$$

Then it takes the form $xu' + u = f(u)$ and can be solved by separation of variables:

$$\int \frac{du}{f(u) - u} = \int \frac{dx}{x} \equiv \ln x + C.$$

2.1.7 Equations of the type $y' = f\left(\frac{ax+by+c}{kx+ly+m}\right)$

This type can be transformed to the standard homogeneous equation (2.29) by a change of variables:

$$\bar{y} = ax + by + c, \quad \bar{x} = kx + ly + m, \tag{2.30}$$

provided that $al - bk \neq 0$. Then $\bar{x}$ and $\bar{y}$ are functionally independent and

$$\frac{d\bar{y}}{d\bar{x}} = \frac{(a + by')dx}{(k + ly')dx} \equiv \frac{a + by'}{k + ly'}.$$

Hence, the equation in question becomes:

$$\frac{d\bar{y}}{d\bar{x}} = \frac{a + bf(\bar{y}/\bar{x})}{k + lf(\bar{y}/\bar{x})}.$$

If $al - bk = 0$, then $k/a = l/b = \lambda$ and the equation has the form

$$y' = f\left(\frac{ax + by + c}{\lambda(ax + by) + m}\right).$$

After the substitution $\bar{y} = ax + by$ it takes the integrable form:

$$\frac{d\bar{y}}{dx} = h(\bar{y}) \equiv a + bf\left(\frac{\bar{y} + c}{\bar{y} + m}\right).$$

2.1.8 The Riccati equation

The general *Riccati equation* [2.7] is an equation (2.1) with quadratic nonlinearity:

$$y' = P(x) + Q(x)y + R(x)y^2. \tag{2.31}$$

Note that equation (1.32) has this form with constant coefficients P, Q and R.

The distinctive property of the Riccati equation is that, unlike an arbitrary equation (2.1), it admits a nonlinear superposition. Namely, the anharmonic ratio of any four solutions y_1, y_2, y_3, y_4 of the Riccati equation is constant:

$$\frac{y_4 - y_2}{y_4 - y_1} : \frac{y_3 - y_2}{y_3 - y_1} = C. \tag{2.32}$$

Consequently, given any *three* distinct particular solutions $y_1(x), y_2(x), y_3(x)$ of (2.31), its general solution y is expressed rationally in terms of y_1, y_2, y_3 and an arbitrary constant C upon solving (with respect to y) the equation

$$\frac{y - y_2(x)}{y - y_1(x)} : \frac{y_3(x) - y_2(x)}{y_3(x) - y_1(x)} = C. \tag{2.33}$$

Definition. A change of variables, $(x, y) \mapsto (\bar{x}, \bar{y})$, is called an *equivalence transformation* of the Riccati equation if any equation of the form (2.31) is transformed into an equation of the same type with possibly different coefficients. Equations related by an equivalence transformation are said to be *equivalent*.

Theorem. The general equivalence transformation of the Riccati equation comprises of

(i) changes of the independent variable

$$\bar{x} = \phi(x), \qquad \phi'(x) \neq 0, \tag{2.34}$$

(ii) linear-rational transformations of the dependent variable

$$\bar{y} = \frac{\alpha(x)y + \beta(x)}{\gamma(x)y + \delta(x)}, \qquad \alpha\delta - \beta\gamma \neq 0. \tag{2.35}$$

Exercise 1. Find the transformation of the coefficients of the Riccati equation under changes of the independent variable (2.34).

Solution. According to the chain rule of differentiation,

$$\frac{dy}{dx} = \phi'(x)\frac{dy}{d\bar{x}},$$

equation (2.31) is written:

$$\phi'(x)\frac{dy}{d\bar{x}} = P(x) + Q(x)y + R(x)y^2.$$

Upon division by $\phi'(x)$ and substitution $x = \phi^{-1}(\bar{x})$, it becomes

$$\frac{dy}{d\bar{x}} = \overline{P}(\bar{x}) + \overline{Q}(\bar{x})y + \overline{R}(\bar{x})y^2,$$

where

$$\overline{P}(\overline{x}) = \frac{P(\phi^{-1}(\overline{x}))}{\phi'(\phi^{-1}(\overline{x}))}, \quad \overline{Q}(\overline{x}) = \frac{Q(\phi^{-1}(\overline{x}))}{\phi'(\phi^{-1}(\overline{x}))}, \quad \overline{R}(\overline{x}) = \frac{R(\phi^{-1}(\overline{x}))}{\phi'(\phi^{-1}(\overline{x}))}. \tag{2.36}$$

Exercise 2. Show that any Riccati equation is equivalent to the *canonical* one:

$$y' + y^2 = P(x). \tag{2.37}$$

Solution. The linear substitution $\overline{y} = -R(x)y$ (a particular transformation (2.35)) reduces (2.31) to the form $\overline{y}' + \overline{y}^2 = \overline{Q}(x)\overline{y} + \overline{P}(x)$. On the other hand, one can annul the coefficient Q in (2.31) without changing $R(x)$ by means of a substitution of the form $\overline{y} = y + \beta(x)$ with an appropriately chosen function $\beta(x)$. Combining these two substitutions, one ultimately reaches the canonical form (2.37).

The Riccati equation is not, in general, integrable by quadratures. However, it reduces to a linear equation of the second order by the substitution

$$y = u'/u. \tag{2.38}$$

After this substitution, e.g., the canonical equation (2.37) becomes $u'' = P(x)u$.

2.1.9 The Bernoulli equation

The Bernoulli equation [2.8]

$$y' + P(x)y = Q(x)y^n, \quad n \neq 0, \neq 1, \tag{2.39}$$

reduces to the linear equation

$$z' + (1-n)P(x)z = (1-n)Q(x)$$

by the substitution $z = y^{1-n}$.

2.2 Higher-order equations

2.2.1 Integration of the equation $y^{(n)} = f(x)$

The solution of the equation

$$y^{(n)} = f(x)$$

is similar to that of (1.9). Namely, consecutive integration yields:

$$y^{(n-1)} = \int f(x)dx + C_1, \quad y^{(n-2)} = \int dx \int f(x)dx + C_1 x + C_2, \quad \ldots,$$

$$y = \int dx \int dx \ldots \int f(x)dx + C_1 \frac{x^{n-1}}{(n-1)!} + C_2 \frac{x^{n-2}}{(n-2)!} + \cdots + C_{n-1}x + C_n,$$

where $C_1, \ldots, C_n$ are arbitrary constants. The meaning of the multiple indefinite integral is explained in Section 1.2.1.

Example. Consider an equation of the fourth order:

$$y^{iv} = x.$$

The successive stages of the calculation are as follows:

$$y''' = \frac{x^2}{2!} + C_1, \quad y'' = \frac{x^3}{3!} + C_1 x + C_2, \quad y' = \frac{x^4}{4!} + C_1 \frac{x^2}{2!} + C_2 x + C_3,$$

and ultimately,

$$y = \frac{x^5}{5!} + C_1 \frac{x^3}{3!} + C_2 \frac{x^2}{2!} + C_3 x + C_4.$$

2.2.2 Equations $F(y, y', \ldots, y^{(n)}) = 0$. Reduction of order

Let an nth-order equation not explicitly contain the independent variable x:

$$F(y, y', \ldots, y^{(n)}) = 0. \tag{2.40}$$

Then its order is reduced to $n - 1$ by regarding y' as a new variable p depending on y. Indeed, the chain rule yields:

$$y' = p(y), \quad y'' = p\frac{dp}{dy}, \quad y''' = p\frac{d}{dy}\left(p\frac{dp}{dy}\right), \ldots.$$

Hence the given equation takes the form

$$\Phi\left(y, p, \ldots, \frac{d^{n-1}p}{dy^{n-1}}\right) = 0. \tag{2.41}$$

Let equation (2.41) be solved and let its general solution be given, e.g., implicitly:

$$\Psi(y, p, C_1, \ldots, C_{n-1}) = 0. \tag{2.42}$$

Upon solving it with respect to $p = dy/dx$, one obtains a first-order equation with separated variables (and hence integrable by quadrature, see Section 2.1.2):

$$\frac{dy}{dx} = \psi(y, C_1, \ldots, C_{n-1})$$

with $n - 1$ arbitrary constants. Consequently, the integration of an nth-order equation (2.40) is reduced to that of equation (2.41) of the order $n - 1$. Thus we reduced by one the order of the original differential equation in accordance with the following definition.

Definition. The order of an equation $F(x, y, y', \ldots, y^{(n)}) = 0$ is said to be reduced to $n - s$ if it is transformed to the form $\Phi(t, u, u', \ldots, u^{(n-s)}) = 0$, where $u = u(t), 0 < s < n$, by a substitution $t = t(x, y, \ldots, y^{(k)}), u = u(x, y, \ldots, y^{(l)})$, $k, l \leq s$, such that any equation $\Psi(t(x, y, \ldots, y^{(k)}), u(x, y, \ldots, y^{(l)})) = 0$ is integrable by quadratures.

2.2. HIGHER-ORDER EQUATIONS

2.2.3 General and intermediate integrals

The *general integral* of an nth-order differential equation

$$F(x, y, y', \ldots, y^{(n)}) = 0 \tag{2.43}$$

is an equation

$$\Psi(x, y, C_1, \ldots, C_n) = 0 \tag{2.44}$$

defining the whole set of solutions to the equation (2.43) so that any particular solution is obtained by assigning to arbitrary constants $C_1, \ldots, C_n$ definite values, and conversely, every function $y = y(x)$ defined implicitly by (2.44) for arbitrarily chosen values of these constants solves the differential equation (2.43). In other words, the elimination of the constants C_i from the general integral (cf. Section 1.2.2) leads to the differential equation (2.43). Examples are provided, e.g., by (1.7), (1.9) (see also the solution formula in Section 2.2.1), (2.6) and (2.28).

An *intermediate integral* of equation (2.43) has the form

$$\Psi(x, y, y', \ldots, y^{(s)}, C_{s+1}, \ldots, C_n) = 0. \tag{2.45}$$

It is a consequence of the given equation (2.43) and provides a general relation between its solutions and $n - s$ arbitrary constants C. The elimination of the constants C_i from equation (2.45) and its $n - s$ differential consequences (see the Notation in Section 1.2.2),

$$D_x \Psi = 0, \ldots, D_x^{n-s} \Psi = 0,$$

leads to equation (2.43). The formula (2.42), $\Psi(y, y', C_1, \ldots, C_{n-1}) = 0$, provides an example of an intermediate integral with $s = 1$.

The integration of an intermediate integral (2.45) regarded as an sth order differential equation, provides s additional constants of integration and hence yields the general integral of the original equation (2.43). Thus, knowledge of an intermediate integral (2.45) immediately reduces the order of an nth-order equation to $n - s$ (cf. Section 2.2.2).

An intermediate integral involving a single arbitrary constant:

$$\Psi(x, y, y', y'', \ldots, y^{(n-1)}, C) = 0,$$

is termed a *first integral*. For example, the first step in the integration of the nth-order equation in Section 2.2.1 gives a first integral, $y^{(n-1)} - \int f(x) \mathrm{d}x - C_1 = 0$, and further steps yield intermediate integrals involving two, three, etc. constants.

2.2.4 $F(x, y, y', \ldots, y^{(n)}) = 0$ with F a total derivative

The equations mentioned in the title are of the form:

$$F(x, y, y', \ldots, y^{(n)}) \equiv D_x \Phi = 0, \tag{2.46}$$

where $\Phi = \Phi(x, y, y', \ldots, y^{(n-1)})$ and hence (see (1.14))

$$F(x, y, y', \ldots, y^{(n)}) \equiv \frac{\partial \Phi}{\partial x} + \frac{\partial \Phi}{\partial y} y' + \cdots + \frac{\partial \Phi}{\partial y^{(n-1)}} y^{(n)}.$$

One can readily integrate equation (2.46) once to obtain its first integral:

$$\Phi(x, y, y', \ldots, y^{(n-1)}) = C. \tag{2.47}$$

In particular, equations (2.46) of the first order are exact equations discussed in Section 2.1.4. Indeed, invoking (2.26), an exact equation (2.2) can be written in the form

$$\frac{\partial \Phi}{\partial x} + \frac{\partial \Phi}{\partial y} y' = 0.$$

Example 1. The equation $y'' - (x + x^2)e^y y' - (1 + 2x)e^y = 0$ has the form (2.46):

$$y'' - (x + x^2)e^y y' - (1 + 2x)e^y \equiv D_x(y' - (x + x^2)e^y) = 0. \tag{2.48}$$

Hence, the first integral (2.47): $y' = (x + x^2)e^y + C_1$. The substitution $y = -\ln z$ reduces it to the linear equation $z' + C_1 z + x + x^2 = 0$, whence upon integration

$$z = \left(C_2 - \int (x + x^2) e^{C_1 x} dx\right) e^{-C_1 x}.$$

Hence, the solution of the equation (2.48) is obtained by quadrature:

$$y = C_1 x - \ln \left| C_2 - \int (x + x^2) e^{C_1 x} dx \right|.$$

Example 2. Equation (1.21) has the form (2.46):

$$y'' + \frac{1}{x}(y' + y'^3) \equiv D_x\left(\frac{xy'}{\sqrt{1 + y'^2}}\right) = 0.$$

2.2.5 The equations $y^{(n)} = f(y^{(n-1)})$ and $y^{(n)} = f(y^{(n-2)})$

These equations admit a reduction of order in agreement with Section 2.2.2.
Indeed, the equation

$$y^{(n)} = f(y^{(n-1)})$$

is reduced, by the substitution $z = y^{(n-1)}$, to an equation of the first order, $z' = f(z)$. It is readily integrable. Its general integral $\int (1/f(z)) dz = x + C_1$, upon solving with respect to z, yields $z = \psi(x + C_1)$. Invoking the definition of z, one arrives at the equation

$$y^{(n-1)} = \psi(x + C_1)$$

integrated in Section 2.2.1.

2.2. HIGHER-ORDER EQUATIONS

The equation
$$y^{(n)} = f(y^{(n-2)})$$
is reduced, by the substitution $z = y^{(n-2)}$, to an equation of the second order, $z'' = f(z)$. It is a particular case of the equations discussed in Section 2.2.2 and can be easily solved by the substitution $z' = p(z)$. The resulting equation (2.41) has the form $pp' = f(z)$ and yields upon integration: $p^2 = 2\int f(z)\mathrm{d}z + C_1$, or

$$\left(\frac{\mathrm{d}z}{\mathrm{d}x}\right)^2 = 2\int f(z)\mathrm{d}z + C_1, \tag{2.49}$$

whence
$$\int \frac{\mathrm{d}z}{\pm\sqrt{2\int f(z)\mathrm{d}z + C_1}} = x + C_2.$$

Solving it with respect to z and invoking the definition of z, one arrives at an integrable equation of the type discussed in Section 2.2.1:

$$y^{(n-2)} = \psi(x, C_1, C_2).$$

2.2.6 Equations $F(x, y^{(k)}, y^{(k+1)}, \ldots, y^{(n)}) = 0$

The order of the equation given in the title is reduced to $n - k$ by the substitution $z = y^{(k)}$. Indeed, the equation in question becomes $F(x, z, z', \ldots, z^{(n-k)}) = 0$. Provided that its general integral $\Psi(x, z, C_1, \ldots, C_{n-k}) = 0$ is known, one obtains an intermediate integral of the original equation:

$$\Psi(x, y^{(k)}, C_1, \ldots, C_{n-k}) = 0.$$

After solving with respect to $y^{(k)}$, it has the form discussed in Section 2.2.1.

2.2.7 Linear equations. Method of variation of parameters

The general nth-order linear equation has the form

$$L_n[y] \equiv y^{(n)} + a_1(x)y^{(n-1)} + \ldots + a_{n-1}(x)y' + a_n(x)y = f(x), \tag{2.50}$$

where $a_i(x)$ and $f(x)$ are continuous functions in a neighborhood of a generic point x. The term *linear* refers to the fundamental property,

$$L_n[C_1 y_1 + C_2 y_2] = C_1 L_n[y_1] + C_2 L_n[y_2], \tag{2.51}$$

peculiar to the nth order differential operator

$$L_n = D^n + a_1 D^{n-1} + \cdots + a_{n-1} D + a_n,$$

where $D = \mathrm{d}/\mathrm{d}x$. Accordingly, L_n is termed a *linear differential operator*. Equation (2.50) is said to be *homogeneous* if $f(x) = 0$, and *non-homogeneous* otherwise (cf. Section 2.1.1).

The *linear superposition principle* follows from the property (2.51) and states that the general solution of the homogeneous equation, $L_n[y] = 0$,

$$y^{(n)} + a_1(x)y^{(n-1)} + \cdots + a_{n-1}(x)y' + a_n(x)y = 0, \qquad (2.52)$$

is given by the *linear superposition* of n linearly independent particular solutions:

$$y = C_1 y_1(x) + \cdots + C_n y_n(x), \qquad (2.53)$$

where $C_1, \ldots, C_n$ are arbitrary constants. Consequently, any set $y_1(x), \ldots, y_n(x)$ of n linearly independent solutions is termed a *fundamental system* for (2.52).

Theorem. Let a fundamental system of solutions for the homogeneous equation (2.52), $L_n[y] = 0$, be known. Then the general solution to the non-homogeneous equation (2.50), $L_n[y] = f(x)$, can be obtained by quadratures.

Proof. The solution can be obtained by the general method of *variation of parameters* due to Lagrange. Namely, just as in the case of equations of the first order (Section 2.1.1), one replaces the constants C_i in (2.53) by functions $u_i(x)$:

$$y = u_1(x)y_1(x) + \cdots + u_n(x)y_n(x).$$

Lagrange's method [2.9] provides the following relations for determining the unknown functions $u_i(x)$:

$$y_1 \frac{du_1}{dx} + \cdots + y_n \frac{du_n}{dx} = 0,$$

$$y_1' \frac{du_1}{dx} + \cdots + y_n' \frac{du_n}{dx} = 0,$$

$$\cdots \cdots \cdots \cdots \cdots \cdots \cdots \cdots \cdots$$

$$y_1^{(n-2)} \frac{du_1}{dx} + \cdots + y_n^{(n-2)} \frac{du_n}{dx} = 0,$$

$$y_1^{(n-1)} \frac{du_1}{dx} + \cdots + y_n^{(n-1)} \frac{du_n}{dx} = f(x). \qquad (2.54)$$

Since $y_1, \ldots, y_n$ are known and they are linearly independent, equations (2.54) can be solved with respect to the derivatives of the unknown functions and written in the form integrable by quadrature:

$$\frac{du_k}{dx} = \psi_k(x), \quad k = 1, \ldots, n.$$

2.2.8 Linear equations with constant coefficients

The general solution of the homogeneous equation with constant coefficients,

$$y^{(n)} + a_1 y^{(n-1)} + \cdots + a_{n-1} y' + a_n y = 0, \quad a_1, \ldots, a_n = \text{const.}, \qquad (2.55)$$

2.2. HIGHER-ORDER EQUATIONS

was given by Euler [2.10]. He introduced particular solutions of the form [2.11]

$$y = e^{\lambda x}, \quad \lambda = \text{const.},$$

and reduced the nth-order differential equation (2.55) to an auxiliary algebraic equation of the nth degree,

$$P_n[\lambda] \equiv \lambda^n + a_1 \lambda^{n-1} + \cdots + a_{n-1} \lambda + a_n = 0. \tag{2.56}$$

The polynomial $P_n(\lambda)$ is known as the *characteristic polynomial* for (2.52). Accordingly, the auxiliary algebraic equation (2.56) is termed the *characteristic equation*. Let $\lambda_1, \ldots, \lambda_n$ be the roots of the characteristic equation (2.56). Then the particular solutions $e^{\lambda_1 x}, \ldots, e^{\lambda_n x}$ provide a fundamental system. According to the superposition principle (2.53), the general solution of the equation (2.55) with constant coefficients is given by the linear combination

$$y = C_1 e^{\lambda_1 x} + \cdots + C_n e^{\lambda_n x}, \tag{2.57}$$

provided that the roots $\lambda_1, \ldots, \lambda_n$ are distinct.

Let (2.56) have a complex root $\lambda = \alpha + \beta i$, and hence its complex conjugate $\overline{\lambda} = \alpha - \beta i$ as well, where $i = \sqrt{-1}$. Invoking Euler's relation in the theory of complex numbers, the corresponding solutions are written:

$$y = e^{\alpha x}(\cos \beta x + i \sin \beta x), \quad \overline{y} = e^{\alpha x}(\cos \beta x - i \sin \beta x).$$

Since any linear combination of solutions is again a solution, one can replace the complex solutions by the real ones: $y_1 = (y + \overline{y})/2$ and $y_2 = (y - \overline{y})/(2i)$. Thus, the pair of conjugate complex roots provide two distinct real solutions:

$$y_1(x) = e^{\alpha x} \cos \beta x, \quad y_2(x) = e^{\alpha x} \sin \beta x. \tag{2.58}$$

Example 1. The equation (1.47) for free ($f(t) = 0$) oscillations of a damped system, upon setting $l/m = 2b$, $k/m = c$, is written:

$$\frac{d^2 y}{dt^2} + 2b \frac{dy}{dt} + cy = 0, \quad b, c > 0,$$

where b and c are positive constants. Consider the case of a small damping force so that $b^2 < c$. Then the characteristic polynomial $P_2(\lambda) = \lambda^2 + 2b\lambda + c$ has the complex zeroes $\lambda = -b \pm i\sqrt{c - b^2}$. Accordingly, the fundamental system of solutions (2.58) has the form

$$y_1(t) = e^{-bt} \cos\left(\sqrt{c - b^2}\, t\right), \quad y_2(t) = e^{-bt} \sin\left(\sqrt{c - b^2}\, t\right).$$

Hence, the general solution (2.57) is

$$y = e^{-bt} \left(C_1 \cos\left(\sqrt{c - b^2}\, t\right) + C_2 \sin\left(\sqrt{c - b^2}\, t\right)\right).$$

Example 2. For equation (1.52), $u^{iv} = \alpha^4 u$, $\alpha = $ const., the characteristic equation $\lambda^4 - \alpha^4 = 0$ has four distinct roots, $\lambda_1 = \alpha$, $\lambda_2 = -\alpha$, $\lambda_3 = \alpha i$, and $\lambda_4 = -\alpha i$. Thus, one arrives at formula (1.53) for the general solution.

Consider now the case when multiple roots occur. For example, let λ_1 be repeated s times. Then the corresponding solution is given by

$$y_1 = (C_1 + C_2 x + \cdots + C_s x^{s-1})e^{\lambda_1 x} \tag{2.59}$$

with arbitrary constants C. Taking into account all multiple roots, one obtains the following modification of formula (2.57) for the general solution:

$$y = q_1(x)e^{\lambda_1 x} + \ldots + q_r(x)e^{\lambda_r x}. \tag{2.60}$$

Here the $q'(x)$s are polynomials with arbitrary coefficients, of degrees equal to the order of multiplicity of the corresponding root λ.

Remark. In the case of complex $\lambda_1 = \alpha_1 + \beta_1 i$, the right-hand side of expression (2.59) (and hence the first term of (2.60)) should be replaced by

$$(C_1 + \cdots + C_s x^{s-1})e^{\alpha_1 x}\cos(\beta_1 x) + (C_{s+1} + \cdots + C_{2s} x^{s-1})e^{\alpha_1 x}\sin(\beta_1 x).$$

Example 3. For the equation $y^{iv} + 2y'' + y = 0$, the characteristic equation $\lambda^4 + 2\lambda^2 + 1 = 0$ has repeated imaginary roots, $\lambda = i$ and $\overline{\lambda} = i$. Hence, the general solution is $y = (C_1 + C_2 x)\cos x + (C_3 + C_4 x)\sin x$.

Exercise. Solve equation (1.47) for undamped oscillations ($l = 0$) of a particle subject to the periodic external force $f(t) = \sin t$. Let, for the sake of brevity, $m = k = 1$.

Solution. The equation to be solved is $y'' + y = \sin t$. The homogeneous equation $y'' + y = 0$ has the fundamental system of solutions $y_1 = \cos t$, $y_2 = \sin t$ (Example 1). According to Theorem 2.2.7, we take $y = u_1(t)\cos t + u_2(t)\sin t$, where the unknown functions $u_1(t)$ and $u_2(t)$ are to be determined from the relations (2.54) with $n = 2$:

$$y_1 u_1' + y_2 u_2' = 0, \quad y_1' u_1' + y_2' u_2' = f,$$

where the prime denotes differentiation. In our case, these equations are written

$$u_1'\cos t + u_2'\sin t = 0, \quad u_2'\cos t - u_1'\sin t u_1' = \sin t.$$

Solving with respect to u_1' and u_2', one obtains $u_1' = -\sin^2 t$, $u_2' = \cos t \sin t$, whence upon integration $u_1 = C_1 - (t - \cos t \sin t)/2$, $u_2 = C_2 - (\cos^2 t)/2$. Hence the solution

$$y = -\frac{1}{2}t\cos t + C_1 \cos t + C_2 \sin t, \quad C_1, C_2 = \text{const}.$$

2.3 Systems of first-order linear equations

Euler's method discussed above applies also to the general system of linear homogeneous equations of the first order with constant coefficients [2.12]:

$$(ED + A)\mathbf{y} = 0. \tag{2.61}$$

2.3. SYSTEMS OF FIRST-ORDER LINEAR EQUATIONS

Here $y = (y^1, \ldots, y^n)$ denotes the dependent variables (for the vector notation, see the footnote in Section 1.4.5), $D = d/dx$, $A = \| a_{kj} \|$ is a constant $n \times n$ matrix so that $Ay = \sum_{j=1}^{n} a_{kj} y^j$, and E is a unit $n \times n$ matrix. Euler's formula for particular solutions is now written

$$y = e^{\lambda x} l, \quad \lambda = \text{const}. \tag{2.62}$$

Here $l = (l^1, \ldots, l^n)$ is an unknown constant vector to be determined from equation (2.61). Substitution of (2.62) into (2.61) yields:

$$(A + \lambda E)l = 0. \tag{2.63}$$

The system of linear homogeneous equations (2.63) has a solution $l \neq 0$ if and only if the determinant $|A + \lambda E| = \det(A + \lambda E)$ vanishes. Subsequently, the *characteristic polynomial* P_n and the *characteristic equation* for the system (2.61) are defined by

$$P_n(\lambda) \equiv |A + \lambda E| = 0. \tag{2.64}$$

Let the characteristic equation (2.64) have distinct roots, $\lambda_1, \ldots, \lambda_n$. Then one obtains precisely n linearly independent solutions $l_{(1)}, \ldots, l_{(n)}$ of (2.63), and hence the following *fundamental system* of solutions of the system (2.61):

$$y_{(1)} = e^{\lambda_1 x} l_{(1)}, \ \ldots, \ y_{(n)} = e^{\lambda_n x} l_{(n)}. \tag{2.65}$$

The general solution to the system (2.61) is given by

$$y = C_1 e^{\lambda_1 x} l_{(1)} + \cdots + C_n e^{\lambda_n x} l_{(n)}. \tag{2.66}$$

Let equation (2.64) have a complex root, $\lambda = \alpha + \beta i$ (and hence its complex conjugate). Then the one obtains from equation (2.63) a complex vector, $l = p + iq$. The corresponding solution (2.62) splits into two real solutions:

$$y_{(1)} = e^{\alpha x}(p \cos \beta x - q \sin \beta x), \quad y_{(2)} = e^{\alpha x}(p \sin \beta x + q \cos \beta x). \tag{2.67}$$

Exercise. Solve the system

$$\frac{dy^1}{dt} = y^2, \quad \frac{dy^2}{dt} = -y^1. \tag{2.68}$$

Solution. Here $y = (y^1, y^2)$ and $a_{11} = a_{22} = 0$, $a_{12} = -1$, $a_{21} = 1$. The characteristic equation has a pair of complex roots, $\lambda = i$, $\overline{\lambda} = -i$. The solution of (2.63) yields $l = p + iq$ with $p = (1, 0)$, $q = (0, 1)$. The formulae (2.67) provide a fundamental system consisting of $y_{(1)} = (y^1_{(1)}, y^2_{(1)})$ and $y_{(2)} = (y^1_{(2)}, y^2_{(2)})$, where $y^1_{(1)} = \cos t$, $y^2_{(1)} = -\sin t$ and $y^1_{(2)} = \sin t$, $y^2_{(2)} = \cos t$, respectively. Hence, the general solution has the form

$$y^1 = C_1 \cos t + C_2 \sin t, \quad y^2 = C_2 \cos t - C_1 \sin t. \tag{2.69}$$

Alternative solution. Differentiating the first equation of (2.68) and substituting dy^2/dt from the second equation, one reduces the problem to integration of a single second-order equation, $d^2y^1/dt^2 + y^1 = 0$. Its fundamental system is $y^1_{(1)} = \cos t$, $y^1_{(2)} = \sin t$ (see Exercise in Section 2.2.8), and by Section 2.2.8, $y^1 = C_1 \cos t + C_2 \sin t$. The function y^2 is obtained by differentiation, $y^2 = dy^1/dt$, i.e. $y^2 = C_2 \cos t - C_1 \sin t$.

Problems

2.1. Solve the equations (i) $y' - \frac{y}{x} = x$; (ii) $y' - y\cos x = 0$; (iii) $y' - y\cos x = x$;
(iv) $y' + \tan x = \cos x$; (v) $y' - \frac{y}{x} = A\ln(x^2) + 2Bx^2 + C$, where $A, B, C = $ const.

2.2. Integrate the equation (1.40), $d\tau/dt = k[8 - \tau - 3(t-t_0)/2]$.

2.3. Prove that formula (2.7) is unaltered when $y_0(x)$, given in Section 2.1.1, is replaced by any other particular solution $\bar{y}_0(x) \neq $ const. of equation (2.4).

2.4. Integrate the equation $x(y^2 - l^2)^\alpha dx + y(x^2 - k^2)^\beta dy = 0$, where k, l, α, β are arbitrary real constants.

2.5. Derive the nonlinear superposition (2.14) and prove its invariance under a change of a particular solution $y_0(x)$.

2.6. Derive formula (2.23):
$$d(M\,dx + N\,dy) = \left(\frac{\partial N}{\partial x} - \frac{\partial M}{\partial y}\right) dx \wedge dy.$$

2.7. Check that the function $\Phi(x, y)$ defined by (2.27) solves equations (2.26).

2.8. Check that the following equation is exact and integrate it:
$$\left(\frac{1}{x} - \frac{y^2}{x^2}\right) dx + \frac{2y}{x} dy = 0.$$

2.9. Find the Riccati equation obtained from $y' = e^{-x} - e^x y^2$ by the substitution $\bar{x} = e^x$.

2.10. Find the transformation of the coefficients of the Riccati equation (2.31) under linear-rational transformations (2.35).

2.11*. Prove Theorem 2.1.8.

2.12. Transform the general Riccati equation to the canonical form (2.37).

2.13*. A change of variables, $(x, y) \mapsto (\bar{x}, \bar{y})$, is called an *equivalence transformation* of the general linear equation if any equation of the form (2.3) is transformed into an equation of the same type with possibly different coefficients. Equations related by an equivalence transformation are said to be *equivalent*.

Show that the most general equivalence transformation of the linear equation (2.3) comprises changes of the independent variable (2.34) and non-homogeneous linear transformations of the dependent variable, $\bar{y} = \alpha(x)y + \beta(x)$, $\alpha(x) \neq 0$.

2.14. Integrate the Riccati equation (2.31) with constant coefficients P, Q, and R, e.g. equations (1.32) and (1.58).

2.15*. Derive equations (2.54) (i) for $n = 2$; (ii) for arbitrary n.

2.16. Solve (i) the equation $y'' + \omega^2 y = 0$ of free harmonic oscillations, where $\omega \neq 0$ is an arbitrary constant and $y = y(t)$; (ii) the equation $y'' + \omega^2 y = f(t)$ of undamped oscillations of a particle subject to an arbitrary external force $f(t)$.

2.17. Solve the equation $y''' + y = 0$, where $y = y(x)$.

Chapter 3

General properties of solutions

Existence theorems furnish the core of the general theory of differential equations. Furthermore, the classical existence and uniqueness theorem for systems of ordinary differential equations of the first order plays a central role in the theory of Lie groups, their invariants and invariant equations.

3.1 Existence and uniqueness theorems

3.1.1 Classical solutions. Cauchy's problem

Throughout the book the following classical definition of solutions of differential equations is assumed.

Definition. A function $y = \phi(x)$, defined in a neighborhood of x_0 and continuously differentiable n times, is said to be a solution of a differential equation $F(x, y, y', \ldots, y^{(n)}) = 0$ if $F(x, \phi(x), \phi'(x), \ldots, \phi^{(n)}(x)) = 0$ identically in x from an interval $(x_0 - \varepsilon, x_0 + \varepsilon)$, $\varepsilon > 0$. Since any function $y = y(x)$ represents a curve in the (x, y) plane, solutions of ordinary differential equations are also termed *integral curves*.

The first systematic investigations on the existence of solutions of differential equations are due to Cauchy [3.1]. Note that, e.g. in the case of equation (1.6)

$$\frac{dy}{dx} = f(x)$$

with continuous $f(x)$, one can readily obtain the solution that assumes a given value y_0 at $x = x_0$, by means of formula (1.7). The solution of this *initial value problem* is unique, is defined in a neighborhood of the point x_0 and is given by

$$y(x) = y_0 + \int_{x_0}^{x} f(t)dt.$$

Cauchy extended this result by proving the existence of solutions of the initial value problem for the general first-order equation [3.2]:

$$\frac{dy}{dx} = f(x,y), \quad y|_{x=x_0} = y_0, \tag{3.1}$$

where $f(x,y)$ is a continuous function in a neighborhood of the point (x_0, y_0) in the (x,y) plane; the notation $|_{x=x_0}$ means evaluated at $x = x_0$ (cf. (1.31)). Consequently, initial value problems are often referred to as the *Cauchy problem*.

Thus, Cauchy's result states the existence of integral curves passing through any given point (x_0, y_0). However, the solution need not be unique if only the continuity of the right-hand side, $f(x,y)$, is required. For example, the initial value problem

$$\frac{dy}{dx} = 2\sqrt{|y|}, \quad y|_{x=x_0} = 0, \tag{3.2}$$

has two solutions, $y = 0$ and $y = |x - x_0|(x - x_0)$.

Therefore, Cauchy continued his investigations and proved the existence and uniqueness theorem for *analytic* differential equations. His proof is based on the ingenious *method of majorants* (which he called the *calculus of limits*) [3.3] and is applicable to both ordinary and partial differential equations.

An alternative approach, known as the *method of successive approximations* [3.4], furnishes a simple proof of the existence and uniqueness theorem for ordinary differential equations satisfying the so-called *Lipschitz condition*. This approach enjoys widespread acceptance in contemporary university texts.

3.1.2 The method of successive approximations

Let $f(x,y)$ be a single-valued continuous function of two variables defined in a rectangular domain $|x - x_0| \leq a$, $|y - y_0| \leq b$, where $a, b > 0$.

Definition. We say that a function $f(x,y)$ satisfies the Lipschitz condition if there exists a positive constant K such that

$$|f(x,y_1) - f(x,y_2)| \leq K|y_1 - y_2| \tag{3.3}$$

for any points (x, y_1) and (x, y_2) within the preceding domain.

Theorem. *Let the right-hand side of the differential equation* (3.1), $f(x,y)$, *satisfy the Lipschitz condition* (3.3). *Then the initial value problem* (3.1) *has one and only one solution* $y = \phi(x)$ *defined in a neighborhood of* x_0.

Proof. Since $f(x,y)$ is continuous in a closed domain, it is bounded, i.e. $|f(x,y)| \leq M$ when $|x - x_0| \leq a$, $|y - y_0| \leq b$. Let h be the smaller of a and b/M, and let x and y vary within the intervals $x_0 - h \leq x \leq x_0 + h$ and $y_0 - b \leq y \leq y_0 + b$, respectively.

Consider the sequence of functions $\{y_n(x)\}$ defined by

$$y_1(x) = y_0 + \int_{x_0}^{x} f(t, y_0) dt, \quad y_{n+1}(x) = y_0 + \int_{x_0}^{x} f(t, y_n(t)) dt, \; n = 1, 2, \ldots. \tag{3.4}$$

3.1. EXISTENCE AND UNIQUENESS THEOREMS

The terms of the sequence $\{y_n\}$ are contained in the interval $(y_0 - b, y_0 + b)$ and hence the process of successive approximations (3.4) can be continued indefinitely. The proof is accomplished by induction. Indeed, $|y_1 - y_0| = |\int_{x_0}^{x} f(t, y_0) dt| \leq M|x - x_0| \leq Mh \leq b$. Furthermore, the inductive assumption $|y_n - y_0| \leq b$ implies $|f(t, y_n(t))| \leq M$, whence $|y_{n+1} - y_0| = |\int_{x_0}^{x} f(t, y_n(t)) dt| \leq Mh \leq b$.

Let us prove now that the sequence (3.4) is absolutely and uniformly convergent and that its limit $y(x) = \lim_{n \to \infty} y_n(x)$ is a continuous function. The formula

$$y_n(x) = y_0 + \sum_{s=1}^{n} [y_s(x) - y_{s-1}(x)]$$

reduces the convergence problem of the sequence $\{y_n\}$ to that of the series

$$\sum_{s=1}^{\infty} [y_s(x) - y_{s-1}(x)]. \tag{3.5}$$

For the latter, one obtains by induction the following estimation of its terms:

$$|y_s(x) - y_{s-1}(x)| \leq \frac{MK^{s-1}}{s!} |x - x_0|^s, \quad s = 1, 2, \ldots. \tag{3.6}$$

Indeed, it was verified above that $|y_1 - y_0| \leq M|x - x_0|$. Hence, (3.6) is true for $s = 1$. Further, the Lipschitz condition (3.3) yields

$$|y_{s+1}(x) - y_s(x)| \leq \left| \int_{x_0}^{x} |f(t, y_s(t)) dt - f(t, y_{s-1}(t))| dt \right|$$

$$\leq K \left| \int_{x_0}^{x} |y_s(t) - y_{s-1}(t)| dt \right|.$$

Hence, using the inductive assumption that (3.6) is true for s,

$$|y_s(t) - y_{s-1}(t)| \leq \frac{MK^{s-1}}{s!} |t - x_0|^s,$$

one arrives at (3.6) for $s + 1$ (consider separately $x \leq x_0$ and $x \geq x_0$):

$$|y_{s+1}(x) - y_s(x)| \leq \frac{MK^s}{s!} \left| \int_{x_0}^{x} |t - x_0|^s dt \right| = \frac{MK^s}{(s+1)!} |x - x_0|^{s+1}.$$

Let us replace, in (3.6), $|x - x_0|$ by its maximal value h and consider the series

$$\sum_{s=1}^{\infty} u_s \equiv \sum_{s=1}^{\infty} \frac{MK^{s-1} h^s}{s!}. \tag{3.7}$$

According to the *ratio test*,

$$\lim_{s \to \infty} \frac{u_{s+1}}{u_s} = \lim_{s \to \infty} \frac{s! MK^s h^{s+1}}{(s+1)! MK^{s-1} h^s} = \lim_{s \to \infty} \frac{Kh}{s+1} = 0 < 1,$$

the series (3.7) converges. Since the absolute value of each term of (3.5) is less than the corresponding term of the convergent positive series (3.7), the series (3.5) is absolutely and uniformly convergent. Hence so is (3.4). Its limit

$$y(x) = \lim_{n \to \infty} y_n(x) \qquad (3.8)$$

is a continuous function in the interval $(x_0 - h, x_0 + h)$.

Invoking (3.4), formula (3.8) is written

$$y(x) = y_0 + \int_{x_0}^{x} f(t, \lim_{n \to \infty} y_n(t)) \mathrm{d}t = y_0 + \int_{x_0}^{x} f(t, y(t)) \mathrm{d}t.$$

Since the integral of a continuous function can be differentiated with respect to its upper limit, it follows that $y'(x) = f(x, y(x))$. Hence, the function $y(x)$ defined be (3.8) is continuously differentiable, solves the differential equation (3.1) and reduces to y_0 at $x = x_0$. Thus, Cauchy's problem (3.1) has a solution given by the function (3.8) with successive approximations (3.4).

To complete the proof of the theorem, it remains to demonstrate the uniqueness of the solution. In other words, one has to show that $Y(x) = y(x)$, where $y(x)$ is given by (3.8), whenever $Y(x)$ solves the problem (3.1). Let $Y(x)$ be any solution. Then

$$\frac{\mathrm{d}Y(t)}{\mathrm{d}t} = f(t, Y(t))$$

identically in the interval $x_0 - h \leq t \leq x_0 + h$. The integration of both sides of this equation from x_0 to x, taking into account the initial condition $Y(x_0) = y_0$, yields the *integral equation*

$$Y(x) = y_0 + \int_{x_0}^{x} f(t, Y(t)) \mathrm{d}t. \qquad (3.9)$$

Consider the interval $x_0 \leq x \leq x_0 + h$ (the interval $x_0 - h \leq x \leq x_0$ is treated similarly). Equations (3.9), (3.4) and the Lipschitz condition (3.3) yield:

$$|Y(x) - y_n(x)| = \left| \int_{x_0}^{x} [f(t, Y(t)) - f(t, y_{n-1}(t))] \mathrm{d}t \right| \leq K \int_{x_0}^{x} |Y(t) - y_{n-1}(t)| \mathrm{d}t.$$

Let $n = 1$, then

$$|Y(x) - y_1(x)| \leq K \int_{x_0}^{x} |Y(t) - y_0| \mathrm{d}t,$$

whence by virtue of $|Y - y_0| \leq b$, one obtains $|Y(x) - y_1(x)| \leq Kb(x - x_0)$. Iterating,

$$|Y(x) - y_2(x)| \leq \frac{1}{2} b K^2 (x - x_0)^2, \quad \ldots$$

one ultimately reaches a sequence of inequalities:

$$|Y(x) - y_n(x)| \leq \frac{K^n (x - x_0)^n}{n!} b.$$

It follows that $Y(x) = \lim_{n \to \infty} y_n(x) = y(x)$, i.e. the solution is unique [3.5].

3.1.3 Systems of first-order equations

Consider the general system of first-order ordinary differential equations:

$$\frac{dy^i}{dx} = f^i(x, y^1, y^2, \ldots, y^n), \quad i = 1, 2, \ldots, n, \tag{3.10}$$

where the functions f^i are continuous in a neighborhood of $x_0, y_0^1, \ldots, y_0^n$. The Lipschitz condition to be imposed on each function f^i is a natural generalization of that for the case of a single dependent variable y, viz.

$$|f^i(x, y_1^1, y_1^2, \ldots, y_1^n) - f^i(x, y_2^1, y_2^2, \ldots, y_2^n)| \leq K_1^i |y_1^1 - y_2^1|$$
$$+ K_2^i |y_1^2 - y_2^2| + \cdots + K_n^i |y_1^n - y_2^n|, \quad i = 1, \ldots, n, \tag{3.11}$$

where $K_1^i, \ldots, K_n^i$ are appropriately chosen constants.

Let us use the vector notation $\boldsymbol{y} = (y^1, \ldots, y^n)$, $\boldsymbol{f} = (f^1, \ldots, f^n)$ and write an initial value problem for (3.10) in the compact form:

$$\frac{d\boldsymbol{y}}{dx} = \boldsymbol{f}(x, \boldsymbol{y}), \quad \boldsymbol{y}|_{x=x_0} = \boldsymbol{y}_0, \tag{3.12}$$

where $\boldsymbol{y}_0 = (y_0^1, \ldots, y_0^n)$. Thus, $\boldsymbol{y}$ is an n-tuple of dependent variables, the ith one of which is denoted by y^i. We will adopt also the usual notation $\boldsymbol{y} \in \mathbb{R}^n$ and will call y^i the ith coordinate of the vector $\boldsymbol{y}$.

The definition of classical solutions (see Section 3.1.1) applies to systems of differential equation as well with the natural replacement of the single variable y and a function F by the vector $\boldsymbol{y}$ and an n-dimensional vector-function $\boldsymbol{F} = (F^1, \ldots, F^n)$, respectively.

Furthermore, the Lipschitz conditions (3.11) can be conveniently written in compact form for the vector-function $\boldsymbol{f}$ after introducing a *norm* ("length") of n-dimensional vectors, e.g.

$$\|\boldsymbol{y}\| = |y^1| + \cdots + |y^n|, \quad \|\boldsymbol{y}_1 - \boldsymbol{y}_2\| = |y_1^1 - y_2^1| + \cdots + |y_1^n - y_2^n|.$$

Definition. Let $\boldsymbol{f}(x, \boldsymbol{y})$ be a single-valued continuous vector-function defined in a domain $|x - x_0| \leq a$, $\|\boldsymbol{y} - \boldsymbol{y}_0\| \leq b$, where $a, b > 0$. The function $\boldsymbol{f}$ satisfies the Lipschitz condition if there exists a positive constant K such that

$$\|\boldsymbol{f}(x, \boldsymbol{y}_1) - \boldsymbol{f}(x, \boldsymbol{y}_2)\| \leq K \|\boldsymbol{y}_1 - \boldsymbol{y}_2\| \tag{3.13}$$

for any values of $(x, \boldsymbol{y}_1)$ and $(x, \boldsymbol{y}_2)$ within the preceding domain.

The method of successive approximations presented in the preceding section can be extended without difficulty to systems of differential equations of the first order. Hence, the following existence and uniqueness theorem is true.

Theorem. Let the function $\boldsymbol{f}(x, \boldsymbol{y})$ satisfy the Lipschitz condition (3.13). Then problem (3.12) has a unique solution defined in a neighborhood of x_0.

Remark. It follows that the general solution of a system of n first-order differential equations (3.10) depends precisely on n arbitrary constants $C_1,\ldots,C_n$, e.g. on arbitrarily chosen initial values $y_0^1,\ldots,y_0^n$ of the dependent variables at $x=x_0$. Accordingly, the general solution to (3.10) is written [3.6]

$$y^i = \phi^i(x,C_1,\ldots,C_n), \quad i=1,2,\ldots,n. \tag{3.14}$$

3.1.4 Higher order equations

A single equation of the nth order,

$$y^{(n)} = f(x,y,y',\ldots,y^{(n-1)}), \tag{3.15}$$

can be rewritten in the form (3.10) by considering the function y and its successive derivatives as new dependent variables, $y^1 = y, y^2 = y',\ldots,y^n = y^{(n-1)}$. Then equation (3.15) is equivalent to the system of n equations of the first order:

$$\frac{dy^1}{dx} = y^2, \quad \frac{dy^2}{dx} = y^3, \quad \ldots, \quad \frac{dy^{(n-1)}}{dx} = y^n, \quad \frac{dy^n}{dx} = f(x,y^1,\ldots,y^n). \tag{3.16}$$

The initial value problem (3.12) for the system (3.16) corresponds to that for the nth-order equation (3.15) with the following initial data:

$$y|_{x=x_0} = y_0, \quad \frac{dy}{dx}\bigg|_{x=x_0} = y_0', \quad \ldots, \quad \frac{d^{n-1}y}{dx^{n-1}}\bigg|_{x=x_0} = y_0^{(n-1)}. \tag{3.17}$$

The Lipschitz condition (3.13) for the system (3.16) is equivalent to that for equation (3.15), viz.

$$|f(x,y_1,y_1',\ldots,y_1^{(n-1)}) - f(x,y_2,y_2',\ldots,y_2^{(n-1)})| \leq K_1|y_1 - y_2|$$
$$+ K_2|y_1' - y_2'| + \cdots + K_n|y_1^{(n-1)} - y_2^{(n-1)}|, \quad K_1,\ldots,K_n = \text{const.} \tag{3.18}$$

Section 3.1.3 provides the following existence and uniqueness theorem.

Theorem. Let the function f in the differential equation (3.15) be continuous in a neighborhood of $x_0, y_0, y_0',\ldots,y_0^{(n-1)}$ and satisfy the Lipschitz condition (3.18). Then the initial value problem (3.15), (3.17) has a unique solution defined in a neighborhood of x_0.

Remark. It follows that the general solution of nth-order differential equations (3.15) depends precisely on n arbitrary constants $C_1,\ldots,C_n$.

3.1.5 Analytic solutions. The method of majorants

Let a function $f(x,y)$ be *locally analytic*, i.e. expandable in a power series in x and y that converges near the point (x_0, y_0) of the (x,y) plane. Let us reduce, for the sake of brevity, the initial point x_0 to zero by merely changing x into $x-x_0$. According to Section 3.1.2, the Cauchy problem (3.1) has a unique solution $y=y(x)$ defined in a neighborhood of the initial point $x=0$.

3.1. EXISTENCE AND UNIQUENESS THEOREMS

Theorem. For the Cauchy problem

$$\frac{dy}{dx} = f(x,y), \quad y|_{x=0} = y_0, \tag{3.19}$$

with a locally analytic function $f(x,y)$, the solution $y = y(x)$ is analytic near zero.

Proof. Let us look for a power series solution:

$$y(x) = y_0 + y'_0 x + y''_0 \frac{x^2}{2!} + \cdots + y_0^{(s)} \frac{x^s}{s!} + \cdots.$$

Its coefficients can be determined from the differential equation (3.19) and those derived from it by repeated differentiations,

$$y' = f(x,y), \quad y'' = D_x f(x,y), \quad y''' = D_x^2 f(x,y), \quad \ldots, \tag{3.20}$$

where D_x is the total derivation (1.14) defined in the notation of Section 1.2.2:

$$D_x = \frac{\partial}{\partial x} + y'\frac{\partial}{\partial y} + y''\frac{\partial}{\partial y'} + \cdots + y^{(s+1)}\frac{\partial}{\partial y^{(s)}} + \cdots.$$

Equations (3.20) can be written in an expanded form, e.g. (cf. (1.12))

$$y'' = \frac{\partial f(x,y)}{\partial x} + \frac{\partial f(x,y)}{\partial y} y'.$$

Setting $x = 0$ and $y = y_0$ in equations (3.20), one calculates consecutively:

$$y'_0 = \left(f\right)_0 \equiv f(0,y_0), \quad y''_0 = \left(D_x f\right)_0 \equiv \frac{\partial f(0,y_0)}{\partial x} + \frac{\partial f(0,y_0)}{\partial y} f(0,y_0), \ldots.$$

Proceeding in this manner, one ultimately constructs the power series

$$y(x) = y_0 + \left(f\right)_0 \frac{x}{1!} + \left(D_x f\right)_0 \frac{x^2}{2!} + \cdots + \left(D_x^{(s-1)} f\right)_0 \frac{x^s}{s!} + \cdots. \tag{3.21}$$

By construction, formula (3.21) provides the unique power series solution to the problem (3.19). Consequently, the question of the existence of an analytic solution is reduced to proving the convergence of the above series.

The crucial idea of the method of majorants [3.3] in proving the existence of analytic solutions is to associate with (3.19) an auxiliary, *majorant problem*:

$$\frac{dY}{dx} = F(x,Y), \quad Y|_{x=0} = Y_0. \tag{3.22}$$

It is obtained by replacing the function $f(x,y)$ in (3.19) by a *dominant function* $F(x,Y) \gg f(x,y)$ and the initial value y_0 by $Y_0 \gg y_0$. The dominance relation $F(x,Y) \gg f(x,y)$ means that $F(x,Y)$ is an analytic function near $x = 0, Y = Y_0$ such that the coefficients for the development of $F(x,Y)$ into power series in

x, Y are positive numbers greater than the absolute values of the corresponding coefficients for the development of $f(x, y)$ in x, y. In particular, the relation $Y_0 \gg y_0$ simply means $Y_0 \geq |y_0|$.

Let the solution $Y = Y(x)$ to the majorant problem be analytic. Then the Taylor expansion for $Y(x)$ has the form (3.21) with y_0 and f replaced by Y_0 and F, respectively. Since $F \gg f$, it follows that $Y(x) \gg y(x)$. Consequently the development (3.21) for $y(x)$ converges whenever that for $Y(x)$ does.

Thus, to complete the proof of the theorem, it remains to prove the existence of an analytic solution of an auxiliary problem (3.22). It seems that the latter problem is not simpler than the first one. Indeed, (3.22) has the same form as the original initial value problem (3.19). The major difference, however, is that the auxiliary problem is not uniquely determined and hence one can choose a 'simple' dominant function F.

Indeed, let $f(x, y)$ be analytic in a domain $|x| \leq a$, $|y - y_0| \leq b$, and let M be the upper limit of $|f(x, y)|$ therein. One can choose

$$F(x, Y) = M \left(1 - \frac{x}{a}\right)^{-1} \left(1 - \frac{Y - Y_0}{b}\right)^{-1} \qquad (3.23)$$

as a dominant function. It is analytic when $|x| < a$, $|Y - Y_0| < b$, and the corresponding majorant problem (3.22) is readily integrable by separation of variables. The integration yields:

$$Y(x) = Y_0 + b - b\sqrt{1 + 2\frac{aM}{b}\ln\left(1 - \frac{x}{a}\right)}. \qquad (3.24)$$

The function $Y(x)$ is analytic in the interval

$$|x| < a\left(1 - e^{-\frac{b}{2aM}}\right),$$

for $|(2aM)/b \ln(1 - x/a)| < 1$ in that interval, and hence the radical is an analytic function of x. This concludes the proof.

3.1.6 Systems of analytic equations of the first order

Let us reduce, in (3.12), the initial point x_0 to the origin by changing x into $x - x_0$ and consider the Cauchy problem

$$\frac{d\boldsymbol{y}}{dx} = \boldsymbol{f}(x, \boldsymbol{y}), \quad \boldsymbol{y}|_{x=0} = \boldsymbol{y}_0. \qquad (3.25)$$

Theorem. If $\boldsymbol{f}(x, \boldsymbol{y})$ is analytic in $x, \boldsymbol{y} = (y^1, \ldots, y^n)$ in a neighborhood of $0, \boldsymbol{y}_0$, then the solution $\boldsymbol{y} = \boldsymbol{y}(x)$ of the Cauchy problem (3.25) is analytic near $x = 0$.

Proof. The method of majorants used in the case of a single equation can be readily extended to the system (3.25) as follows. Let $\boldsymbol{f}(x, \boldsymbol{y})$ be analytic in the

3.1. EXISTENCE AND UNIQUENESS THEOREMS

domain $|x| < a$, $\|\boldsymbol{y} - \boldsymbol{y}_0\| < b$, and let M be the upper limit of $\|\boldsymbol{f}(x, \boldsymbol{y})\|$ in the preceding domain. Let us replace the majorant problem (3.22) and the dominant function (3.23), respectively, by

$$\frac{dY^1}{dx} = \cdots = \frac{dY^n}{dx} = F(x, \boldsymbol{Y}), \quad Y^1 = \cdots = Y^n = Y_0 \geq \|\boldsymbol{y}_0\|, \quad (3.26)$$

and by

$$F(x, \boldsymbol{Y}) = M\left(1 - \frac{x}{a}\right)^{-1} \left(1 - \frac{Y^1 - Y_0^1}{b}\right)^{-1} \cdots \left(1 - \frac{Y^n - Y_0^n}{b}\right)^{-1}. \quad (3.27)$$

The functions $Y^k(x)$, having their derivatives equal and all assuming the same value Y_0 at $x = 0$, will be identical near the origin. Hence, the majorant problem (3.26)–(3.27) reduces to the problem

$$\frac{dY}{dx} = M\left(1 - \frac{x}{a}\right)^{-1} \left(1 - \frac{Y - Y_0}{b}\right)^{-n} \quad (3.28)$$

for a single variable Y. Whence, upon integration,

$$Y(x) = Y_0 + b - b\left[1 + (n+1)\frac{aM}{b} \ln\left(1 - \frac{x}{a}\right)\right]^{\frac{1}{n+1}}. \quad (3.29)$$

This concludes the proof, since the function (3.29) is analytic in the interval

$$|x| < a\left(1 - e^{-\frac{b}{(n+1)aM}}\right).$$

One can easily prove a similar theorem for the higher order equation (3.15) with analytic f, e.g. by reducing (3.15) to the system (3.16).

3.1.7 Example: The exponential map

Consider the Cauchy problem

$$\frac{dy^i}{dt} = \xi^i(\boldsymbol{y}), \quad y^i|_{t=0} = x^i, \quad i = 1, \ldots, n, \quad (3.30)$$

where $\xi^i(\boldsymbol{y})$ are analytic functions of $\boldsymbol{y} = (y^1, \ldots, y^n)$ and do not involve the independent variable t. Note that the initial value of $\boldsymbol{y}$ at $t = 0$ is denoted by $\boldsymbol{x} = (x^1, \ldots, x^n)$. The development (3.21) of the analytic solution is written

$$y^i = x^i + \frac{t}{1!}\left(\frac{dy^i}{dt}\right)_0 + \frac{t^2}{2!}\left(\frac{d^2 y^i}{dt^2}\right)_0 + \cdots + \frac{t^s}{s!}\left(\frac{d^s y^i}{dt^s}\right)_0 + \cdots, \quad (3.31)$$

where the zero means evaluated at $t = 0$. For any function $h = h(\boldsymbol{y})$, the chain rule of differentiation and the differential equation (3.30) yield

$$\left(\frac{dh}{dt}\right)_0 = \left(\frac{\partial h}{\partial y^k}\right)_0 \left(\frac{dy^k}{dt}\right)_0 = \xi^k \frac{\partial h}{\partial x^k}, \quad \text{or} \quad \left(\frac{dh}{dt}\right)_0 = X(h),$$

where X is the linear partial differential operator of the first order defined by

$$X = \xi^k(x)\frac{\partial}{\partial x^k}. \tag{3.32}$$

It is evident that $X(x^i) = \xi^i(x)$. Consequently, one obtains from equations (3.30) by repeated differentiations:

$$\left(\frac{dy^i}{dt}\right)_0 = \xi^i(x) = X(x^i), \quad \left(\frac{d^2y^i}{dt^2}\right)_0 = \left(\frac{d\xi^i}{dt}\right)_0 = X(\xi^i) = X^2(x^i).$$

Proceeding in this manner, one ultimately obtains the general formula:

$$\left(\frac{d^sy^i}{dt^s}\right)_0 = X^s(x^i), \quad s = 1, 2, 3, \ldots.$$

Hence, the development (3.31) takes the form

$$y^i = x^i + \frac{t}{1!}X(x^i) + \frac{t^2}{2!}X^2(x^i) + \cdots + \frac{t^s}{s!}X^s(x^i) + \cdots. \tag{3.33}$$

Thus, we proved the following theorem.

Theorem. Let e^{tX} be the operator defined by the power series expansion:

$$e^{tX} = 1 + \frac{t}{1!}X + \frac{t^2}{2!}X^2 + \cdots + \frac{t^s}{s!}X^s + \cdots. \tag{3.34}$$

The solution of the Cauchy problem (3.30) is given, for sufficiently small t, by

$$y^i = e^{tX}(x^i) \equiv x^i + tX(x^i) + \frac{t^2}{2!}X^2(x^i) + \cdots + \frac{t^s}{s!}X^s(x^i) + \cdots, \quad i = 1, \ldots, n. \tag{3.35}$$

The formula (3.35) is known as the exponential map, the basis for this designation being that it maps an initial point $x \in \mathbb{R}^n$ into new positions $y = y(t) \in \mathbb{R}^n$.

Exercise. Calculate the exponential map for the Cauchy problem

$$\frac{dy^1}{dt} = y^2, \quad \frac{dy^2}{dt} = -y^1; \quad y^1|_{t=0} = x^1, \quad y^2|_{t=0} = x^2.$$

Solution (cf. Exercise in Section 2.3). Here the differential operator (3.32) has the form

$$X = x^2\frac{\partial}{\partial x^1} - x^1\frac{\partial}{\partial x^2}.$$

Therefore $X(x^1) = x^2$ and $X(x^2) = -x^1$ whence, iterating, $X^2(x^1) \equiv X(X(x^1)) = X(x^2) = -x^1$, $X^2(x^2) = X(-x^1) = -x^2, \ldots$, one ultimately arrives at

$$X^{2s-1}(x^1) = (-1)^{s-1}x^2, \quad X^{2s}(x^1) = (-1)^s x^1, \quad s = 1, 2, \ldots,$$
$$X^{2s-1}(x^2) = (-1)^s x^1, \quad X^{2s}(x^2) = (-1)^s x^2, \quad s = 1, 2, \ldots.$$

Consequently, formula (3.33) yields:

$$y^1 = x^1\left(1 - \frac{t^2}{2!} + \frac{t^4}{4!} - \cdots\right) + x^2\left(t - \frac{t^3}{3!} + \frac{t^5}{5!}X^2(x^i) - \cdots\right) = x^1\cos t + x^2\sin t,$$

$$y^2 = x^2\left(1 - \frac{t^2}{2!} + \frac{t^4}{4!} - \cdots\right) - x^1\left(t - \frac{t^3}{3!} + \frac{t^5}{5!}X^2(x^i) - \cdots\right) = x^2\cos t - x^1\sin t.$$

3.2 Equations of the first order not solved for y'

In the previous sections of this chapter, the differential equations were supposed to be solved with respect to the highest derivatives involved, e.g. equations of the first order were solved for y'. Here we will consider several types of equations of the first order where y' is related implicitly to x and y:

$$F(x, y, y') = 0. \tag{3.36}$$

3.2.1 Equations algebraic in y and y'

Definition. An equation (3.36) is said to be *algebraic* in the variables y and y' if F is a polynomial in y':

$$F(x,y,y') \equiv A_0(x,y)y'^n + A_1(x,y)y'^{n-1} + \cdots + A_n(x,y) = 0, \tag{3.37}$$

where the functions $A(x, y)$ are assumed to be rational fractions, $A = P/Q$ with polynomials P and Q in y whose coefficients are continuous functions of x. It is supposed that $A_0 \neq 0$. Equation (3.37) is then termed a first-order differential equation of the nth degree.

Remark. One can assume $A(x, y)$ to be polynomials by multiplying equation (3.37) by the least common multiple of Q's in $A_i = P_i/Q_i$, $i = 0, 1, \ldots, n$.

Let the algebraic equation $A_0 p^n + A_1 p^{n-1} + \cdots + A_n = 0$ associated with (3.37) have n distinct real[1] roots $p_1, \ldots, p_n$. Then the equation (3.37) splits into n equations solved with respect to y':

$$y' = f_1(x, y), \ y' = f_2(x, y), \ \ldots, \ y' = f_n(x, y). \tag{3.38}$$

By the definition of implicit functions, the right-hand sides of equations (3.38) are continuous and have a bounded partial derivative with respect to y in a domain where $|\partial F/\partial y'| \geq \varepsilon > 0$. Hence, each equation (3.38) satisfies the Lipschitz condition (see Problem 3.1) and has a unique solution (Section 3.1.2). A point (x_0, y_0) within the preceding domain will be referred to as a *regular point*. We conclude that the differential equation (3.37) of the nth degree has precisely n integral curves (branches) passing through any regular point (x_0, y_0).

Example 1. The equation of the second degree,

$$y'^2 - 2xyy' + 2xy - 1 = 0, \tag{3.39}$$

splits into two equations, $y' = 1$ and $y' = 2xy - 1$. Accordingly, it has two distinctly different families of solutions:

$$y = x + C_1 \quad \text{and} \quad y = e^{x^2} \left(C_2 - \int_{x_0}^{x} e^{-t^2} dt \right). \tag{3.40}$$

---

[1] We are not interested in complex roots because, in this book, differential equations are considered in the real domain. A discussion of ordinary differential equations in the complex domain is to be found, e.g., in Part II of Ince's book [2.1] (i).

Example 2. The equation of the second degree,

$$y'^2 + y^2 - 1 = 0, \tag{3.41}$$

splits into two real equations, $y' = \sqrt{1-y^2}$ and $y' = -\sqrt{1-y^2}$, satisfying the Lipschitz condition provided that $|y| < 1$. Accordingly, one obtains two families of solutions determined in the strip $-1 < y < 1$ of the (x,y) plane:

$$y = \sin(x + C_1) \quad \text{and} \quad y = \sin(-x + C_2), \tag{3.42}$$

where

$$-\frac{\pi}{2} < x + C_1 < \frac{\pi}{2} \quad \text{and} \quad -\frac{\pi}{2} < -x + C_1 < \frac{\pi}{2}.$$

If one extends the open domain $-1 < y < 1$ by adding its boundaries $y = \pm 1$ (i.e. the singular lines where the Lipschitz condition is violated) and notes that $\sin(-x + C) = \sin(x + \pi + C)$, one can encapsulate the two branches (3.42) into a one-parameter family of solutions:

$$y = \sin(x + C), \quad -\infty < x < \infty. \tag{3.43}$$

Unlike (3.41), equation (3.39) has two distinctly different families of solutions (3.40), one of them being a simple linear function whereas the other is transcendental. This is the chief difference between Examples 1 and 2, if solutions are not confined to the vicinity of regular points only but are embraced globally in the whole domain of existence. One, apparent, reason for this difference is based on the fact that the left-hand side of equation (3.39) is *reducible*, that is to say decomposable into factors of the same algebraic type (3.37),

$$y'^2 - 2xyy' + 2xy - 1 = (y' - 1)(y'y - 2xy + 1),$$

whereas (3.41) is *irreducible*, i.e. not decomposable into factors of this type. To discern the true nature of the situation, one has, however, to handle the problem in the framework of the analytic theory of differential equations.

3.2.2 Equations $x = f(y')$ and $y = f(y')$. Parametric solutions

First-order equations involving only one of the variables x or y, i.e. $F(x, y') = 0$ or $F(y, y') = 0$, were integrated by quadrature in Section 1.2.1 provided that they can be solved for y'. It may happen, however, that it is easier to solve them with respect to x or y. Then one can use a parametric integration.

Let us consider first an equation soluble for x. Let it has the solved form:

$$x = f(y'). \tag{3.44}$$

Denoting $p = y'$, the equation is rewritten as the system

$$x = f(p), \quad \frac{dy}{dx} = p.$$

3.2. EQUATIONS OF THE FIRST ORDER NOT SOLVED FOR Y'

Upon differentiating the first equation, $dx = f'(p)dp$, the second equation is rewritten $dy = pf'(p)dp$, whence

$$y = \int pf'(p)dp + C = g(p) + C. \tag{3.45}$$

Hence one arrives at a parametric representation of the solution:

$$x = f(p), \quad y = g(p) + C. \tag{3.46}$$

One can proceed similarly with equations soluble for y. Let

$$y = f(y'). \tag{3.47}$$

Using the above notation $p = y'$ and then expressing $dy = f'(p)dd$ from (3.47), one obtains

$$dx = \frac{dy}{p} = \frac{f'(p)}{p}dp,$$

whence

$$x = \int \frac{f'(p)dp}{p} + C = h(p) + C. \tag{3.48}$$

Hence one arrives at a parametric representation of the solution:

$$x = h(p) + C, \quad y = f(p). \tag{3.49}$$

An alternative method is to transform (3.47) to (3.44) by interchanging x and y.

Example 1. The equation $y' = e^{x-y'}$ is solvable for x:

$$x = y' + \ln y'.$$

Here $f(p) = p + \ln p$, and (3.45) yields $g(p) \equiv \int pf'(p)dp = \frac{1}{2}p^2 + p$. Hence the parametric representation (3.46) of the solution has the form

$$x = p + \ln p, \quad y = \frac{1}{2}p^2 + p + C.$$

Example 2. The equation $y = y' + \ln y'$ has the form (3.47). The formulae (3.48)–(3.49) yield:

$$x = \ln p - \frac{1}{p} + C, \quad y = p + \ln p.$$

The solutions (3.46) and (3.49) can be written in the explicit form $y = \phi(x)$ provided that one can eliminate the parameter p. See, e.g. Example 1.

3.1.3 Lagrange's equation

This is a first-order equation linear in x and y:

$$y = x\phi(y') + \psi(y'). \tag{3.50}$$

The derived equation, using the notation $p = y'$, is written:

$$p = \phi(p) + [x\phi'(p) + \psi'(p)]\frac{\mathrm{d}p}{\mathrm{d}x}. \tag{3.51}$$

Regarding x as a function of p, one obtains a linear differential equation (2.3),

$$[\phi(p) - p]\frac{\mathrm{d}x}{\mathrm{d}p} + x\phi'(p) + \psi'(p) = 0, \tag{3.52}$$

provided that $\phi(p) - p \neq 0$. Its solution is given by two quadratures (formula (2.6)) and has the form $x = C\omega(p) + \mu(p)$. Hence the solution of Lagrange's equation is given parametrically:

$$x = C\omega(p) + \mu(p), \quad y = x\phi(p) + \psi(p). \tag{3.53}$$

The method fails when the coefficient $\phi(p) - p$ in equation (3.52) vanishes. Let $\phi(p) - p = 0$ for one or more particular values $p = p_1, p_2, \ldots$. Then integration, e.g., of the equation $y' = p_1$ yields $y = p_1 x + C$. Invoking the condition $p_1 = \phi(p_1)$ and substituting into equation (3.50), one obtains $C = \psi(p_1)$. Hence, one arrives at the following integral curves:

$$y = \phi(p_s)x + \psi(p_s), \quad p_s = \text{const.}, \quad s = 1, 2, \ldots. \tag{3.54}$$

Unlike the general solution (3.53) which depends on an arbitrary continuous parameter C, the straight lines (3.54) are *isolated solutions*. They are not included, in general, into the one-parameter family of solutions (3.53).

The case when $\phi(p) - p$ vanishes identically is considered in the next section.

3.1.4 Clairaut's equation. The envelope of the general integral as a singular solution

The equation has the form

$$y = xp + \psi(p), \tag{3.55}$$

where $p = y'$. Differentiation of both sides of (3.55) yields

$$\left(x + \psi'(p)\right)\frac{\mathrm{d}p}{\mathrm{d}x} = 0. \tag{3.56}$$

By equating to zero the second factor in the left-hand side of (3.56), $\mathrm{d}p/\mathrm{d}x = 0$ or $p = C$, and invoking (3.55), one arrives at the solution which depends on an arbitrary constant C:

$$y = Cx + \psi(C). \tag{3.57}$$

3.2. EQUATIONS OF THE FIRST ORDER NOT SOLVED FOR Y'

The second possibility provided by equation (3.56) is

$$x + \psi'(p) = 0. \tag{3.58}$$

Equations (3.58) and (3.55) define a parametric representation of an isolated solution distinctly different from any particular straight line (3.57). Furthermore, it is readily seen that (3.58) is the derived equation (3.55) with respect to the parameter p. The elimination of p between equations (3.58)–(3.55) will lead therefore to the *envelope* [3.7] of the family of lines (3.57).

Thus, *the general solution of Clairaut's equation (3.55) comprises the one-parameter family of straight lines (3.57) and the envelope of this family defined by (3.58) and (3.55).*

Example 1. For the differential equation

$$y = xy' + \sqrt{b^2 + a^2 y'^2},$$

formula (3.57) provides the general solution:

$$y = Cx + \sqrt{b^2 + a^2 C^2}.$$

The envelope of these straight lines is obtained by eliminating the parameter p from the two equations:

$$y = xp + \sqrt{b^2 + a^2 p^2}, \quad x + \frac{a^2 p}{\sqrt{b^2 + a^2 p^2}} = 0.$$

The substitution of x from the second equation into the first one yields

$$y = \frac{b^2}{\sqrt{b^2 + a^2 p^2}}, \quad x = -\frac{a^2 p}{\sqrt{b^2 + a^2 p^2}}.$$

Hence, the envelope is furnished by the ellipse:

$$\frac{x^2}{a^2} + \frac{y^2}{b^2} = 1.$$

Example 2. For the differential equation

$$y = xy' + 2\sqrt{-y'},$$

formula (3.57) provides the general solution:

$$y = Cx + 2\sqrt{-C}, \quad C < 0.$$

The envelope is furnished by the hyperbola $y = 1/x$ obtained by eliminating p between the two equations

$$y = xp + 2\sqrt{-p}, \quad x - \frac{1}{\sqrt{-p}} = 0.$$

Let (x_0, y_0) be any point on the envelope of the family of straight lines (3.57). According to the above discussions, there are two distinctly different integral curves passing through (x_0, y_0), namely the envelope itself and a straight line obtained from (3.57) by assigning to C a particular value. Hence, the uniqueness of the solution to the Cauchy problem is violated near (x_0, y_0). Bearing this in mind, we will say that the envelope is a locus of *singular points*, or that it is a *singular solution* for the Clairaut equation (3.55). The general definition of singular solutions is as follows.

Definition. Given a differential equation, its singular solution [3.8] is an integral curve such that, for any point (x_0, y_0) on this curve, there are at least two different solutions of the differential equation passing through (x_0, y_0).

Example 3. For equation (3.41), the straight lines $y = \pm 1$ are singular solutions and furnish the envelope of the sine curves (3.43).

Example 4. The equation $y' = 3y^{2/3}$ is solved by $y = (x + C)^3$ and by $y = 0$, the latter being its singular solution.

Remark. In both Examples 3 and 4, the singular solutions are loci of singular points where the Lipschitz condition is violated, namely where the right-hand sides of the differential equations under consideration have unbounded derivatives with respect to y. The violation of the Lipschitz condition does not, however, provide a sufficient condition for singular solutions. Indeed, the y-derivative of the right-hand side of the equation $y' = 3y^{2/3} + K$, $K \neq 0$, is unbounded on the line $y = 0$, but the function $y = 0$ does not solve the differential equation and hence it is not a singular solution.

3.2 Global behavior of solutions

3.2.1 Singularities of first-order equations

The existence theorems of Section 3.1 describe the local behavior of integral curves in the vicinity of ordinary points. A global investigation of solutions in the whole domain of their existence requires an analysis of the behavior of integral curves near *singular points* at which the conditions for existence and uniqueness of solutions cease to hold.

The classification of singularities [3.9] is displayed here by presenting typical situations for the simple equation integrable by elementary methods:

$$\frac{dy}{dx} = \frac{ax + by}{kx + ly}, \quad al - bk \neq 0. \tag{3.59}$$

The origin $O = (0, 0)$ is its *isolated singular point*. The diverse ways in which the solutions of equation (3.59) may behave near the singular point O are combined in a few distinctly different types. It turns out that the behavior of integral curves depends on the sign of $(b - k)^2 + 4al$ and $al - bk$. Accordingly, the origin is qualified as a *node, saddle point, focus* or *center* of the integral curves.

3.3. GLOBAL BEHAVIOR OF SOLUTIONS

Note that an arbitrary linear substitution $\bar{x} = \alpha x + \beta y$, $\bar{y} = \gamma x + \delta y$ is an *equivalence transformation* of equation (3.59) in the sense of Definition 2.1.8. Furthermore, the behavior of integral curves is stable under these equivalence transformations. Therefore, one can eliminate equivalent equations. In this way, one arrives at the standard representations of different types of equations (3.59) given below. Note that the equation presented in Fig. 3.2 is a particular case of that in Fig. 3.1. Furthermore, the center presented in Fig. 3.6 is the limiting case of the focus (Fig. 3.5) when $b = 0$.

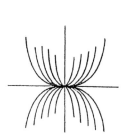
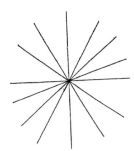

Figure 3.1 Node: $\dfrac{dy}{dx} = a\dfrac{y}{x}$, $a > 0$.   Figure 3.2 Node: $\dfrac{dy}{dx} = \dfrac{y}{x}$.

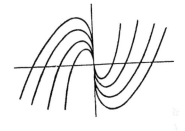

Figure 3.3 Saddle: $\dfrac{dy}{dx} = a\dfrac{y}{x}$, $a < 0$.   Figure 3.4 Node: $\dfrac{dy}{dx} = \dfrac{x + by}{bx}$.

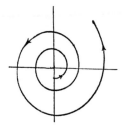
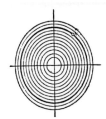

Figure 3.5 Focus: $\dfrac{dy}{dx} = \dfrac{ax + by}{bx - ay}$.   Figure 3.6 Center: $\dfrac{dy}{dx} = -\dfrac{x}{y}$.

3.2.2 Equations in the complex domain. Movable and fixed singular points

The discussion of analytic solutions (Section 3.1.5) reveals that a natural path for the investigation of the true nature of solutions, including their singularities, is the extension of differential equations to the complex domain [3.10]. Indeed, the method of majorants immediately applies to equations

$$\frac{dw}{dz} = f(z, w) \qquad (3.60)$$

with the complex variables z and w and a complex-valued function $f(z, w)$. Then the existence theorem of Section 3.1.5 can be formulated as follows.

Let the function $f(z, w)$ be analytic in a neighborhood of (z_0, w_0). Then equation (3.60) has a unique solution which reduces to w_0 for $z = z_0$ and which is analytic in a circle in the complex z-plane with the center at z_0 and with the radius provided by Theorem 3.1.5.

The proof of the theorem enables one to calculate the analytic solution only in the interior of the circle of convergence of the power series (3.21). However, according to the theory of analytic functions of complex variables, knowledge of these local developments suffices to determine analytic solutions in the whole domain of their existence. The possibility of an analytic continuation is one of the advantages gained from the extension to the complex domain. If the general solution of (3.60) is given in a closed form, it virtually determines the solution, together with its singularities, in the whole domain of existence.

Example 1. The right-hand side of the equation $dw/dz = w/z^2$ is analytic in the neighborhood of (z_0, w_0), $z_0 \neq 0$. The general solution of the equation is $w = Ce^{-1/z}$, where C is an arbitrary complex constant. It provides the integral curve $w = w_0 e^{\frac{1}{z_0} - \frac{1}{z}}$ passing through (z_0, w_0) with the singularity at $z = 0$.

Example 2. Similarly, one can find the general solution of the equation $dw/dz = w^2$ and single out the integral curve $w = w_0/[1 - (z - z_0)w_0]$ passing through any (z_0, w_0). This solution has the singularity, namely the pole $z = z_0 + 1/w_0$.

Example 3. The integral curve of the equation $dw/dz = w^3$ passing through (z_0, w_0) is $w = w_0/\sqrt{1 - 2(z - z_0)w_0^2}$. It has the branch point $z = z_0 + 1/(2w_0^2)$.

Example 4. The integral curve of the equation $dw/dz = -z/w$ passing through (z_0, w_0) is $w = \sqrt{w_0^2 + z_0^2 - z^2}$. It has the branch points $z = \pm\sqrt{w_0^2 + z_0^2}$.

These examples manifest that differential equations of the first order (3.60) may have two distinctly different types of singularities.

Definition. A singular point is said to be *fixed* if its position in the z-plane is independent of the initial values (z_0, w_0) (see Example 1), and *movable* if it moves over the z-plane as the initial values are varied (see Examples 2-4).

3.3. GLOBAL BEHAVIOR OF SOLUTIONS

Consider, in particular, equations (3.60) with f a rational function of w:

$$\frac{dw}{dz} = f(z,w) \equiv \frac{g(z,w)}{h(z,w)}, \qquad (3.61)$$

where g and h are polynomials in w whose coefficients are analytic functions of z. It is supposed that g and h have no common polynomial (in w) divisor. Then the only movable singularities of solutions of (3.61) are *poles* or *branch points*, and hence they cannot have movable essential singularities [3.10]. The crux of the matter is whether or not equation (3.61) has solutions with movable branch points. The chief result is as follows.

Theorem. The most general equation (3.61) that has no solutions with movable branch points is the Riccati equation (2.31):

$$\frac{dw}{dz} = P(z) + Q(z)w + R(z)w^2. \qquad (3.62)$$

Proof. The integral curve $w = \phi(z)$ of equation (3.61) that approaches w_0 when z approaches z_0 has z_0 as its branch point if $h(z_0, w_0) = 0$ while $g(z_0, w_0)$ is not zero [3.11]. Therefore, in order that any solution of equation (3.61) be free of movable branch points, it is necessary that there should be no value $w = w_0$ for which the equation $h(z_0, w_0) = 0$ has solutions z_0 depending on w_0. This is possible only if the denominator h in (3.61) does not contain w. Hence, the right-hand side f of the desired equation is a polynomial in w:

$$f(z,w) = a_0(z) + a_1(z)w + \cdots + a_s(z)w^s.$$

Furthermore, to investigate the behavior of solutions at $w = \infty$, let us substitute $w = 1/W$. Then equation (3.61) becomes $dW/dz = -W^2 f(z, W^{-1})$, or

$$\frac{dW}{dz} = -\Big[a_0(z)W^2 + a_1(z)W + a_2(z) + a_3(z)W^{-1} + \cdots + a_s(z)W^{2-s}\Big].$$

For the same reason as above, the right-hand side of this equation must be a polynomial in W. It follows that $a_3(z) = \cdots = a_s(z) \equiv 0$. Thus, the necessary condition for the nonappearance of movable branch points is that (3.61) reduces to the Riccati equation (3.62). It can be verified by inspection that the Riccati equation has no movable branch points. This completes the proof.

Remark 1. Let m and n be the degrees (with respect to w) of the numerator and denominator in (3.61), respectively: $m = \deg g$ and $n = \deg h$. Then the integrals of equation (3.61) have no *movable poles* if and only if $m - 2 < n$ [3.12].

Remark 2. It follows from the above Theorem and Remark 1 that the only type of equation that has no *movable singular points of any kind* is the linear equation.

Problems

3.1. Let a function $f(x,y)$ have a bounded partial derivative with respect to y in the domain defined in Section 3.1.2. Prove that $f(x,y)$ satisfies the Lipschitz condition.

3.2. Show that the function $f(x,y) = |y|$ is not differentiable with respect to y at $y = 0$ and that, nevertheless, it satisfies the Lipschitz condition.

3.3. Show that the right-hand side of equation (3.2), $f(x,y) = 2\sqrt{|y|}$, does not satisfy the Lipschitz condition in any domain containing a segment of the line $y = 0$.

3.4*. Let the functions $f^1(x, y^1, \ldots, y^n), \ldots, f^n(x, y^1, \ldots, y^n)$ and their partial derivatives with respect to $y^1, \ldots, y^n$ be continuous in a neighborhood of $(x_0, y_0^1, \ldots, y_0^n)$. Prove that the vector-function $\boldsymbol{f} = (f^1, \ldots, f^n)$ satisfies the Lipschitz condition (3.12).

3.5. Prove that conditions (3.11) and (3.13) are equivalent, i.e. that one can be derived from the other by an appropriate choice of constants K and K_j^i, $i,j = 1, \ldots, n$.

3.6. Prove that the function $Y(x)$ defined by (3.29) is analytic in the interval
$$|x| < a\left(1 - e^{-b/((n+1)aM)}\right).$$

3.7. Find the exponential map for the following equations:

(i) $\dfrac{d\boldsymbol{y}}{dt} = \boldsymbol{y}$, $\boldsymbol{y}|_{t=0} = \boldsymbol{x}$, where $\boldsymbol{x}, \boldsymbol{y} \in \mathbb{R}^n$;

(ii) $\dfrac{dy^1}{dt} = y^1$, $\dfrac{dy^2}{dt} = ky^2$, $y^1|_{t=0} = x^1$, $y^2|_{t=0} = x^2$, where $k = \text{const}$.

3.8. Solve the system
$$\dfrac{dy^1}{dt} = y^2, \quad \dfrac{dy^2}{dt} = y^1, \quad y^1|_{t=0} = x^1, \quad y^2|_{t=0} = x^2.$$

3.9. Show that any point (x_0, y_0) is a regular point for equation (3.39). Find the integral curves passing through (x_0, y_0).

3.10. Show that (x_0, y_0), where $|y_0| < 1$, is a regular point for equation (3.41). Find the integral curves (3.42) passing through (x_0, y_0).

3.11. Use the global solution formula (3.43) to find two branches of the integral curve passing through (x_0, y_0) (cf. Problem 3.10).

3.12. Integrate the equation $x = y' + e^{y'}$.

3.13. Integrate Lagrange's equation $3(y + xy') = 4y'^{3/2}$ assuming $y' > 0$.

3.14. Solve Clairaut's equations (i) $y = xy' + 1/y'$, (ii) $y = xy' + y' - y'^2$.

3.15. Integrate the differential equations given in Figs. 3.1–3.6.

3.16. Examine for singular points the following Riccati equation in the complex domain:
$$\frac{dw}{dz} + w^2 - \frac{2}{z^2} = 0.$$

Chapter 4

Partial differential equations of the first order

An acquaintance with the elements of the theory of first-order partial differential equations is a prerequisite for Lie's theory. This chapter contains basic classical devices for solving a single equation (linear and nonlinear) as well as systems of linear homogeneous equations with one dependent variable [4.1]. The coefficients of all equations are assumed to be locally analytic.

4.1 Notation

4.1.1 Equations of the first order and classical solutions

Let $x = (x^1, \ldots, x^n)$ be $n \geq 2$ independent variables and u a single dependent variable. Denote by $p = (p_1, \ldots, p_n)$ the partial derivatives, $p_i = \partial u / \partial x^i$.

It was mentioned in Section 1.1 that equations in which the number of independent variables is greater than one are termed partial differential equations. An equation is said to be of the first order if the partial derivatives of highest order that occur are of order one. A single partial differential equation of the first order with one dependent variable is written $F(x^1, \ldots, x^n, u, p_1, \ldots, p_n) = 0$, or

$$F(x, u, p) = 0. \tag{4.1}$$

The following definition of classical solutions of partial differential equations is similar to that for ordinary differential equations given in Definition 3.1.1.

Definition. A function $u = \phi(x)$, defined and continuously differentiable in a neighborhood of a point $x_0 = (x_0^1, \ldots, x_0^n)$, is said to be a solution (or integral) of the partial differential equation (4.1) if the substitution $u = \phi(x), p_i = \partial \phi(x) / \partial x^i$ converts (4.1) into an identity in a neighborhood of the point x_0.

In the case of two independent variables, we will use the following notation. The independent variables are denoted by x, y, the dependent one by u, and its first derivatives by $p = \partial u/\partial x$ and $q = \partial u/\partial y$. Equation (4.1) is then written

$$F(x,y,u,p,q) = 0. \tag{4.2}$$

A solution $u = \phi(x,y)$ of the differential equation (4.2) defines a surface in the three-dimensional space with the Cartesian coordinates x, y, and u, and therefore it is termed, in the classical literature, an *integral surface*.

4.1.2 Linear, quasi-linear and nonlinear equations

The standard form of the *linear* partial differential equation of the first order is

$$\xi^1(\boldsymbol{x})p_1 + \cdots + \xi^n(\boldsymbol{x})p_n + c(\boldsymbol{x})u = f(\boldsymbol{x}), \tag{4.3}$$

or (cf. the equation (2.3)) [4.2]

$$\xi^1(\boldsymbol{x})\frac{\partial u}{\partial x^1} + \cdots + \xi^n(\boldsymbol{x})\frac{\partial u}{\partial x^n} + c(\boldsymbol{x})u = f(\boldsymbol{x}), \tag{4.4}$$

where $\xi^i(\boldsymbol{x}), c(\boldsymbol{x})$ and $f(\boldsymbol{x})$ are given functions of the independent variables. In particular, equation (4.3) with $c(\boldsymbol{x}) = 0$ and $f(\boldsymbol{x}) = 0$,

$$\xi^1(\boldsymbol{x})p_1 + \cdots + \xi^n(\boldsymbol{x})p_n = 0, \tag{4.5}$$

is termed a *homogeneous linear* equation owing to the fact that its left-hand side is a *linear form* in $\boldsymbol{p}$, i.e., there is no term not involving a derivative and that the derivatives p_i occur in the first power only (cf., however, equation (2.4)).

The general *quasi-linear* equation of the first order is [4.3]

$$\xi^1(\boldsymbol{x},u)p_1 + \cdots + \xi^n(\boldsymbol{x},u)p_n = g(\boldsymbol{x},u), \tag{4.6}$$

where ξ^i and g are given functions of both the independent and dependent variables. Since (4.6) is linear in $\boldsymbol{p}$, it is sometimes called the *non-homogeneous linear* equation (see, e.g., [2.1] (ii)–(iv)). However, equation (4.6) is nonlinear because the unknown function u is introduced into its coefficients. Consequently, equation (4.6) is also termed, more properly, *quasi-linear* instead of linear [4.4]. This modern nomenclature, reflecting the peculiar attributes of equation (4.6), is adopted in the present text.

Equations (4.1) different from (4.3) and (4.6) are termed *nonlinear* partial differential equations of the first order.

When several equations of the form (4.1) are given instead of a single one, they furnish a *system* (known also as a *simultaneous system*) of partial differential equations of the first order. Differential equations composing a simultaneous system may be linear, quasi-linear or nonlinear.

4.2 Integration of linear equations

4.2.1 First integrals of systems of ordinary differential equations

Consider a system of ordinary differential equations of the first order (3.10) with $n-1$ dependent variables:

$$\frac{dy^i}{dx} = f^i(x, y^1, y^2, \ldots, y^{n-1}), \quad i = 1, \ldots, n-1. \tag{4.7}$$

According to Theorem 3.1.3, its general solution has the form (3.14),

$$y^i(x) = \phi^i(x, C_1, \ldots, C_{n-1}), \quad i = 1, \ldots, n-1,$$

whence, upon solving with respect to the constants of integration C_i,

$$\psi_i(x, y^1, y^2, \ldots, y^{n-1}) = C_i, \quad i = 1, \ldots, n-1. \tag{4.8}$$

In terms of Section 2.2.3, the relations (4.8) provide the *general integral* of the system (4.7). The left-hand side of each relation (4.8) reduces to a constant when $y^1, y^2, \ldots, y^{n-1}$ are replaced by the coordinates $y^1(x), y^2(x), \ldots, y^{n-1}(x)$ of any solution of the system (4.7). For this reason every single relation in (4.8) is known as a *first integral* of the system (4.7).

Example 1. Consider the system

$$\frac{dx}{dt} = x, \quad \frac{dy}{dt} = y.$$

Integration yields the general solution

$$x = C_1 e^t, \quad y = C_2 e^t.$$

Upon solving with respect to the constants of integration, one obtains two first integrals:

$$xe^{-t} = C_1, \quad ye^{-t} = C_2.$$

The set of first integrals (4.8) is not the only possible representation of the general solution. Indeed, any relation $\Psi(\psi_1, \ldots, \psi_{n-1}) = C$ is a first integral, and hence one can replace the functions ψ by any $n-1$ functionally independent functions $\Psi_i(\psi_1, \ldots, \psi_{n-1})$, $i = 1, \ldots, n-1$. Therefore, it is useful to have the definition of first integrals independent on the general integral (4.8).

Definition 1. Given a system (4.7), its first integral is a relation of the form

$$\psi(x, y^1, y^2, \ldots, y^{n-1}) = C$$

satisfied for any solution $y^i = y^i(x)$, $i = 1, \ldots, n-1$, where the function ψ is not identically constant. In other words, the function ψ holds a constant value along each solution with the constant C depending on the solution.

The system (4.7) can be rewritten in the form

$$\frac{\mathrm{d}x}{1} = \frac{\mathrm{d}y^1}{f^1} = \frac{\mathrm{d}y^2}{f^2} = \cdots = \frac{\mathrm{d}y^{n-1}}{f^{n-1}}.$$

Since the denominators can be multiplied by any function distinct from zero, one can rewrite these equations (using the notation $\boldsymbol{x} = (x^1, x^2, \ldots, x^n)$ for the variables $x, y^1, \ldots, y^{n-1}$) in the *symmetric form*:

$$\frac{\mathrm{d}x^1}{\xi^1(\boldsymbol{x})} = \frac{\mathrm{d}x^2}{\xi^2(\boldsymbol{x})} = \cdots = \frac{\mathrm{d}x^n}{\xi^n(\boldsymbol{x})}. \tag{4.9}$$

The term *symmetric* is due to the fact that the form (4.9) of $n-1$ first-order ordinary differential equations does not specify the independent variable, which may be now any of the n variables $x^1, x^2, \ldots, x^n$. A *first integral* of the system (4.9) is given by Definition 1 and is written

$$\psi(\boldsymbol{x}) = C. \tag{4.10}$$

Definition 2. A set of $n-1$ first integrals

$$\psi_k(\boldsymbol{x}) = C_k, \quad k = 1, \ldots, n-1, \tag{4.11}$$

is said to be *independent* if the functions $\psi_k(\boldsymbol{x})$ are *functionally independent*, i.e. if there is no relation of the form $F(\psi_1, \ldots, \psi_{n-1}) = 0$.

Any set of $n-1$ independent first integrals represents the general solution of the system (4.9). Since the general solution of a system of $n-1$ first order equations depends precisely on $n-1$ arbitrary constants (see Remark 3.1.3), one arrives at the following statement.

Theorem. A system of $n-1$ first-order ordinary differential equations (4.9) has $n-1$ independent first integrals (4.11). Any other first integral (4.10) of the system (4.9) is expressible in terms of (4.11), viz.

$$\psi = F(\psi_1, \ldots, \psi_{n-1}). \tag{4.12}$$

Example 2. Consider the system

$$\frac{\mathrm{d}x}{yz} = \frac{\mathrm{d}y}{xz} = \frac{\mathrm{d}z}{xy}.$$

It is equivalently rewritten as

$$\frac{\mathrm{d}x}{yz} = \frac{\mathrm{d}y}{xz}, \quad \frac{\mathrm{d}y}{xz} = \frac{\mathrm{d}z}{xy},$$

or, multiplying through by z and x, respectively,

$$\frac{\mathrm{d}x}{y} = \frac{\mathrm{d}y}{x}, \quad \frac{\mathrm{d}y}{z} = \frac{\mathrm{d}z}{y}.$$

4.2. INTEGRATION OF LINEAR EQUATIONS

Rewriting them in the form $x dx - y dy = 0$ and $y dy - z dz = 0$ and integrating, one arrives at the following two independent first integrals:

$$\psi_1 \equiv x^2 - y^2 = C_1, \quad \psi_2 \equiv y^2 - z^2 = C_2.$$

Alternatively, the system in question can be written in the form

$$\frac{dx}{y} = \frac{dy}{x}, \quad \frac{dx}{z} = \frac{dz}{x}.$$

Then one arrives at the first integrals

$$\psi_1 \equiv x^2 - y^2 = C_1, \quad \psi_3 \equiv x^2 - z^2 = C_3,$$

and hence one obtains three different first integrals, $\psi_1 = C_1, \psi_2 = C_2,$ and $\psi_3 = C_3$. However, they are not independent. Indeed, e.g. $\psi_3 = \psi_1 + \psi_2$, in accordance with the above theorem.

4.2.2 Homogeneous linear partial differential equations

Lemma 1. A function $\psi(x) = \psi(x^1, \ldots, x^n)$ provides a first integral (4.10) of the system (4.9) if and only if it solves the partial differential equation

$$\xi^1(x) \frac{\partial \psi}{\partial x^1} + \cdots + \xi^n(x) \frac{\partial \psi}{\partial x^n} = 0. \tag{4.13}$$

Proof. Let a function $\psi(x)$ provide a first integral. Since $\psi(x) = $ const. for every solution $x = (x^1, \ldots, x^n)$ of the system (4.9), the *directional differential* $d\psi$, taken along any integral curve of equations (4.9), vanishes. In other words,

$$d\psi \equiv \frac{\partial \psi}{\partial x^1} dx^1 + \cdots + \frac{\partial \psi}{\partial x^n} dx^n = 0 \tag{4.14}$$

whenever the differential $dx = (dx^1, \ldots, dx^n)$ is proportional to the vector $\xi = (\xi^1, \ldots, \xi^n)$, viz. $dx = \lambda \xi$ where λ is the common value of the ratios dx^i/ξ^i in (4.9). Consequently, substituting $dx^i = \lambda \xi^i$ in (4.14), one arrives at equation (4.13). The latter is satisfied at points x belonging to any integral curve of the system (4.9). But, according to the existence theorem in Section 3.1.3, integral curves pass through any point. Hence, equation (4.13) is satisfied identically in a neighborhood of any generic point x.

Conversely, let $\psi(x)$ be a solution of the partial differential equation (4.13). Since the left-hand side of (4.13) is the directional derivative, $\xi \cdot \nabla \psi$, of ψ in the direction ξ, it follows that the *directional differential* $d\psi$ along any integral curve of the system (4.9) vanishes. Hence, $\psi(x)$ holds a constant value along any integral curve, i.e., $\psi(x) = C$ is a first integral of (4.9).

Lemma 2. Consider the partial differential operator of the first order:

$$X = \xi^1(x) \frac{\partial}{\partial x^1} + \cdots + \xi^n(x) \frac{\partial}{\partial x^n}. \tag{4.15}$$

Let x'^i be new independent variables defined by an invertible transformation

$$x'^i = \varphi^i(x), \quad i = 1, \ldots, n. \tag{4.16}$$

Then the operator (4.15) is written in the new variables in the form

$$\overline{X} = X(\varphi^1)\frac{\partial}{\partial x'^1} + \cdots + X(\varphi^n)\frac{\partial}{\partial x'^n}, \tag{4.17}$$

where

$$X(\varphi^i) = \xi^1\frac{\partial \varphi^i}{\partial x^1} + \cdots + \xi^n(x)\frac{\partial \varphi^i}{\partial x^n}.$$

Proof. The chain rule for the partial derivatives yields:

$$\frac{\partial}{\partial x^i} = \sum_{k=1}^{n} \frac{\partial \varphi^k}{\partial x^i}\frac{\partial}{\partial x'^k}.$$

One can easily verify that the substitution of the above expressions into the operator (4.15) transforms it to the form (4.17).

Theorem. The general solution to the homogeneous linear partial differential equation (4.5),

$$X(u) \equiv \xi^1(x)\frac{\partial u}{\partial x^1} + \cdots + \xi^n(x)\frac{\partial u}{\partial x^n} = 0, \tag{4.18}$$

is given by the formula

$$u = F\left(\psi_1(x), \ldots, \psi_{n-1}(x)\right), \tag{4.19}$$

where F is an arbitrary function of $n-1$ variables and

$$\psi_1(x) = C_1, \quad \ldots, \quad \psi_{n-1}(x) = C_{n-1}$$

is a set of $n-1$ independent first integrals of the system (4.9) associated with (4.18) (the characteristic system of the equation (4.18), see Sections 4.3.4 and 4.4.4):

$$\frac{dx^1}{\xi^1(x)} = \frac{dx^2}{\xi^2(x)} = \cdots = \frac{dx^n}{\xi^n(x)}. \tag{4.20}$$

Proof. The function u defined by (4.19) solves equation (4.18). For, $X(\psi_1) = 0, \ldots, X(\psi_{n-1}) = 0$ by Lemma 1, and the equation $X(u) = 0$ follows from the obvious relation

$$X\left(F(\psi_1, \ldots, \psi_{n-1})\right) = \frac{\partial F}{\partial \psi_1}X(\psi_1) + \cdots + \frac{\partial F}{\partial \psi_{n-1}}X(\psi_{n-1}).$$

Let us verify that any solution of equation (4.18) has the form (4.19). We introduce new independent variables by

$$x'^1 = \psi_1(x), \ldots, x'^{n-1} = \psi_{n-1}(x), \, x'^n = \phi(x),$$

4.2. INTEGRATION OF LINEAR EQUATIONS

where $\psi_1(\boldsymbol{x}), \ldots, \psi_{n-1}(\boldsymbol{x})$ are the left-hand sides of $n-1$ independent first integrals of the characteristic system (4.20), and $\phi(\boldsymbol{x})$ is any function that is functionally independent of $\psi_1(\boldsymbol{x}), \ldots, \psi_{n-1}(\boldsymbol{x})$. According to Lemma 1, $X(\psi_1) = \cdots = X(\psi_{n-1}) = 0$, whereas $X(\phi) \neq 0$. Now one can use Lemma 2 to reduce equation (4.18) to the form

$$X(u) = X(\phi)\frac{\partial u}{\partial x'^n} = 0,$$

whence $\partial u/\partial x'^n = 0$. Therefore, the general solution is an arbitrary function of $x'^1, \ldots, x'^{n-1}$, in accordance with formula (4.19).

4.2.3 Non-homogeneous equations

We will discuss here the integration of non-homogeneous linear equations (4.4) of the type (see also Section 4.3.1)

$$\xi^1(\boldsymbol{x})\frac{\partial u}{\partial x^1} + \cdots + \xi^n(\boldsymbol{x})\frac{\partial u}{\partial x^n} = f(\boldsymbol{x}). \qquad (4.21)$$

Invoking the notation (4.15), equation (4.21) is written

$$X(u) = f(\boldsymbol{x}). \qquad (4.22)$$

Theorem. Given a particular solution $u = \varphi(\boldsymbol{x})$ of the non-homogeneous equation $X(u) = f(\boldsymbol{x})$, its general solution is obtained by adding to $\varphi(\boldsymbol{x})$ the general solution of the corresponding homogeneous equation $X(u) = 0$ (cf. also Remark 2.1.1.). Thus, by virtue of Theorem 4.2.2, the general solution of equation (4.21) is written:

$$u = \varphi(\boldsymbol{x}) + F\left(\psi_1(\boldsymbol{x}), \ldots, \psi_{n-1}(\boldsymbol{x})\right), \qquad (4.23)$$

where $\varphi(\boldsymbol{x})$ is any particular solution of (4.21), $\psi_1(\boldsymbol{x}), \ldots, \psi_{n-1}(\boldsymbol{x})$ are the left-hand sides of any set of $n-1$ independent first integrals of the system of ordinary differential equations (4.20), and F is an arbitrary function.

Proof. Let $X(\varphi(\boldsymbol{x})) = f(\boldsymbol{x})$. By setting $u = v + \varphi(\boldsymbol{x})$, one obtains

$$X(u) = X(v) + X(\varphi(\boldsymbol{x})) = X(v) + f(\boldsymbol{x}).$$

It follows that $X(u) = f(\boldsymbol{x})$ if and only if v solves the homogeneous equation (4.18), $X(v) = 0$. Hence, after we have replaced v in $u = \varphi(\boldsymbol{x}) + v$ by its expression (4.19), $v = F(\psi_1(\boldsymbol{x}), \ldots, \psi_{n-1}(\boldsymbol{x}))$, we arrive at formula (4.23).

This theorem reduces the problem of integration of a non-homogeneous linear partial differential equation (4.21) to that of the associated system of ordinary differential equations (4.20), provided that a particular solution $\varphi(\boldsymbol{x})$ of (4.21) is known. In general, it is not a simple matter to find a solution $\varphi(\boldsymbol{x})$. However, one can easily arrive at a desired particular solution in special cases, e.g. in the following case.

Exercise. Solve an equation (4.21) where one of the ξ^i's and the function f depend upon the single variable x^i, e.g.

$$\xi^1(x^1)\frac{\partial u}{\partial x^1} + \xi^2(x^1,\ldots,x^n)\frac{\partial u}{\partial x^2} + \cdots + \xi^n(x^1,\ldots,x^n)\frac{\partial u}{\partial x^n} = f(x^1). \qquad (4.24)$$

Solution. One readily obtains a particular solution by letting $u = \varphi(x^1)$. Indeed, (4.24) yields the ordinary differential equation

$$\xi^1(x^1)\frac{\mathrm{d}\varphi}{\mathrm{d}x^1} = f(x^1),$$

whence the solution is obtained by quadrature (see Notation 1.2.1):

$$\varphi(x^1) = \int \frac{f(x^1)}{\xi^1(x^1)}\mathrm{d}x^1.$$

The general solution is provided now by formula (4.23).

Example 1. The equation with the independent variables x and y,

$$x^2\frac{\partial u}{\partial x} + xy\frac{\partial u}{\partial y} = 1,$$

has the form (4.24) with $\xi^1 = x$ and $f = 1$. Consequently, one can look for a particular solution of the form $u = \varphi(x)$. Then the equation in question reduces to the ordinary differential equation $x^2 \mathrm{d}\varphi/\mathrm{d}x = 1$, whence one obtains (ignoring the additive constant of integration) the particular solution $\varphi = -1/x^2$. The associated system (4.20),

$$\frac{\mathrm{d}x}{x^2} = \frac{\mathrm{d}y}{xy},$$

has the first integral $y/x = C$ (cf. Problem 4.1). Thus, the general solution (4.23) is written

$$u = -\frac{1}{x} + F\left(\frac{y}{x}\right).$$

Example 2. Consider the equation

$$y\frac{\partial u}{\partial x} - x\frac{\partial u}{\partial y} = y.$$

Upon dividing by y, it takes the form (4.24) with $\xi^1 = 1$ and $f = 1$. Consequently, assuming $u = \varphi(x)$, one obtains from $\mathrm{d}\varphi/\mathrm{d}x = 1$ a particular solution $\varphi = x$. Since the general solution of the corresponding homogeneous equation is given by $v = F(x^2 + y^2)$ (see Problem 4.3(ii)), the general solution of the non-homogeneous equation has the form:

$$u = x + F\left(x^2 + y^2\right).$$

To integrate arbitrary non-homogeneous equations (4.4), one can use the methods discussed in Sections 4.3.2 and 4.3.3.

4.3 Quasi-linear equations

4.3.1 Laplace's method

Laplace [4.3] integrated the equation with two independent variables,

$$\alpha(x,y)\frac{\partial u}{\partial x} + \beta(x,y)\frac{\partial u}{\partial y} = g(x,y,u), \tag{4.25}$$

by reducing it to an ordinary differential equation by means of an appropriate change of independent variables. Both coefficients α and β are supposed to be different from zero, since otherwise (4.25) is an ordinary differential equation. Equation (4.25) is rewritten in new variables $x' = x$ and $y' = \psi(x,y)$ in the form:

$$\alpha \frac{\partial u}{\partial x'} + \left(\alpha \frac{\partial \psi}{\partial x} + \beta \frac{\partial \psi}{\partial y}\right)\frac{\partial u}{\partial y'} = g(x,y,u).$$

Letting $\psi(x,y)$ be a solution (different from constant) of the homogeneous linear equation associated with (4.25),

$$\alpha(x,y)\frac{\partial \psi}{\partial x} + \beta(x,y)\frac{\partial \psi}{\partial y} = 0, \tag{4.26}$$

one arrives (invoking that $x = x$) at the ordinary differential equation:

$$\alpha(x,y)\frac{\partial u}{\partial x} = g(x,y,u), \tag{4.27}$$

where y should be expressed in terms of x and y' by solving $y' = \psi(x,y)$ with respect to y. Thus, (4.25) is reduced to the ordinary differential equation of the first order (4.27). One can use an alternative change of variables, $x' = \psi(x,y)$ and $y' = y$, to obtain, instead of (4.27), the following equation:

$$\beta(x,y)\frac{\partial u}{\partial y} = g(x,y,u), \tag{4.28}$$

where x should be expressed in terms of x' and y from $x' = \psi(x,y)$.

Example. Consider the equation

$$y\frac{\partial u}{\partial x} - x\frac{\partial u}{\partial y} = 1. \tag{4.29}$$

To solve the homogeneous equation (4.26),

$$y\frac{\partial \psi}{\partial x} - x\frac{\partial \psi}{\partial y} = 0, \tag{4.30}$$

we use the characteristic equation $dx/y = -dy/x$, or $xdx + ydy = 0$. Integration yields the first integral $x^2 + y^2 = $ const. Hence, $\psi = x^2 + y^2$ solves equation (4.30) (see Lemma 1 in Section 4.2.2). Consequently, the change of variables

$x' = x^2 + y^2$ and $y' = y$ transforms (4.29) to the form (4.28), $x\partial u/\partial y = -1$, or, upon substitution of $x = \sqrt{x' - y^2}$,

$$\frac{\partial u}{\partial y} = -\frac{1}{\sqrt{x' - y^2}}.$$

After integration with respect to y, one readily obtains the general solution of this equation:

$$u = -\arcsin(y/\sqrt{x'}) + F(x').$$

Using the elementary formula $\arcsin t = \arctan(t/\sqrt{1-t^2})$ and returning to the original variables x and y, one arrives at the general solution to equation (4.29):

$$u = -\arctan(y/x) + F\left(x^2 + y^2\right). \tag{4.31}$$

4.3.2 Extension of Laplace's method to many variables

Here, Laplace's method is extended to equations of the form

$$\xi^1(\boldsymbol{x})\frac{\partial u}{\partial x^1} + \cdots + \xi^n(\boldsymbol{x})\frac{\partial u}{\partial x^n} = g(\boldsymbol{x}, u) \tag{4.32}$$

with many variables $\boldsymbol{x} = (x^1, \ldots, x^n)$. Invoking the operator (4.15), equation (4.32) is written (cf. (4.22))

$$X(u) = g(\boldsymbol{x}, u).$$

Lemma. Given an operator X, (4.15), one can find variables x'^i by solving the homogeneous equation $X(u) = 0$, such that X reduces to the "one-dimensional" form, e.g.

$$X = \xi(\boldsymbol{x}')\frac{\partial}{\partial x'^n}, \tag{4.33}$$

with an arbitrary coefficient $\xi(\boldsymbol{x}')$.

Proof. It follows from Section 4.2.2 that the required variables can be determined by the equations

$$x'^1 = \psi_1(\boldsymbol{x}), \ \ldots, \ x'^{n-1} = \psi_{n-1}(\boldsymbol{x}), \ x'^n = \phi(\boldsymbol{x}), \tag{4.34}$$

where $\psi_1(\boldsymbol{x}), \ldots, \psi_{n-1}(\boldsymbol{x})$ are any $n-1$ functionally independent solutions of the homogeneous equation, viz. $X(\psi_1) = \cdots = X(\psi_{n-1}) = 0$, and $\phi(\boldsymbol{x})$ is functionally independent of $\psi_1(\boldsymbol{x}), \ldots, \psi_{n-1}(\boldsymbol{x})$. Indeed, in these variables, the operator X is written (see Lemma 2 in Section 4.2.2 and Theorem 4.2.2):

$$X = X(\phi)\frac{\partial}{\partial x'^n}.$$

It has the form (4.33) with $\xi(\boldsymbol{x}') = X(\phi(\boldsymbol{x}))$, where in the right-hand side $\boldsymbol{x}$ has to be expressed in terms of the new variables $\boldsymbol{x}'$ by inverting (4.34).

4.3. QUASI-LINEAR EQUATIONS

Definition. Variables (4.34) in which X has the "one-dimensional" form (4.33), are termed *semi-canonical variables* [4.5] for the operator X.

Theorem. Let $\xi^n(\boldsymbol{x}) \neq 0$. Then there exist semi-canonical variables (4.34), such that (4.32) is written as the ordinary differential equation of the first order:

$$\xi^n(\boldsymbol{x})\frac{\partial u}{\partial x'^n} = g(\boldsymbol{x}, u). \qquad (4.35)$$

Proof. It suffices to choose the following semi-canonical variables (4.34):

$$x'^1 = \psi_1(\boldsymbol{x}), \ \ldots, \ x'^{n-1} = \psi_{n-1}(\boldsymbol{x}), \ x'^n = x^n. \qquad (4.36)$$

Remark. After we have expressed the variables $\boldsymbol{x}$ in (4.35) in terms of the new variables by inverting (4.36), equation (4.35) becomes an ordinary differential equation with an independent variable x'^n, where the remaining variables $x'^1, \ldots, x'^{n-1}$ are regarded as parameters. Consequently, the constant of integration C will depend on these parameters, $C = F(x'^1, \ldots, x'^{n-1})$.

Example. Consider the non-homogeneous linear equation

$$x^1\frac{\partial u}{\partial x^1} + \cdots + x^n\frac{\partial u}{\partial x^n} = 1.$$

Assuming $x^n \neq 0$ and integrating the homogeneous equation, one readily obtains the solution $\psi_1 = x^1/x^n$, $\psi_2 = x^2/x^n, \ldots, \psi_{n-1} = x^{n-1}/x^n$ (see Problem 4.3(i)), and hence the semi-canonical variables (4.36):

$$x'^1 = x^1/x^n, \ \ldots, \ x'^{n-1} = x^{n-1}/x^n, \ x'^n = x^n.$$

The ordinary differential equation (4.35) is written

$$x'^n \frac{\partial u}{\partial x'^n} = 1$$

and has the general solution $u = \ln|x'^n| + F(x'^1, x'^2, \ldots, x'^{n-1})$. Whence, returning to the original variables,

$$u = \ln|x^n| + F\left(x^1/x^n, x^2/x^n, \ldots, x^{n-1}/x^n\right).$$

4.3.3 Reduction to a homogeneous linear equation

The general quasi-linear equation (4.6) with n independent variables,

$$\xi^1(\boldsymbol{x}, u)\frac{\partial u}{\partial x^1} + \cdots + \xi^n(\boldsymbol{x}, u)\frac{\partial u}{\partial x^n} = g(\boldsymbol{x}, u), \qquad (4.37)$$

in particular, an arbitrary non-homogeneous linear equation (4.4), can be reduced to a homogeneous linear equation with $n + 1$ variables as follows.

Let us define u as a function of $\boldsymbol{x} = (x^1, \ldots, x^n)$ implicitly by

$$V(x^1, \ldots, x^n, u) = 0 \tag{4.38}$$

and treat V as an unknown function of $n+1$ variables, $x^1, \ldots, x^n$ and u. Let

$$D_i = \frac{\partial}{\partial x^i} + p_i \frac{\partial}{\partial u} \tag{4.39}$$

be the operator of *total differentiation* with respect to x^i (cf. Notation 1.2.2). It follows from (4.38), upon total differentiation, the equation

$$D_i V \equiv \frac{\partial V}{\partial x^i} + p_i \frac{\partial V}{\partial u} = 0,$$

whence

$$p_i = -\frac{\partial V}{\partial x^i} \bigg/ \frac{\partial V}{\partial u}, \quad i = i, \ldots, n. \tag{4.40}$$

Replacing p_i in equation (4.6),

$$\xi^1(\boldsymbol{x}, u) p_1 + \cdots + \xi^n(\boldsymbol{x}, u) p_n = g(\boldsymbol{x}, u),$$

by expressions (4.40), one obtains the homogeneous linear equation

$$X(V) \equiv \xi^1(\boldsymbol{x}, u) \frac{\partial V}{\partial x^1} + \cdots + \xi^n(\boldsymbol{x}, u) \frac{\partial V}{\partial x^n} + g(\boldsymbol{x}, u) \frac{\partial V}{\partial u} = 0 \tag{4.41}$$

for an unknown function V of $n+1$ variables $x^1, \ldots, x^n$ and u. Applying Theorem 4.2.2 to the linear equation (4.41), one ultimately arrives at the following effective method for solving quasi-linear equations.

Theorem. Let

$$\psi_1(\boldsymbol{x}, u) = C_1, \quad \ldots, \quad \psi_n(\boldsymbol{x}, u) = C_n$$

be a set of n independent first integrals of the system of equations (called the characteristic system for the quasi-linear equation (4.37), see Section 4.3.4):

$$\frac{dx^1}{\xi^1(\boldsymbol{x}, u)} = \frac{dx^2}{\xi^2(\boldsymbol{x}, u)} = \cdots = \frac{dx^n}{\xi^n(\boldsymbol{x}, u)} = \frac{du}{g(\boldsymbol{x}, u)}. \tag{4.42}$$

Then the general solution to (4.41) is given by

$$V(\boldsymbol{x}, u) = F(\psi_1(\boldsymbol{x}, u), \ldots, \psi_n(\boldsymbol{x}, u)), \tag{4.43}$$

with an arbitrary function F of n variables. Consequently, the solution of the quasi-linear equation (4.37) is defined implicitly by (4.38), $V(\boldsymbol{x}, u) = 0$. Provided that $\partial V/\partial u \neq 0$, the solution can be written explicitly, $u = \phi(\boldsymbol{x})$.

Example 1. Let us apply the method of this section to equation (4.29),

$$y \frac{\partial u}{\partial x} - x \frac{\partial u}{\partial y} = 1.$$

4.3. QUASI-LINEAR EQUATIONS

Here $g(x, y, u) = 1$ and hence the characteristic system (4.42) is written

$$\frac{dx}{y} = -\frac{dy}{x} = \frac{du}{1}.$$

We have to find two independent first integrals of this system. The first equation, $x dx + y dy = 0$, yields $x^2 + y^2 = a^2 = \text{const}$. By virtue of this relation, the second equation is rewritten

$$du + \frac{dy}{\sqrt{a^2 - y^2}} = 0,$$

whence, upon integration, $u + \arcsin(y/a) = C$, or $u + \arctan(y/x) = C$. Hence, the two independent first integrals have the form

$$\psi_1 \equiv x^2 + y^2 = C_1, \quad \psi_2 \equiv u + \arctan(y/x) = C_2.$$

Therefore, the general solution of the corresponding equation (4.41),

$$y\frac{\partial V}{\partial x} - x\frac{\partial V}{\partial y} + \frac{\partial V}{\partial u} = 0,$$

is given by formula (4.43):

$$V = F\left(x^2 + y^2, u + \arctan(y/x)\right).$$

Hence, equation (4.38) is written

$$F\left(x^2 + y^2, u + \arctan(y/x)\right) = 0.$$

Under the assumption $\partial V/\partial u$, the solution formula (4.31) is obtained by solving the latter equation with respect to u.

Example 2. Consider the quasi-linear equation

$$\frac{\partial u}{\partial t} + u\frac{\partial u}{\partial x} = 0.$$

The homogeneous linear equation (4.41) for the function $V(t, x, u)$ is

$$X(V) \equiv \frac{\partial V}{\partial t} + u\frac{\partial V}{\partial x} = 0.$$

The characteristic system (4.42) can be written formally as

$$\frac{dt}{1} = \frac{dx}{u} = \frac{du}{0},$$

where the last term simply means that this system has the first integral $u = C_1$ (it follows from Lemma 1 of Section 4.2.2, since $X(u) = 0$). By virtue of this first integral, the characteristic system reduces to $dx - C_1 dt = 0$, whence $x - C_1 t = C_2$. We have two first integrals, $u = C_1$ and $x - tu = C_2$. Thus, $V = F(u, x - tu)$, and the solution of the equation in question is given implicitly by (4.38):

$$F(u, x - tu) = 0, \quad \text{or} \quad u = f(x - tu).$$

4.3.4 Integral surfaces as loci of characteristic curves

Characteristics play a central role in the whole theory of differential equations. The general notion of characteristic curves [4.6] is susceptible to a vivid geometric description in the case of quasi-linear equations (4.6) with two independent variables:
$$\xi^1(x,y,u)p + \xi^2(x,y,u)q = g(x,y,u), \tag{4.44}$$
where $p = \partial u/\partial x$ and $q = \partial u/\partial y$ (in the notation of Section 4.1.1).

Recall the elementary facts from the geometry of surfaces. Consider a surface given in the form $u = \phi(x,y)$ and a point $P(x,y,u)$ on the surface. Let $p = \partial \phi/\partial x$ and $q = \partial \phi/\partial y$. The equation of the *tangent plane* to the surface at P is
$$U - u = p(X - x) + q(Y - y), \tag{4.45}$$
where (X, Y, U) are points on the tangent plane. The equation of a straight line in the direction of a given vector (a^1, a^2, a^3) is
$$\frac{X-x}{a^1} = \frac{Y-y}{a^2} = \frac{U-u}{a^3}, \tag{4.46}$$
where (X, Y, U) are points on the straight line. If the line is perpendicular to the tangent plane to the surface at P, it is called the *normal line* at P. The equation of the normal line to the surface is
$$\frac{X-x}{p} = \frac{Y-y}{q} = \frac{U-u}{-1}. \tag{4.47}$$

Consequently, the line represented by (4.46) lies in the tangent plane (4.45) if the vectors (a^1, a^2, a^3) and $(p, q, -1)$ are orthogonal, i.e. if $a^1 p + a^2 q - a^3 = 0$.

Let now $u = \phi(x,y)$ be any integral surface of the partial differential equation (4.44). In the above geometric language, the relation (4.44) means that the tangent plane (4.45) to the integral surface contains the straight line passing through the point of contact $P(x,y,u)$ in the direction of the vector (ξ^1, ξ^2, g):
$$\frac{X-x}{\xi^1(x,y,z)} = \frac{Y-y}{\xi^2(x,y,z)} = \frac{U-u}{g(x,y,z)}. \tag{4.48}$$

Definition. Curves which are tangent at each of their points $P(x,y,u)$ to the straight line (4.48) are termed *characteristic curves*, or simply *characteristics*, of the quasi-linear equation (4.44).

Since the tangent vector to a curve is directed along the vector (dx, dy, du), the characteristic curves are determined by the system (cf. equations (4.42))
$$\frac{dx}{\xi^1(x,y,u)} = \frac{dy}{\xi^2(x,y,u)} = \frac{du}{g(x,y,u)} \tag{4.49}$$
obtained from (4.48) by merely letting (X, Y, U) tend to (x, y, u).

4.3. QUASI-LINEAR EQUATIONS

This definition and Theorem 4.3.3 disclose the geometric nature of integral surfaces as a loci of characteristic curves.

Theorem. Every integral surface of the quasi-linear equation (4.44) is generated by a one-parameter family of characteristic curves.

Proof [4.7]. Consider two independent first integrals of the system (4.49),

$$\psi_1(x, y, u) = a, \quad \psi_2(x, y, u) = b. \tag{4.50}$$

Equations (4.50) define a two-parameter family of characteristic curves. Let us single out a one-parameter family by subjecting the parameters a and b to any relation, $F(a, b) = 0$. Eliminating a and b from the latter relation and from equations (4.50), one obtains a surface given by

$$F\left(\psi_1(x, y, u), \psi_2(x, y, u)\right) = 0.$$

This surface, generated by the one-parameter family of characteristics, has the form (4.38), (4.43), and hence it is an integral surface for equation (4.44). Since F is an arbitrary function, any integral surface can be obtained by this construction.

The characteristics of the general quasi-linear equation (4.37) are determined by the *characteristic system* (4.42). Theorem 4.3.3 extends to this case as follows: A function $u = \phi(x^1, \ldots, x^n)$ is a solution of the equation (4.37) if and only if it is formed by a family of characteristics depending on $n - 1$ parameters.

In the particular case of equation (4.44), where $g = 0$ and the coefficients ξ^1 and ξ^2 do not depend on u, i.e. in the case of the homogeneous linear equation

$$\xi^1(x, y)\frac{\partial u}{\partial x} + \xi^2(x, y)\frac{\partial u}{\partial y} = 0, \tag{4.51}$$

the characteristics are given by equations (4.50) of the form $\psi(x, y) = a, u = b$, where the first relation, $\psi(x, y) = a$, provides a first integral of the equation

$$\frac{dx}{\xi^1(x, y)} = \frac{dy}{\xi^2(x, y)}. \tag{4.52}$$

It follows that the characteristic curves are merely cuts of cylinders protracted along the u-axis, with directrices $\psi(x, y) = a$, by planes $u = $ const. parallel to the xy-plane. Consequently, it suffices to consider only projections of the characteristic curves to the xy-plane. Then one arrives at plane curves, called the *characteristics* of equation (4.51), determined by the ordinary differential equation (4.52).

A similar treatment of the case of many variables provides the characteristic system (4.20) for the general homogeneous linear equation (4.18).

Example. For the equation (Problem 4.3(ii), see also the Example 4.3.1)

$$y\frac{\partial u}{\partial x} - x\frac{\partial u}{\partial y} = 0,$$

the characteristics are concentric circles $x^2 + y^2 = C$.

4.4 Nonlinear equations

4.4.1 Complete, general and singular integrals

Consider the general equation of the first order (4.2),

$$F(x, y, u, p, q) = 0, \tag{4.53}$$

with two independent variables. Its solution involves, in general, an arbitrary function (see, e.g., the solution formulas (4.19), (4.23) and (4.43) with $n = 2$). The fundamental result due to Lagrange [4.8] states, however, that it suffices to know a solution depending on two parameters only. Then all other integrals of (4.53) can be derived from such a solution through differentiation and elimination of the parameters. Precise definitions and propositions are formulated as follows.

Definition 1. The *complete integral* of equation (4.53) is a solution depending on two arbitrary constants:

$$u = \phi(x, y, a, b). \tag{4.54}$$

This means that the relation (4.53) becomes an identity in x, y, a, b, whenever u and p, q are replaced by $u = \phi(x, y, a, b)$ and

$$p = \phi_x(x, y, a, b), \quad q = \phi_y(x, y, a, b), \tag{4.55}$$

respectively, where $\phi_x = \partial \phi / \partial x$ and $\phi_y = \partial \phi / \partial y$. Furthermore, it is assumed that elimination of the parameters a and b from the three relations (4.54) and (4.55) leads precisely to equation (4.53).

Theorem. Given a complete integral (4.54), let the parameters a and b undergo an arbitrary relation $b = \sigma(a)$. Let

$$u = f_\sigma(x, y) \tag{4.56}$$

be the envelope of the one-parameter family of integral surfaces,

$$u = \phi(x, y, a, \sigma(a)). \tag{4.57}$$

Then (4.56) is an integral surface for equation (4.53). The subscript σ in (4.56) indicates that the solution depends upon choice of the function σ.

Proof. The envelope (4.56) of the family of surfaces (4.57) is obtained by eliminating the parameter a from equation (4.57) and the equation (see [3.7]):

$$\left[\frac{\partial \phi(x, y, a, b)}{\partial a} + \frac{\partial \phi(x, y, a, b)}{\partial b} \sigma'(a) \right]_{b=\sigma(a)} = 0. \tag{4.58}$$

By definition, the envelope has the same p and q along the curve of contact with the enveloped surface. Consequently, the envelope (4.56) of integral surfaces (4.57) is also an integral surface of equation (4.53).

4.4. NONLINEAR EQUATIONS

Definition 2. The *general integral* is the set of all particular solutions (4.57) obtained for all possible relations $b = \sigma(a)$ between the two parameters. Hence, the general integral involves, indirectly, an arbitrary function, viz. $\sigma(a)$.

Definition 3. The *singular integral* is the envelope of the family of integral surfaces (4.54) depending on two parameters. It is obtained by eliminating a and b from equation (4.54) and the equations

$$\frac{\partial \phi(x,y,a,b)}{\partial a} = 0, \quad \frac{\partial \phi(x,y,a,b)}{\partial b} = 0, \qquad (4.59)$$

provided that this elimination is possible.

For equations (4.1) with any number of variables, the name *complete integral* is given to a solution involving as many parameters as there are independent variables. Hence,

Definition 4. The complete integral of the equation

$$F(x^1, \ldots, x^n, u, p_1, \ldots, p_n) = 0 \qquad (4.60)$$

is a solution

$$u = \phi(x^1, \ldots, x^n, a_1, \ldots, a_n) \qquad (4.61)$$

containing n arbitrary parameters a_i and such that elimination of the parameters from equation (4.61) and the equations

$$p_i = \frac{\partial \phi(x^1, \ldots, x^n, a_1, \ldots, a_n)}{\partial x^i}, \quad i = 1, \ldots, n, \qquad (4.62)$$

leads precisely to the differential equation (4.60).

4.4.2 Completely integrable systems. The Lagrange-Charpit method

According to the preceding section, one can find all solutions of equation (4.53) by calculating its complete integral. To solve the latter problem, one can use, e.g., Lagrange and Charpit's method [4.9]. The method is based on the following notion of completely integrable systems.

Let u be an unknown function of n variables $\boldsymbol{x} = (x^1, \ldots, x^n)$, and let $f_1(\boldsymbol{x}, u), \ldots, f_n(\boldsymbol{x}, u)$ be given functions of $x^1, \ldots, x^n$ and u. Consider the system of partial differential equations of the first order,

$$\frac{\partial u}{\partial x^1} = f_1(\boldsymbol{x}, u), \quad \ldots, \quad \frac{\partial u}{\partial x^n} = f_n(\boldsymbol{x}, u), \qquad (4.63)$$

and define an associated 1-form ω by

$$\omega = f_1(\boldsymbol{x}, u) \mathrm{d}x^1 + \cdots + f_n(\boldsymbol{x}, u) \mathrm{d}x^n. \qquad (4.64)$$

Definition. The system (4.63) is said to be *completely integrable* if, for any solution $u = u(\boldsymbol{x})$ of (4.63), the form ω is *exact*, specifically:

$$f_1(\boldsymbol{x}, u) \mathrm{d}x^1 + \cdots + f_n(\boldsymbol{x}, u) \mathrm{d}x^n = \mathrm{d}u. \qquad (4.65)$$

According to Section 2.1.3, ω is exact if and only if it is closed. Hence, the condition of complete integrability is $d\omega|_{(4.63)} = 0$. Invoking the exterior differentiation formula (2.17) (see also Problem 2.6), one arrives at the following.

Theorem. The system (4.64) is completely integrable if and only if the following $n(n-1)/2$ equations are satisfied identically in $x^1, \ldots, x^n, u$:

$$\frac{\partial f_i}{\partial x^k} + \frac{\partial f_i}{\partial u} f_k = \frac{\partial f_k}{\partial x^i} + \frac{\partial f_k}{\partial u} f_i, \quad i, k = 1, \ldots, n. \tag{4.66}$$

The crucial idea of Lagrange and Charpit's method is to find an auxiliary differential equation,

$$\Phi(x, y, u, p, q) = a, \quad a = \text{const.}, \tag{4.67}$$

such that equations (4.53) and (4.67) can be solved in the form $p = f_1(x, y, u, a)$, $q = f_2(x, y, u, a)$, to provide a completely integrable system (4.63):

$$\frac{\partial u}{\partial x} = f_1(x, y, u, a), \quad \frac{\partial u}{\partial y} = f_2(x, y, u, a). \tag{4.68}$$

The general solution of this system is obtained by integration of ordinary differential equations (cf. Section 2.1.4) and contains an arbitrary constant of integration, b. Thus, upon solving the system (4.68), one obtains a complete integral $u = \phi(x, y, a, b)$ of equation (4.53).

Construction of the auxiliary equation (4.67) requires the following calculations. The values of the partial derivatives $\partial p/\partial y$, $\partial p/\partial u$, $\partial q/\partial x$ and $\partial q/\partial u$, are obtained from (4.53) and (4.67) by differentiation with respect to x and y and elimination. The test for complete integrability is obtained by substituting these values in the integrability condition (4.66),

$$\frac{\partial p}{\partial y} + q\frac{\partial p}{\partial u} = \frac{\partial q}{\partial x} + p\frac{\partial q}{\partial u},$$

and can be written explicitly in terms of the functions F and Φ as the following linear partial differential equation in five independent variables, x, y, u, p, q [4.10]:

$$P\frac{\partial \Phi}{\partial x} + Q\frac{\partial \Phi}{\partial y} + (pP + qQ)\frac{\partial \Phi}{\partial u} - (X + pU)\frac{\partial \Phi}{\partial p} - (Y + qU)\frac{\partial \Phi}{\partial q} = 0. \tag{4.69}$$

Here the functions X, Y, U, P and Q are defined by

$$X = \frac{\partial F}{\partial x}, \quad Y = \frac{\partial F}{\partial y}, \quad U = \frac{\partial F}{\partial u}, \quad P = \frac{\partial F}{\partial p}, \quad Q = \frac{\partial F}{\partial q}.$$

To integrate equation (4.69), one needs first integrals of the characteristic system:

$$\frac{dx}{P} = \frac{dy}{Q} = \frac{du}{pP + qQ} = -\frac{dp}{X + pU} = -\frac{dq}{Y + qU}. \tag{4.70}$$

Moreover, the method requires a knowledge of one first integral (4.67) only.

4.4. NONLINEAR EQUATIONS

Exercise. Find a complete integral and investigate the singular and general integrals of the nonlinear equation

$$\frac{\partial u}{\partial x}\frac{\partial u}{\partial y} + x\frac{\partial u}{\partial x} + y\frac{\partial u}{\partial y} - u = 0. \tag{4.71}$$

Solution. Let us apply the Lagrange-Charpit method. Here $F = pq + xp + yq - u$, whence

$$X = p, \ Y = q, \ U = -1, \ P = x + q, \ Q = y + p.$$

It follows that $X + pU = Y + qU = 0$, and hence equation (4.69) is written:

$$(x+q)\frac{\partial \Phi}{\partial x} + (y+p)\frac{\partial \Phi}{\partial y} + (xp + yq + 2pq)\frac{\partial \Phi}{\partial u} = 0.$$

One of its simple solutions is, e.g., $\Phi = p$. Consequently, we take the first integral (4.67) in the form $p = a$. The equations

$$p = a, \quad u + xp + yq + pq$$

yield $p = a$, $q = (u - ax)/(y + a)$. Hence, the completely integrable system (4.68) has the form:

$$\frac{\partial u}{\partial x} = a, \quad \frac{\partial u}{\partial y} = \frac{u - ax}{y + a}.$$

Integrating the first equation and substituting the resulting formula $u = ax + v(y)$ into the second one, we obtain $v'(y) = v/(y + a)$. Integration yields $v = b(y + a)$. Thus, one arrives at the following *complete integral* for equation (4.71):

$$u = ax + by + ab. \tag{4.72}$$

This complete integral comprises a two-parameter family of planes. These planes envelop the singular integral obtained by eliminating a and b from the relations (see Definition 3 in Section 4.4.1)

$$u = ax + by + ab, \ x + b = 0, \ y + a = 0.$$

Hence, the *singular integral* is the hyperboloid:

$$u = -xy.$$

The *general integral* is the envelope of the one-parameter family of planes obtained be setting $b = \sigma(a)$ in the complete integral (4.72). This envelope is represented parametrically by equations (4.57)-(4.58) containing an arbitrary function $\sigma(a)$:

$$u = ax + (y + a)\sigma(a), \quad x + \sigma(a) + (y + a)\sigma'(a) = 0. \tag{4.73}$$

It is clearly tangent to the hyperboloid $u = -xy$ representing the singular integral.

Particular solutions can be obtained by specifying $\sigma(a)$ in (4.73). Let us take, e.g., $\sigma(a) = 1/a$. Then equations (4.73) are written

$$u = 1 + ax + \frac{y}{a}, \quad x - \frac{y}{a^2} = 0.$$

Expressing the parameter a from the second equation and substituting its value $a = \sqrt{y/x}$ into the first equation, one obtains a particular solution of equation (4.71):

$$u = 1 + 2\sqrt{xy}.$$

4.4.3 Solution of Cauchy's problem via complete integrals

The Cauchy problem for equation (4.53), $F(x, y, u, p, q) = 0$, is that of determining an integral surface passing through a given curve γ (cf. Section 3.1.1). Cauchy's problem has, in general, a unique solution just as in the case of ordinary differential equations.

Let an *initial curve* γ be given parametrically by

$$x = x_0(s), \quad y = y_0(s), \quad u = u_0(s),$$

and let a complete integral (4.54), $u = \phi(x, y, a, b)$, of equation (4.53) be known. we introduce the function

$$W(s, a, b) = u_0(s) - \phi(x_0(s), y_0(s), a, b) \tag{4.74}$$

obtained by considering the complete integral on the initial curve γ. Elimination of the parameter s from the equations

$$W(s, a, b) = 0, \quad \frac{\partial W(s, a, b)}{\partial s} = 0 \tag{4.75}$$

provides a relation $b = \sigma(a)$, i.e. a one-parameter family of integral surfaces (4.57). The envelope of this family passes through the curve γ by construction. Moreover, it is an integral surface by Theorem 4.4.1. Hence, it provides the solution of the Cauchy problem in question.

Example. Let us consider again equation (4.71), $pq + xp + yq - u = 0$, and find its integral surface passing through the parabola $x = 0, u = y^2$.

We can use the complete integral obtained in Exercise 4.4.2, $u = ax + by + ab$. The initial curve can be written parametrically, $x = 0, y = s, u = s^2$, and formula (4.74) yields $W(s, a, b) = s^2 - bs - ab$. Consequently, equations (4.75) yield:

$$s^2 - bs - ab = 0, \quad 2s - b = 0,$$

whence eliminating s one obtains $b = -4a$. The solution of the Cauchy problem is obtained now by eliminating a from the equations

$$u = ax - 4ay - 4a^2, \quad x - 4y - 8a = 0$$

and has the form $u = (x - 4y)^2/16$.

4.4.4 Monge's theory of characteristics

Monge [4.11] amplified the basis of the contemporary theory of partial differential equations by providing a visible geometric picture of Lagrange's theory. The fundamental concept of characteristics introduced by Monge can be explained in a natural manner in the framework of complete integrals.

Let $V(x, y, u, a, b) = 0$ be an implicit representation of a complete integral of equation (4.53). The general integral is the envelope of the one-parameter family of surfaces $V(x, y, u, a, \sigma(a)) = 0$ and is obtained by eliminating a from

4.4. NONLINEAR EQUATIONS

$$V(x,y,u,a,\sigma(a)) = 0, \quad \frac{\partial V}{\partial a} + \frac{\partial V}{\partial b}\sigma'(a) = 0, \qquad (4.76)$$

where $\sigma(a)$ is an arbitrary function. Consider now these two equations for a given value of a. Then equations (4.76) represent a curve of contact of the envelope with enveloped surfaces. This is precisely what Monge called a *characteristic curve*.

The locus of characteristic curves obtained by varying a is clearly the general integral (cf. Theorem 4.3.4). Furthermore, since the function $\sigma(a)$ is arbitrary, the values of this function and of its derivative $\sigma'(a)$ at any given value of a can be regarded as arbitrary parameters b and c, respectively. Thus, *the set of all characteristic curves of a given equation* (4.53) *depends on three arbitrary parameters, a, b, and c, and is defined by the equations*

$$V(x,y,u,a,b) = 0, \quad \frac{\partial V(x,y,u,a,b)}{\partial a} + \frac{\partial V(x,y,u,a,b)}{\partial b}c = 0. \qquad (4.77)$$

It follows from the definition of the envelope that the envelope and the enveloped surfaces have common tangent planes along the characteristics. The respective values p and q, determining the common tangent planes, are obtained from the relations

$$\frac{\partial V}{\partial x} + p\frac{\partial V}{\partial u} = 0, \quad \frac{\partial V}{\partial y} + q\frac{\partial V}{\partial u} = 0, \qquad (4.78)$$

where $V = V(x,y,u,a,\sigma(a))$. Equations (4.76) and (4.78) define, for any fixed value of a, a characteristic curve and a tangent plane to that curve. This combination of a characteristic curve and the tangent plane to it is referred to as a *characteristic strip* or a *characteristic of the first order*. The latter name is self-explanatory, since characteristics of the first order involve the variables x, y, and u and the first derivatives p and q. Accordingly, given an equation (4.53), its characteristics of the first order are described by the following system of ordinary differential equations (for the notation, see Section 4.4.2):

$$\frac{dx}{P} = \frac{dy}{Q} = \frac{du}{pP+qQ} = -\frac{dp}{X+pU} = -\frac{dq}{Y+qU} = d\tau. \qquad (4.79)$$

Equations (4.79) are identical to the system (4.70). An auxiliary parameter τ is introduced merely for symmetry in the variables x, y, u, p, and q.

For a given differential equation (4.53), the quantities X, Y, U, P, Q are known functions of five variables x, y, u, p, q. Consequently, the characteristics are defined, independently of complete integrals, by the *characteristic equations* (4.79). The latter can be treated as a system (4.7) with dependent variables x, y, u, p, q:

$$\frac{dx}{d\tau} = P, \; \frac{dy}{d\tau} = Q, \; \frac{du}{d\tau} = pP+qQ, \; \frac{dp}{d\tau} = -(X+pU), \; \frac{dq}{d\tau} = -(Y+qU), \quad (4.80)$$

and can be solved, e.g., by using any initial values x_0, y_0, u_0, p_0, q_0 at $\tau = 0$. The solution defines a characteristic of the first order provided that the initial values satisfy the differential equation (4.53), $F(x_0, y_0, u_0, p_0, q_0) = 0$.

The previous notions can be extended without difficulty to equations (4.1) with many independent variables,

$$F(x^1, \ldots, x^n, u, p_1, \ldots, p_n) = 0.$$

In this case, the characteristic equations (4.79) have the form

$$\frac{dx^1}{P_1} = \cdots = \frac{dx^n}{P_n} = \frac{du}{\sum p_i P_i} = -\frac{dp_1}{X_1 + Up_1} = \cdots = -\frac{dp_n}{X_n + Up_n} = d\tau, \quad (4.81)$$

where

$$X_i = \frac{\partial F}{\partial x^i}, \quad U = \frac{\partial F}{\partial u}, \quad P_i = \frac{\partial F}{\partial p_i}.$$

Remark. If (4.1) is the quasi-linear equation (4.6), then $P_i = \xi^i(x, u)$ and $\sum p_i P_i = g(x, u)$. Therefore the first n equations (4.81) reduce to the characteristic system (4.42), and hence they can be integrated independently of the remaining equations of the whole system (4.81). Likewise, in the case of homogeneous linear equations (4.5), one has $P_i = \xi^i(x)$. Consequently, it is sufficient to consider the first $n-1$ equations (4.81) only, thus arriving at the system (4.20).

4.4.5 Example: Light rays as visible characteristics

In everyday life, characteristics are visible, e.g. in the form of narrow beams of light. The mathematical background of this phenomenon is as follows. The propagation of light waves is governed by the wave equation (cf. Section 1.1),

$$\frac{\partial^2 u}{\partial t^2} - c^2 \left(\frac{\partial^2 u}{\partial x^2} + \frac{\partial^2 u}{\partial y^2} + \frac{\partial^2 u}{\partial z^2} \right) = 0, \quad (4.82)$$

where $c = 2.99793 \times 10^{10}$ cm/s is the velocity of light in the vacuum. In the theory of partial differential equations, the characteristics of equation (4.82) are defined by $\Phi(t, x, y, z) = C$, where Φ solves the differential equation, the *characteristic equation* for (4.82):

$$\left(\frac{\partial \Phi}{\partial t}\right)^2 - c^2 \left[\left(\frac{\partial \Phi}{\partial x}\right)^2 + \left(\frac{\partial \Phi}{\partial y}\right)^2 + \left(\frac{\partial \Phi}{\partial z}\right)^2 \right] = 0. \quad (4.83)$$

The characteristics of the nonlinear partial differential equation of the first order (4.83) are called *bicharacteristics* for the wave equation (4.82). They are determined by the equations (4.81) written, in this case, in the form:

$$\frac{dx}{\Phi_x} = \frac{dy}{\Phi_y} = \frac{dz}{\Phi_z} = -c^2 \frac{dt}{\Phi_t} = d\tau,$$

where the subscripts denote the partial derivatives, e.g. $\Phi_t = \partial \Phi / \partial t$.

Bicharacteristics transport light disturbances. Consequently, they are visible as light rays. Therefore bicharacteristics are often termed *rays*.

4.4. NONLINEAR EQUATIONS

4.4.6 Cauchy's method

Consider an alternative approach to the solution of the Cauchy problem. This approach, due to Cauchy, is independent of Lagrange's theory of complete integrals. Cauchy's method reduces the Cauchy problem for partial differential equations (4.1) to that for ordinary differential equations – characteristic equations (4.81).

Consider equations with two independent variables. Let

$$\begin{aligned} x &= f_1(\tau; x_0, y_0, u_0, p_0, q_0), \\ y &= f_2(\tau; x_0, y_0, u_0, p_0, q_0), \\ u &= \phi(\tau; x_0, y_0, u_0, p_0, q_0), \\ p &= \psi_1(\tau; x_0, y_0, u_0, p_0, q_0), \\ q &= \psi_2(\tau; x_0, y_0, u_0, p_0, q_0) \end{aligned} \quad (4.84)$$

be the solution of equations (4.80) which assumes the initial values x_0, y_0, u_0, p_0, q_0 at $\tau = 0$. Let the initial point (x_0, y_0, u_0) circumscribe an *initial curve* γ:

$$x_0 = x_0(s), \quad y_0 = y_0(s), \quad u_0 = u_0(s). \quad (4.85)$$

The crucial idea of the method is to assign p_0 and q_0 as functions $p_0(s)$ and $q_0(s)$ of the parameter s. However, unlike (4.85), $p_0(s)$ and $q_0(s)$ cannot be arbitrary functions. They should satisfy the equation $F(x_0, y_0, u_0, p_0, q_0) = 0$ and the *tangency condition*, $du_0 = p_0 dx_0 + q_0 dy_0$. Dividing the latter relation by ds, one obtains the following two equations for determining $p_0(s)$ and $q_0(s)$:

$$F(x_0, y_0, u_0, p_0, q_0) = 0, \quad \frac{du_0}{ds} = p_0 \frac{dx_0}{ds} + q_0 \frac{dy_0}{ds}. \quad (4.86)$$

Summing up, we formulate Cauchy's method as follows (see Problem 4.12).

Let γ be a given curve (4.85). The integral surface of equation (4.53), $F(x, y, u, p, q) = 0$, passing through γ is defined by the first three equations (4.84), $x = f_1$, $y = f_2$, and $u = \phi$, where x_0, y_0, u_0 are replaced by the functions (4.85), and p_0 and q_0 by $p_0(s)$ and $q_0(s)$ obtained from the algebraic equations (4.86).

Cauchy's method can be extended to equations (4.1), $F(\boldsymbol{x}, u, \boldsymbol{p}) = 0$, with many independent variables, $\boldsymbol{x} = (x^1, \ldots, x^n)$. Then equations (4.80) are naturally replaced by (4.81), while the relations (4.85) and (4.86) are replaced by their respective multi-dimensional extensions, viz.

$$\boldsymbol{x}_0 = \boldsymbol{x}_0(s_1, \ldots, s_{n-1}), \quad u_0 = u_0(s_1, \ldots, s_{n-1}),$$

and

$$F(\boldsymbol{x}_0, u_0, \boldsymbol{p}_0) = 0, \quad \frac{\partial u_0}{\partial s_\nu} = \boldsymbol{p}_0 \cdot \frac{\partial \boldsymbol{x}_0}{\partial s_\nu}, \quad \nu = 1, \ldots, n-1.$$

4.4.7 Hamilton's equations – characteristics of the Hamilton-Jacobi equation

The Hamilton-Jacobi equation is one of the most important examples of nonlinear first-order partial differential equations. It is a particular case of equation (4.1) when it does not contain the dependent variable. The equation has the form:

$$\frac{\partial S}{\partial t} + H\left(t, x^1, \ldots, x^n, \frac{\partial S}{\partial x^1}, \ldots, \frac{\partial S}{\partial x^n}\right) = 0, \tag{4.87}$$

or $S_t + H(t, \boldsymbol{x}, \boldsymbol{p}) = 0$, where $\boldsymbol{p} = (p_1, \ldots, p_n)$ with $p_i = \partial S/\partial x^i$. Here the independent variables are time t and the coordinates $\boldsymbol{x} = (x^1, \ldots, x^n)$. The function $H(t, \boldsymbol{x}, \boldsymbol{p})$ is termed the Hamiltonian of a mechanical system.

The characteristic equations (4.81) for (4.87) are written:

$$dt = \frac{dx^1}{\partial H/\partial p_1} = \cdots = \frac{dx^n}{\partial H/\partial p_n} = \frac{dS}{S_t + \sum p_i \frac{\partial H}{\partial p_i}} = -\frac{dp_1}{\partial H/\partial x^1} = \cdots = -\frac{dS_t}{\partial H/\partial t}.$$

These equations do not contain the variable S, and therefore one can integrate independently the equations not involving dS and dS_t, i.e. Hamilton's equations:

$$\frac{dx^i}{dt} = \frac{\partial H}{\partial p_i}, \quad \frac{dp_i}{dt} = -\frac{\partial H}{\partial x^i}, \quad i = 1, \ldots, n. \tag{4.88}$$

We summarize: *the canonical equations (4.88) of mechanics are characteristic equations for the Hamilton-Jacobi equation. In other words, a mechanical system moves along the characteristics of the Hamilton-Jacobi equation.*

Furthermore, since the equation (4.87) does not contain the variable S, the complete integral (4.61) can be taken with an additive constant, i.e. in the form $S = \phi(t, \boldsymbol{x}, a_1, \ldots, a_n) + a_{n+1}$. Consequently, the complete integral is usually identified with its principal part, viz. $S = \phi(t, \boldsymbol{x}, a_1, \ldots, a_n)$. The following theorem due to Jacobi [4.12] states that the general solution of Hamilton's equations of motion can be obtained from the complete integral of the Hamilton-Jacobi equation.

Given a complete integral, $S = \phi(t, \boldsymbol{x}, a_1, \ldots, a_n)$, of the equation (4.87), the general solution of the canonical equations (4.88) are given by the relations

$$\frac{\partial S}{\partial x^i} = p_i, \quad \frac{\partial S}{\partial a_i} = b_i, \quad i = 1, \ldots, n, \tag{4.89}$$

where a_i and b_i provide $2n$ arbitrary constants of the general solution.

4.5 Systems of homogeneous linear equations

4.5.1 Basic notions

Consider the general system of r homogeneous linear partial differential equations for one unknown function $u = u(\boldsymbol{x})$:

$$\sum_{i=1}^{n} \xi_\alpha^i(\boldsymbol{x}) p_i = 0, \quad \alpha = 1, \ldots, r, \tag{4.90}$$

4.5. SYSTEMS OF HOMOGENEOUS LINEAR EQUATIONS

where $x = (x^1, \ldots, x^n)$ are the independent variables and $p_i = \partial u/\partial x^i$ are the partial derivatives. After introducing r differential operators of the form (4.15),

$$X_\alpha = \xi_\alpha^1(x)\frac{\partial}{\partial x^1} + \cdots + \xi_\alpha^n(x)\frac{\partial}{\partial x^n}, \quad \alpha = 1, \ldots, r, \tag{4.91}$$

the system (4.90) is written in the compact form (cf. equation (4.18)):

$$X_1(u) = 0, \ \ldots\ , X_r(u) = 0. \tag{4.92}$$

Equations (4.90) have a trivial solution $u = $ const., which is of no interest to us. Furthermore, it is apparent that any equation of the system can be multiplied by a function of x. Therefore, if u solves s equations $X_\alpha(u) = 0$, $\alpha = 1, \ldots, s \leq r$, it also satisfies their linear combination with arbitrary variable coefficients, viz. $\sum_{\alpha=1}^{s} \lambda^\alpha(x) X_\alpha(u) = 0$. This motivates the following definitions.

Definition 1. Differential operators $X_1, \ldots, X_s$ are said to be *connected* if there exist functions $\lambda^\alpha(x)$, not all zero, such that

$$\lambda^1(x) X_1 + \cdots + \lambda^s(x) X_s = 0, \tag{4.93}$$

this being satisfied as an operator identity in a neighborhood of a generic x. If the relation (4.93) implies $\lambda^1 = \cdots = \lambda^s = 0$, we say that the operators $X_1, \ldots, X_s$ are *unconnected*. In the latter case, the corresponding differential equations $X_1(u) = 0, \ldots, X_s(u) = 0$ are said to be *independent*.

Definition 2. Let Z_α be r linear combinations of the operators (4.91):

$$Z_\alpha = \sum_{\beta=1}^{r} h_\alpha^\beta(x) X_\beta, \quad \alpha = 1, \ldots, r,$$

with variable coefficients $h_\alpha^\beta(x)$ whose determinant, $|h_\alpha^\beta(x)|$, is not zero. The system of linear homogeneous equations $Z_1(u) = 0, \ldots, Z_r(u) = 0$ has the same set of solutions as the original system (4.92). Consequently, these two systems, as well as the operators X_α and Z_α, are said to be *equivalent*.

Given any r operators (4.91), the number of unconnected ones is determined by the following well-known algebraic lemma.

Lemma. The number r_* of unconnected operators (4.91) is equal to the rank of the $r \times n$ matrix of their coefficients $\xi_\alpha^i(x)$:

$$r_* = \mathrm{rank}\left\|\xi_\alpha^i(x)\right\|, \tag{4.94}$$

where α and i denote rows and columns, respectively. The number r_* is the same for equivalent operators Z_α.

According to this lemma, any system of r homogeneous linear equations can be replaced by a system of r_* independent equations. It is clear that more than n

equations cannot be independent. Furthermore, if $r = n$, and if operators (4.91) are unconnected (i.e. $r_* = r = n$), then the determinant of the coefficients $\xi_\alpha^i(x)$ is not zero. In this case, the linear homogeneous *algebraic* equations (4.90) with respect to $p = (p_1, \ldots, p_n)$ yield $p = 0$, and hence the solution of the *differential* equations (4.92) is trivial, $u = \text{const}$. Thus, a necessary condition for the existence of non-trivial solutions is $r_* < n$.

The following examples manifest typical situations that may occur when dealing with systems (4.92). In particular, they show that the condition $r_* < n$ alone is *not sufficient* for the existence of non-trivial solutions.

Example 1. Consider the following operators in the $x = (x, y, z)$ space:

$$X_1 = z\frac{\partial}{\partial y} - y\frac{\partial}{\partial z}, \quad X_2 = x\frac{\partial}{\partial z} - z\frac{\partial}{\partial x}, \quad X_3 = y\frac{\partial}{\partial x} - x\frac{\partial}{\partial y}.$$

Here

$$\|\xi_\alpha^i(x)\| = \begin{vmatrix} 0 & z & -y \\ -z & 0 & x \\ y & -x & 0 \end{vmatrix}$$

and the formula (4.94) yields $r_* = 2$. Accordingly, the operators X_1, X_2, X_3 are connected, $xX_1 + yX_2 + zX_3 = 0$, but any two of them are unconnected. Hence, to solve the three equations, $X_1(u) = 0, X_2(u) = 0, X_3(u) = 0$, it suffices to solve any two of them, e.g. $X_1(u) = 0, X_2(u) = 0$:

$$z\frac{\partial u}{\partial y} - y\frac{\partial u}{\partial z} = 0, \quad x\frac{\partial u}{\partial z} - z\frac{\partial u}{\partial x} = 0.$$

These two simultaneous equations can be integrated successively using the methods of Section 4.2.2. One readily obtains the general solution of the first equation in the form $u = v(x, \rho)$, where $\rho = \sqrt{y^2 + z^2}$. The second equation is written

$$\frac{z}{\rho}\left(x\frac{\partial v}{\partial \rho} - \rho\frac{\partial v}{\partial x}\right) = 0.$$

Since $z \neq 0$ at a generic point (x, y, z), we arrive at the following single equation for the function v of two variables:

$$x\frac{\partial v}{\partial \rho} - \rho\frac{\partial v}{\partial x} = 0.$$

Its solution can be written in the form $v = \phi(\sqrt{x^2 + \rho^2})$ with an arbitrary function ϕ of one variable. Hence the general solution of the original system is

$$u = \phi\left(\sqrt{x^2 + y^2 + z^2}\right).$$

Example 2. Consider the system of two equations:

$$X_1(u) \equiv z\frac{\partial u}{\partial y} - y\frac{\partial u}{\partial z} = 0, \quad X_2(u) \equiv y\frac{\partial u}{\partial x} + z\frac{\partial u}{\partial y} = 0.$$

4.5. SYSTEMS OF HOMOGENEOUS LINEAR EQUATIONS

It is evident that these equations are independent since they involve differentiations in different variables. Proceeding as in the previous example, one obtains the solution of the first equation in the form $u = v(x, \rho)$, where $\rho = \sqrt{y^2 + z^2}$. Substituting this expression into the second equation and dividing by y/ρ, we get

$$\frac{\partial v}{\partial x} + \frac{z}{\rho}\frac{\partial v}{\partial \rho} = 0.$$

Since v does not involve the variable z *explicitly*, it follows that $\partial v/\partial \rho = 0$ and hence $\partial v/\partial x = 0$. Consequently, $u = v = $ const. Thus, the equations in question do not have a non-trivial solution, even though $r_* = r = 2$ is less than the number $n = 3$ of independent variables.

Example 3. In the space of four variables $x = (x, y, z, t)$, consider the system of two equations:

$$X_1(u) \equiv z\frac{\partial u}{\partial y} - y\frac{\partial u}{\partial z} = 0, \quad X_2(u) \equiv \frac{\partial u}{\partial x} + t\frac{\partial u}{\partial y} + y\frac{\partial u}{\partial t} = 0.$$

The first equation yields $u = v(x, t, \rho)$, where again $\rho = \sqrt{y^2 + z^2}$. Now the second equation is written

$$\frac{\partial v}{\partial x} + y\left(\frac{\partial v}{\partial t} + \frac{t}{\rho}\frac{\partial u}{\partial \rho}\right) = 0.$$

Since v does not involve the variable y explicitly, the latter equation *splits* into two equations:

$$\frac{\partial v}{\partial x} = 0, \quad \rho\frac{\partial v}{\partial t} + t\frac{\partial v}{\partial \rho} = 0.$$

It follows that $v = \phi(\rho^2 - t^2)$, and hence

$$u = \phi(y^2 + z^2 - t^2).$$

To discern the true nature of the situations exhibited by these examples, we present a general approach to integration of any equations (4.90) by reducing them to a complete system.

4.5.2 Complete systems

Any solution u of the system (4.92) also satisfies the second-order equations $X_\alpha(X_\beta(u)) = 0$ for any values of the indices α and β. Consequently, u solves the following equations of the first order:

$$X_\alpha(X_\beta(u)) - X_\beta(X_\alpha(u)) \equiv \sum_{i=1}^{n}\left(X_\alpha(\xi_\beta^i) - X_\beta(\xi_\alpha^i)\right)\frac{\partial u}{\partial x^i} = 0.$$

Hence, one can state that u annuls, together with the operators (4.91), all their *commutators* defined as follows.

Definition 1. The commutator of any two operators (4.91), X_α and X_β, is the differential operator $[X_\alpha, X_\beta]$ of the first order defined by

$$[X_\alpha, X_\beta] = X_\alpha X_\beta - X_\beta X_\alpha,$$

or in the following equivalent form exhibiting the coefficients explicitly:

$$[X_\alpha, X_\beta] = \sum_{i=1}^n \left(X_\alpha(\xi_\beta^i) - X_\beta(\xi_\alpha^i) \right) \frac{\partial}{\partial x^i}. \tag{4.95}$$

Thus, any solution of equations (4.92) solves the equations $[X_\alpha, X_\beta](u) = 0$ as well. We have the following alternatives: either *some of the commutators* (4.95) *are independent of the original operators* (4.91), or *the commutators* (4.95) *are linear combinations with variable coefficients of the operators* (4.91). The latter case means that the combined set of operators (4.91), (4.95) is connected.

In the first case, one should consider an extended system of differential equations of the first order obtained by combining (4.91) with all independent commutators. Then one can apply the above operations to this new system. Proceeding in this manner, one ultimately reaches the second case and hence arrives at what is called a complete system.

Definition 2. Let (4.92) be a system of independent equations. It is called a *complete system* if all commutators (4.95) are *dependent* on the operators (4.91):

$$[X_\alpha, X_\beta] = \sum_{\gamma=1}^r h_{\alpha\beta}^\gamma(x) X_\gamma. \tag{4.96}$$

If $h_{\alpha\beta}^\gamma(x) = 0$, i.e. if all commutators of the operators (4.91) vanish, we have a particular case of a complete system known as a *Jacobian system*.

The system of equations $X_1(u) = 0, X_2(u) = 0$ in Example 1 of the preceding section is complete since $[X_1, X_2] = X_3 = -(x/z)X_1 - (y/z)X_2$. On the other hand, the systems considered in Examples 2 and 3 are not complete (see Problems 4.13 and 4.14). Consequently, the process of their solution gave rise to new equations, and the corresponding complete systems were self-generated.

The theory of complete systems is based on the following properties.

Lemma 1. The commutator (4.95) is invariant under changes of variables (4.16). Namely, let $x'^i = \varphi^i(x), i = 1, \ldots, n$, be new variables and let $\overline{X}_\alpha$ denote the operators X_α written in the variables x'^i (see formula (4.17)). Then

$$\overline{[X_\alpha, X_\beta]} = [\overline{X}_\alpha, \overline{X}_\beta]. \tag{4.97}$$

Invoking (4.96), we conclude that a system (4.92) is complete (or Jacobian) independently on the choice of variables.

Proof. According to Lemma 2 of Section 4.2.2, $X(u) = \overline{X}(u')$, where $\overline{X}$ denotes the operator (4.17) obtained from X by a change of variables (4.16), u is an arbitrary function of x and u' the same function expressed in the new variables x'.

4.5. SYSTEMS OF HOMOGENEOUS LINEAR EQUATIONS

Since u is arbitrary, we also have $X_\alpha(X_\beta(u)) = \overline{X}_\alpha\left(\overline{X}_\beta(u')\right)$, and consequently, $[X_\alpha, X_\beta](u) = [\overline{X}_\alpha, \overline{X}_\beta](u')$. Invoking the identity $[X_\alpha, X_\beta](u) = \overline{[X_\alpha, X_\beta]}(u')$, we arrive at the operator equation (4.97).

Lemma 2. If the system (4.92) is complete, then any equivalent system $Z_1(u) = 0, \ldots, Z_r(u) = 0$ (see Section 4.5.1, Definition 2) is also complete.

Proof. The commutator of the operators $Z_\alpha = \sum_{\sigma=1}^r h_\alpha^\sigma(\boldsymbol{x}) X_\sigma$ and $Z_\beta = \sum_{\nu=1}^r h_\beta^\nu(\boldsymbol{x}) X_\nu$ can be represented as a sum of terms of the form

$$h_\alpha^\sigma h_\beta^\nu [X_\sigma, X_\nu] + h_\alpha^\sigma X_\sigma(h_\beta^\nu) X_\nu - h_\beta^\nu X_\nu(h_\alpha^\sigma) X_\sigma.$$

It follows from (4.96) that each of these expressions is a linear combination of the operators $X_1, \ldots, X_r$. Since the two sets of operators, X_α and Z_α, are equivalent, all the commutators $[Z_\alpha, Z_\beta]$ can be expressed linearly in terms of $Z_1, \ldots, Z_r$.

Lemma 3. Any complete system (4.92) is equivalent to a Jacobian system.

Proof. By Definition 2 of a complete system, equations (4.92) are independent. Hence they can be solved with respect to r derivatives. We can assume without loss of generality that the system is solved for the $x^1, \ldots, x^r$ derivatives:

$$Z_1(u) \equiv \frac{\partial u}{\partial x^1} + \cdots = 0, \quad Z_r(u) \equiv \frac{\partial u}{\partial x^r} + \cdots = 0, \qquad (4.98)$$

where the dots denote terms not containing the $x^1, \ldots, x^r$- derivatives. The system (4.98) is complete since it is equivalent to a complete system (4.92). It is evident that the commutators $[Z_\alpha, Z_\beta](u)$ may involve only the derivatives $\partial u / \partial x^{r+1}, \ldots, \partial u / \partial x^n$, whereas the operators Z_α given by (4.98) contain differentiations with respect to $x^1, \ldots, x^r$. Consequently, $[Z_\alpha, Z_\beta]$ can be linear combinations of Z_α only if $[Z_\alpha, Z_\beta] = 0$. Hence, (4.98) is a Jacobian system.

4.5.3 Integration of complete systems

According to the preceding section, *every system of homogeneous linear partial differential equations can be converted into a complete system by adding all independent commutators*. The following theorem gives a practical guide for integrating a complete system of independent equations (4.92), $X_\alpha(u) = 0$ ($\alpha = 1, \ldots, r$).

Theorem. A complete system of r independent equations (4.92) has $n - r$ functionally independent solutions, $u = \psi_1(\boldsymbol{x}), \ldots, u = \psi_{n-r}(\boldsymbol{x})$. The general solution is an arbitrary function of these $n - r$ particular solutions:

$$u = F(\psi_1(\boldsymbol{x}), \ldots, \psi_{n-r}(\boldsymbol{x})). \qquad (4.99)$$

Proof. The proof consists of r consecutive integrations, where each step requires the integration of a single equation only.

In the first step, integrating one of the equations (4.92), e.g. $X_1(u) = 0$, one obtains $n - 1$ independent solutions $u = \psi_1(\boldsymbol{x}), \ldots, u = \psi_{n-1}(\boldsymbol{x})$. Then one can introduce semi-canonical variables, $x'^1 = \phi(\boldsymbol{x})$, $x'^2 = \psi_1(\boldsymbol{x})$, ..., $x'^n = \psi_{n-1}(\boldsymbol{x})$ (see Definition 4.3.2) and reduce the operator X_1 to the form $Y_1 = f(\boldsymbol{x})\partial/\partial x'^1$. Consequently, the system (4.92) is equivalent to a new complete system of the following form (see Lemma 3 of Section 4.5.2):

$$Y_1(u) \equiv \frac{\partial u}{\partial x'^1} = 0, \quad Y_\alpha(u) \equiv \frac{\partial u}{\partial x'^\alpha} + \sum_{s=r+1}^{n} \eta_\alpha^s \frac{\partial u}{\partial x'^s} = 0, \quad \alpha = 2, \ldots, r. \quad (4.100)$$

It was shown in the preceding section (see Lemma 3) that (4.100) is a Jacobian system. Hence, all commutators of the operators Y vanish. In particular,

$$[Y_1, Y_\alpha] = \sum_{s=r+1}^{n} \frac{\partial \eta_\alpha^s}{\partial x'^1} \frac{\partial}{\partial x'^s} = 0,$$

whence $\partial \eta_\alpha^s / \partial x'^1 = 0$. It follows that $Y_2(u) = 0, \ldots, Y_r(u) = 0$ form a Jacobian system of $r-1$ equations with $n-1$ independent variables $x'_2, \ldots, x'_n$. By applying the above procedure to this system and continuing in this manner, one ultimately reduces the system (4.92) to a single homogeneous linear equation with $n-r+1$ independent variables. Its integration provides the general solution (4.99).

Remark. If some operators X_γ do not enter in the right-hand sides of the commutator relations (4.96), $[X_\alpha, X_\beta] = \sum_{\gamma=1}^{r} h_{\alpha\beta}^\gamma(\boldsymbol{x})X_\gamma$, then it is advantageous to start the integration of the system (4.92) by first solving the sub-system $X_\gamma(u) = 0$ with those X_γ involved in the right-hand sides of (4.96).

Example. Consider again the equations $X_1(u) = 0, X_2(u) = 0$ from Example 3 of Section 4.5.1. Here

$$X_1 = z\frac{\partial}{\partial y} - y\frac{\partial}{\partial z}, \quad X_2 = \frac{\partial}{\partial x} + t\frac{\partial}{\partial y} + y\frac{\partial}{\partial t}.$$

Their commutator has the form $[X_1, X_2] = X_3$, where

$$X_3 = t\frac{\partial}{\partial z} + z\frac{\partial}{\partial t}.$$

The three equations $X_1(u) = 0, X_2(u) = 0$, and $X_3(u) = 0$ form a complete system since (see Answer to Problem 4.14)

$$[X_1, X_2] = X_3, \quad [X_1, X_3] = \frac{t}{z}X_1 + \frac{y}{z}X_3, \quad [X_2, X_3] = -X_1.$$

According to the above Remark, we solve the equations $X_1(u) = 0, X_3(u) = 0$. The first equation yields $u = v(x, t, \rho)$, where $\rho = \sqrt{y^2 + z^2}$. Then $X_3(u) = 0$ reduces to the equation (cf. Example 3 of Section 4.5.1)

$$t\frac{\partial v}{\partial \rho} + \rho\frac{\partial v}{\partial t} = 0,$$

whence $v = w(x, \lambda)$, where $\lambda = \rho^2 - t^2 = y^2 + z^2 - t^2$. Now the last equation, $X_2(u) = 0$, reduces to $\partial w/\partial x = 0$. Thus, the formula (4.99) is written

$$u = \phi(y^2 + z^2 - t^2).$$

Problems

4.1. Find first integrals and the general solution of the system
$$\frac{dx}{dt} = x^2, \quad \frac{dy}{dt} = xy.$$

4.2. Find a first integral of the equation $dx/(2y) = dy/(3x^2)$.

4.3. Solve the homogeneous linear equations

(i) $x^1 \dfrac{\partial u}{\partial x^1} + \cdots + x^n \dfrac{\partial u}{\partial x^n} = 0;$ (ii) $y \dfrac{\partial u}{\partial x} - x \dfrac{\partial u}{\partial y} = 0;$

(iii) $y \dfrac{\partial u}{\partial x} + x \dfrac{\partial u}{\partial y} = 0;$ (iv) $2y \dfrac{\partial u}{\partial x} + 3x^2 \dfrac{\partial u}{\partial y} = 0.$

4.4. Solve the non-homogeneous linear equation
$$x \frac{\partial u}{\partial x} + 2y \frac{\partial u}{\partial y} - 2z \frac{\partial u}{\partial z} = 1.$$

4.5. Solve the equations

(i) $y \dfrac{\partial u}{\partial x} - x \dfrac{\partial u}{\partial y} = x;$ (ii) $y \dfrac{\partial u}{\partial x} - x \dfrac{\partial u}{\partial y} = yg(x);$ (iii) $y \dfrac{\partial u}{\partial x} - x \dfrac{\partial u}{\partial y} = xh(y),$

where $g(x)$ and $h(y)$ are arbitrary functions.

4.6. Show that $u + \arctan(y/x) = C$ is a first integral of the system $dx/y = -dy/x = du$.

4.7. Solve the equation
$$y \frac{\partial u}{\partial x} - x \frac{\partial u}{\partial y} = x^2.$$

4.8. Solve the following linear equation:
$$x^1 \frac{\partial u}{\partial x^1} + \cdots + x^n \frac{\partial u}{\partial x^n} = \sigma u, \quad \sigma = \text{const.} \neq 0.$$

4.9. Show that the change of variables (4.36) is well defined, i.e. that all functions involved are functionally independent.

4.10. Find a complete integral for the equation $u = xp + yq + \psi(p)$.

4.11. Find a complete integral of the natural generalization of Clairaut's equation (3.55):
$$u = xp + yq + \psi(p, q).$$

4.12. Solve the Cauchy problem from Example 4.4.3 by means of Cauchy's method.

4.13. Show that the system considered in Example 2 of Section 4.5.1 is not complete.

4.14. Find consecutive commutators of the operators X_1 and X_2 from Example 3 in Section 4.5.1. Is the system $X_1(u) = 0, X_2(u) = 0$ complete?

Part II

FUNDAMENTALS OF LIE GROUP ANALYSIS

The aim of this part is to present in a concise form the basic methods that lie at the core of modern group analysis.

The concept of a *group* is closely related to that of invariance or symmetry of mathematical objects – surfaces, functions, differential equations, etc. Indeed, given any object $\mathcal{M}$, the set G of all invertible transformations T leaving the object $\mathcal{M}$ unaltered:

$$T : \mathcal{M} \to \mathcal{M},$$

contains the identity transformation I, the inverse T^{-1} of any transformation $T \in G$ and the product (or composition) $T_1 T_2$ of any transformations $T_1, T_2 \in G$. Then G is called a group, or more precisely, a *symmetry group* of the object $\mathcal{M}$.

The symmetry of algebraic equations is discussed in Galois theory, whereas Lie group theory deals with symmetries of differential equations.

Lie groups, unlike Galois groups, contain infinitely many transformations and depend on continuous parameters. Consequently, Lie's theory deals with continuous groups. The crucial idea of this theory is to employ infinitesimal transformations instead of finite group transformations.

Any Lie group is uniquely determined by its infinitesimal transformations. The latter form a vector space closed under the commutator, i.e. a Lie algebra.

The proper language for the formulation and calculation of symmetry groups of differential equations and their Lie algebras is provided by differential algebra. This language and the space $\mathcal{A}$ of *differential functions* are introduced in Chapter 8 and essentially used throughout the book.

Chapter 5

Gateway to modern group analysis: Historical survey

> If the historical exposition given here is correct, I can claim that I was the first who used the concept of groups for an integration theory of differential equations.
>
> S. Lie, 1895

This chapter contains a short account of important events in the development of modern group analysis of differential equations and provides a panoramic view on the multi-faceted possibilities offered by the philosophy of Lie groups.

5.1 Appearance of groups in the theory of equations

The general theory of groups was the final outcome of about three centuries of effort by renown mathematicians to solve algebraic equations by radicals.

After the solution of cubic equations of a particular form by S. del Ferro (1465-1526) in 1515, N. Tartaglia (1500-1557) and L. Ferrari (1522-1565) gave the solution of general cubic and quartic by radicals. Tartaglia's solution of the cubic was published in 1545 by H. Cardan (1501-1576) in his comprehensive treatise *Artis magnae sive de regulis algebrae liber unus* and became known as *Cardan's solution*.

The solution of the cubic is quite similar to that of the quadratic equation and is presented in this book in Section 6.1.1. Ferrari's solution of the quartic is more complicated [5.1]. About 1770, Euler invented a new method of solving the quartic different from that of Ferrari. Euler believed that the general algebraic equation is solvable by radicals. About the same time, Lagrange critically examined the particular devices used by his predecessors. He found that they succeeded because the solution of each equation of degree 2, 3 or 4 is reducible to that of an equation of lower order. Furthermore, he applied his approach to the quintic and obtained an equation of sixth degree: the degree of the original equation was raised, instead of being reduced as before! It took another 54 years before N.H. Abel (1802-1829) proved in 1824 that the general algebraic equation of degree higher than the fourth is not solvable by radicals. Even though Lagrange missed this general result, he actually took the first steps toward the theory of permutation groups.

The true nature of solvability by radicals was disclosed by Galois [5.2] in 1832. He invented the term *group* and formulated general conditions for solvability of an algebraic equation in terms of an associated group known in the literature as the Galois group. *An algebraic equation is solvable by radicals if and only if its Galois group satisfies a certain structural property.* Groups obeying this property are naturally called *solvable*. The solvability or non-solvability of algebraic equations by radicals is based on the following fundamental theorem of the Galois theory:

The Galois group of the general equation of degree n is solvable if $n \leq 4$ and not solvable if $n \geq 5$. In the latter case, it is solvable for particular equations only.

5.2 Sophus Lie and symmetry analysis of differential equations

The idea of symmetry permeates all mathematical models, in particular those formulated in terms of differential equations. Mathematical tools for revealing and using the symmetry of differential equations are provided by the theory of continuous groups. This theory, unlike the purely algebraic Galois theory, combines algebra, analysis and geometry. It was originated and elaborated by an outstanding mathematician of the nineteenth century, Sophus Lie.

5.2.1 His life story

SOPHUS LIE (**Lie**, Marius Sophus, 1842-1899), was born on the 17 December 1842 in the town of Nordfjordeid (Eid-on-Nordfjord) in Norway. He was the sixth (youngest) child in the family of Johann Hermann Lie, a Lutheran pastor in the vicarage at Eid. Starting in 1857 Sophus Lie studied in Christiania (now Oslo), first in a Grammar school and then (1859–1865) in the University where he attended Sylow's lectures [5.3] on the Galois theory. Among other events of importance that determined the choice of his scientific direction were: study in 1868 of the geometrical works of Chasles [5.4], Poncelet [5.5] and Plücker [5.6], a journey to Germany and France (1869–1870) and meetings with F. Klein [5.7], C. Jordan [5.8], G. Darboux [5.9] and M. Chasles. Felix Klein and Sophus Lie became close friends and long-term partners after their first meeting in 1869. In Berlin, Klein and Lie attended Kummer's world famous seminars where they gave a lecture, and Lie had encouraging discussions with Kummer [5.10].

In 1871 Lie was appointed tutor in the University of Christiania, in the same year submitting his work *On complexes, in particular, line and spherical com-*

5.2. SOPHUS LIE AND SYMMETRY ANALYSIS

plexes, with applications to the theory of partial differential equations for a doctor's degree. This work, essentially based on Plücker's theory of complexes and Monge's geometrical interpretation of partial differential equations [4.11], is a splendid blend of Lie's new ideas with results of his contemporaries in projective geometry, contact transformations and theory of partial differential equations. In 1872 Lie was appointed extraordinary professor of the University of Christiania where he worked during the period 1872–1886.

Lie married Anna Birch in 1874. Sophus and Anna Lie had three children, Marie, Dagny and Herman.

In 1869-70, Lie made a crucial discovery, namely that the majority of the old methods of integration of ordinary differential equations, which until then had seemed artificial and not intrinsically related to one another, could be unified by means of group theory. During 1870-74, Lie investigated the role of a general transformation theory in classical integration methods and developed his ideas on what he called *finite continuous groups* of transformations. During this period he perceived the significance of the new notion in geometry and particularly in the theory of differential equations. He also introduced the notion of infinitesimal contact transformations and developed new integration methods for partial differential equations of the first order, supplementing the classical integration methods of Lagrange (1772), Cauchy (1819) and Jacobi (1837). Lie's approach is based on his transformation theories and Lagrange-Monge's notion of characteristics of nonlinear first-order partial differential equations.

Later, in his survey paper [5.11] containing numerous important historical remarks on the development of differential equations and transformation groups, Lie wrote: *"If the historical exposition given here is correct, I can claim that I was the first who used the conception of groups for an integration theory of differential equations."* Lie begins his paper with the remarkable statement: *"The theory of differential equations is the most important discipline in the whole modern mathematics."* This, together with a general survey of Lie's papers, clearly indicates the focus of his research interests. It is a fact, however, that Lie made great contributions in three different directions: *Continuous groups and Lie algebras, Geometry,* and *Differential equations.* It is clear today that the work of Lie in all these directions merits equally high evaluation.

International recognition came from geometers. In 1897 Sophus Lie was awarded the first N.I. Lobachevskii prize established by the Physical-Mathematical Society of the Imperial Kazan University in 1895 to recognize distinguished works on geometry, especially non-Euclidean geometry. The first three prizes awarded were to the following: 1897 – S. Lie (Nominator: F. Klein), 1900 – W. Killing (Nominator: F. Engel), 1904 – D. Hilbert (Nominator: H. Poincaré). *"The extraordinary significance of Lie's work for the general development of geometry can not be overstated. I am convinced that in years to come it will grow still greater."* So wrote Felix Klein in his nomination of Lie's results on group-theoretical foundations of geometry to receive the prize.

In 1886 Lie succeeded F. Klein in the chair of geometry at the University of

Leipzig where he worked until 1898. In 1898 the Norwegian government created at the University of Christiania a special post on continuous groups for Sophus Lie. Lie accepted this post and returned to Norway in September of 1898. But his health was already bad and he died on 18 February 1899 in Christiania.

The reader can find Lie's original papers in [5.12]. His fundamental treatise [5.13] in three volumes written in collaboration with F. Engel and the series of books [5.14]-[5.16] based on Lie's lectures and prepared for publication together with G. Scheffers, represent works of wide range and great originality.

5.2.2 Symmetry groups, Lie algebras and integration of ordinary differential equations

A set G of invertible point transformations in the (x,y) plane $\mathbb{R}^2$,

$$\overline{x} = f(x,y,a), \quad \overline{y} = g(x,y,a), \tag{5.1}$$

depending on a parameter a is called a one-parameter *continuous group*, if G contains the identity transformation (e.g. $f = x, g = y$ at $a = 0$) as well as the inverse of its elements and their composition.

It is said that the transformations (5.1) form a *symmetry group* of a differential equation $F(x,y,y',\ldots,y^{(n)}) = 0$ if the equation is form invariant, i.e.

$$F(\overline{x},\overline{y},\overline{y}',\ldots,\overline{y}^{(n)}) = 0$$

whenever $F(x,y,y',\ldots,y^{(n)}) = 0$. A symmetry group of a differential equation is also termed a group *admitted* by this equation.

Lie's theory reduces the construction of the largest symmetry group G to the determination of its *infinitesimal transformations*,

$$\overline{x} \approx x + a\xi(x,y), \quad \overline{y} \approx y + a\eta(x,y), \tag{5.2}$$

defined as a linear part (in the group parameter a) in the Taylor expansion of the *finite transformations* (5.1) of G. It is convenient to represent an infinitesimal transformation (5.2) by the linear differential operator

$$X = \xi(x,y)\frac{\partial}{\partial x} + \eta(x,y)\frac{\partial}{\partial y} \tag{5.3}$$

called by Lie the *symbol* of the infinitesimal transformation. In the modern literature, (5.3) is often referred to as the *infinitesimal operator* or the *generator* of the group G. The generator X of the group admitted by a differential equation is also termed an operator admitted by this equation.

An infinitesimal operator admitted by the differential equation satisfies a system of linear partial differential equations termed the *determining equation* of the infinitesimal symmetries. The chief property of the determining equation is that the vector space of its solutions is closed under the commutator, i.e. the

5.2. SOPHUS LIE AND SYMMETRY ANALYSIS

commutator (4.95) $X = [X_1, X_2]$ of infinitesimal symmetries X_1, X_2 is again a symmetry. A vector space closed under the commutator is called a Lie algebra. Hence, the infinitesimal symmetries of a given differential equation form a Lie algebra. The theory of Lie algebras is one of the major fields of mathematics.

An essential feature of a symmetry group G is that it conserves the set of solutions of the differential equation admitting this group. Namely, the symmetry transformations merely permute the integral curves among themselves. It may happen that some of the integral curves are individually unaltered under G. Such integral curves are termed *invariant solutions*.

Example 1. Equation (2.40), $F(y, y', \ldots, y^{(n)}) = 0$, furnishes us with a simple example for exhibiting the symmetry of differential equations. Since this equation does not explicitly contain the independent variable x, it does not alter after any transformation $\bar{x} = x + a$ with an arbitrary parameter a. The latter transformations form a group known as the *group of translations* along the x-axis.

Equation (2.40) is in fact the general nth-order ordinary differential equation admitting the group of translations along the x-axis. Moreover, any one-parameter group reduces, in proper variables, to the group of translations. These new variables, *canonical variables t and u*, are obtained by solving the equations

$$X(t) = 1, \quad X(u) = 0, \tag{5.4}$$

where X is the generator (5.3) of the group G. It follows that an nth-order ordinary differential equation admitting a one-parameter group reduces to the form (2.40) in the canonical variables, and hence one can reduce its order to $n - 1$. In particular, any first-order equation with a known one-parameter symmetry group can be integrated by quadrature using canonical variables. Integration of second-order equations requires two independent infinitesimal symmetries (namely, two-dimensional Lie algebras), etc.

Example 2. The Riccati equation (2.31) of the following particular form:

$$\frac{dy}{dx} + y^2 - \frac{2}{x^2} = 0,$$

admits the one-parameter group of non-homogeneous dilations (scaling transformations) $\bar{x} = xe^a$, $\bar{y} = ye^{-a}$. Indeed, the derivative y' is written in the new variables as $\bar{y}' = y'e^{-2a}$, whence

$$\frac{d\bar{y}}{d\bar{x}} + \bar{y}^2 - \frac{2}{\bar{x}^2} = \left(\frac{dy}{dx} + y^2 - \frac{2}{x^2}\right)e^{-2a} = 0.$$

Hence, the equation is form invariant. The symmetry group has the generator

$$X = x\frac{\partial}{\partial x} - y\frac{\partial}{\partial y}.$$

Solution of equations (5.4) with the above operator X provides the canonical variables, $t = \ln x$ and $u = xy$. In these variables, the original Riccati equation takes the following integrable form:

$$\frac{du}{dt} + u^2 - u - 2 = 0.$$

5.2.3 Group classification of differential equations

The great success in integration using symmetries provided Lie with an incentive to begin the classification of all ordinary differential equations of an arbitrary order in terms of symmetry groups, and thus to describe the whole set of equations integrable by his methods. I give here *in extenso* Lie's own words [5.17]: "*In a short communication to the Scientific Society of Göttingen (3 December 1874), I gave, inter alia, a listing of all continuous transformation groups in two variables x, y, and especially emphasized that this might be made the basis of a classification and rational integration theory of all differential equations $f(x, y, y', \ldots, y^{(m)}) = 0$ admitting a continuous group of transformations. The vast program sketched there I have subsequently carried out in detail.*"

The result of implementing this program, as it applies, e.g., to second-order equations, was sensational. At that time, several hundreds of special types of differential equations of the second order were integrated by *ad hoc* methods. Lie's classification showed that most of these equations can be integrated by a single method furnished by group theory and that they can be simplified by a mere change of variables and sorted into only *four* basic types!

Lie's four canonical equations, admitting two-dimensional Lie algebras, are:

$$y'' = f(y'), \quad y'' = f(x), \quad y'' = \frac{1}{x}f(y'), \quad y'' = f(x)y'. \tag{5.5}$$

Here f is an arbitrary function of its arguments. It is readily seen that each equation (5.5) is integrable by quadratures. These and a rich store of other results on symmetry analysis of all ordinary differential equations are to be found in [5.17]. The result of classification of second-order equations is given in [5.20](i).

In 1881, Lie [5.18] gave a group classification of linear second-order partial differential equations with two independent variables, x and y:

$$a_{11}u_{xx} + 2a_{12}u_{xy} + a_{22}u_{yy} + b_1 u_x + b_2 u_y + cu = 0, \tag{5.6}$$

where the coefficients a_{ij}, b_i, and c are arbitrary functions of x and y. Lie also developed a new method for constructing exact solutions known today as invariant solutions. This paper discloses many of Lie's original ideas on symmetry analysis and can serve as a concise introduction to practical methods of group classification of partial differential equations.

5.2.4 Linearization of second-order equations

The solution of this important problem is due to Lie ([5.17], Part III, § 1). Lie's linearization test states that a second-order ordinary differential equation can be linearized, by a suitable change of variables x and y, if it is cubic in y' :

$$y'' + F_3(x,y)y'^3 + F_2(x,y)y'^2 + F_1(x,y)y' + F(x,y) = 0 \qquad (5.7)$$

and the coefficients F_3, F_2, F_1 and F satisfy certain conditions (see Section 12.3).

5.2.5 Nonlinear superposition

It was mentioned in Sections 2.1.2 and 2.1.8 that the general solution of equations with separated variables and of the Riccati equation can be given by nonlinear superposition formulas. The theory of Lie algebras amplified the basis for an extension of this remarkable property to systems of nonlinear ordinary differential equations [5.19]. The result is as follows (cf. equations (2.9) and (2.31)).

The general system of first-order ordinary differential equations admitting a nonlinear superposition has the form

$$\frac{dx^i}{dt} = T_1(t)\xi_1^i(\boldsymbol{x}) + \cdots + T_r(t)\xi_r^i(\boldsymbol{x}), \quad i=1,\ldots,n, \qquad (5.8)$$

where the operators[1]

$$X_\alpha = \xi_\alpha^i(\boldsymbol{x})\frac{\partial}{\partial x^i}, \quad \alpha = 1,\ldots,r, \qquad (5.9)$$

span a finite-dimensional Lie algebra. Here $\boldsymbol{x} = (x^1,\ldots,x^n)$.

This theory, the Vessiot-Guldberg-Lie theory, is discussed in Section 11.2.

5.3 Tangent transformations

5.3.1 Contact transformations and their applications

In addition to point transformations (5.1), contact (*alias* first-order tangent) transformations have been found to be of use in mechanics, optics and geometry. Furthermore, Lie showed that the theory of partial differential equations of the first order reduces to the theory of groups of contact transformations. This is due to the fact that any first-order partial differential equation (4.1), $F(\boldsymbol{x}, u, \boldsymbol{p}) = 0$, can be mapped into any other equation of the first order, $H(\boldsymbol{x}, u, \boldsymbol{p}) = 0$, by means of a suitable contact transformation.

Contact transformations exist, in general, in the case of an arbitrary number of independent variables, but a single dependent variable only. These transformations involve independent variables $\boldsymbol{x} = (x^1,\ldots,x^n)$, a dependent variable u and its partial derivatives $\boldsymbol{p} = (p_1,\ldots,p_n)$, $p_i = \partial u/\partial x^i$. Thus, a contact transformation maps the points $(\boldsymbol{x}, u, \boldsymbol{p}) \in \mathbb{R}^{2n+1}$ into new positions $(\overline{\boldsymbol{x}}, \overline{u}, \overline{\boldsymbol{p}}) \in \mathbb{R}^{2n+1}$:

$$\overline{\boldsymbol{x}} = f(\boldsymbol{x}, u, \boldsymbol{p}), \quad \overline{u} = g(\boldsymbol{x}, u, \boldsymbol{p}), \quad \overline{\boldsymbol{p}} = h(\boldsymbol{x}, u, \boldsymbol{p}), \qquad (5.10)$$

---

[1] Here and in the rest of the book the common convention of dropping the summation symbol is used for expressions containing repeated super- and subscripts: $\xi_\alpha^i \partial/\partial x^i \equiv \sum_{i=1}^n \xi_\alpha^i \partial/\partial x^i$.

subject to the transformation (5.10) being restricted by the *contact condition*

$$\mathrm{d}\bar{u} - \bar{p}_i\,\mathrm{d}\bar{x}^i = \lambda(\boldsymbol{x}, u, \boldsymbol{p})\left(\mathrm{d}u - p_i\mathrm{d}x^i\right), \tag{5.11}$$

where $\lambda(\boldsymbol{x}, u, \boldsymbol{p})$ is an indeterminate multiplier.

Example. The following particular contact transformation is known as the *Legendre transformation*:

$$\bar{\boldsymbol{x}} = \boldsymbol{p}, \quad \bar{u} = -u + \boldsymbol{x}\cdot\boldsymbol{p}, \quad \bar{\boldsymbol{p}} = \boldsymbol{x}, \tag{5.12}$$

where $\boldsymbol{x}\cdot\boldsymbol{p} = x^i p_i$ is the *scalar product* of vectors $\boldsymbol{x}$ and $\boldsymbol{p}$. One can readily verify that the contact condition (5.11) is satisfied (see Problem 5.2).

The Legendre transformation has various geometrical and mechanical applications. For example, it connects the families of parabolas and hyperbolas given in Section 1.2.2 (Exercises 2 and 4). Indeed, the Legendre transformation (5.12), $t = y', z = -y + xy', z' = x$, reduces the differential equation of hyperbolas (1.19):

$$y''' - \frac{3}{2}\frac{y''^2}{y'} = 0$$

to the linear equation $2tz''' + 3z'' = 0$. Here $z = z(t)$ and z', z'', z''' are the successive derivatives of z with respect to t. By substituting $s = \sqrt{t}$, one arrives at the differential equation of parabolas (1.17):

$$\frac{\mathrm{d}^3 z}{\mathrm{d}s^3} = 0.$$

Contact transformations are of considerable use in the Hamilton-Jacobi theory of integration of Hamilton's equations of motion and are known in mechanics as *canonical transformations*. The latter term reflects the chief property of contact transformations, considered in mechanics, to preserve the canonical form (4.88) of the equations of motion. The crux of the matter is that one can simplify an arbitrary Hamilton-Jacobi equation (4.87), e.g. reduce it to that with $H = 0$ by a suitable canonical transformation, and hence immediately integrate the new canonical equations (4.88). It is clear that the inverse of canonical transformations and their composition are again canonical transformations. Hence, the set of all canonical transformations is a group. Thus, the Hamilton-Jacobi theory reduces mechanical problems to the theory of contact transformation groups.

Furthermore, Lie [5.20] indicated the significance of contact transformations in geometrical optics. Namely, he showed that the famous Huygens' construction of wave fronts in light propagation (restricted here, for the sake of simplicity, to the (x, y) plane) is identical to the statement that the formulae

$$\bar{x} = x + \frac{tp}{\sqrt{1+p^2}}, \quad \bar{y} = y - \frac{t}{\sqrt{1+p^2}}, \quad \bar{p} = p, \tag{5.13}$$

define a one-parameter group of contact transformations, where $p = \mathrm{d}y/\mathrm{d}x$ and the time t is regarded as a group parameter.

In the case of several dependent variables, contact transformations are *trivial*, i.e. they are mere extended point transformations.

5.3.2 Infinitesimal contact transformations

The definition of a one-parameter group of point transformations (5.1) naturally generalizes to contact transformations (5.10) depending upon a parameter a. Lie showed that all one-parameter groups of contact transformations are determined by *infinitesimal contact transformations* (see, e.g., [5.13], vol. 2, Chapter 14):

$$\bar{x}^i \approx x + a\xi^i(\boldsymbol{x},u,\boldsymbol{p}), \quad \bar{u} \approx u + a\eta(\boldsymbol{x},u,\boldsymbol{p}), \quad \bar{p}_i \approx x + a\zeta_i(\boldsymbol{x},u,\boldsymbol{p}), \qquad (5.14)$$

where

$$\xi^i = -\frac{\partial W}{\partial p_i}, \quad \eta = W - p_i\frac{\partial W}{\partial p_i}, \quad \zeta_i = \frac{\partial W}{\partial x^i} + p_i\frac{\partial W}{\partial u} \qquad (5.15)$$

with an arbitrary function $W = W(\boldsymbol{x},u,\boldsymbol{p})$ called by Lie the *characteristic function* of the infinitesimal transformation (5.14). See Section 8.3.7 for details.

5.3.3 Higher order tangent transformations: Lie and Bäcklund's discussion

According to the previous section, the variety of contact transformations is confined to the set of functions $W(\boldsymbol{x},u,\boldsymbol{p})$. This set is adequate for a group theoretic description of first-order partial differential equations (mentioned in Section 5.3.1), but it is not sufficient for tackling higher order equations in a similar way. Therefore, Lie raised the question of the existence of higher order tangent transformations and emphasized their importance for the theory of higher order partial differential equations [5.21].

Independently, A.V. Bäcklund [5.22] encountered the question of whether there are surface transformations for which the second-order tangency (osculation) conditions,

$$du - u_i dx^i = 0, \quad du_i - u_{ij} dx^j = 0,$$

are invariant rather than the first-order tangency condition (5.11). He investigated this question in the more general context of arbitrary finite-order tangent transformations of plane curves and surfaces in a three-dimensional space. However, Bäcklund came to a negative answer to the above question. When the result was published in 1874, Bäcklund learned that the idea underlying the existence of higher order tangent transformations and possible applications in the theory of differential equations had been discussed by Lie. This fact encouraged him to prepare and publish in 1876 a revised and enlarged version [5.23] of his work.

Bäcklund begins with the most general transformations written (in our notation) in the form:

$$\bar{x}^i = f^i(x,u,u_{(1)},u_{(2)},\ldots), \quad \bar{u} = g(x,u,u_{(1)},u_{(2)},\ldots), \quad i=1,\ldots,n, \qquad (5.16)$$

where $u_{(1)} = \{u_i\}$, $u_{(2)} = \{u_{ij}\}$, ... are the sets of partial derivatives of the first, second and higher orders. The transformations (5.16) are extended to all derivatives through differentiation and elimination to obtain

$$\bar{u}_i = h_i(x,u,u_{(1)},u_{(2)},\ldots), \quad \bar{u}_{ij} = h_{ij}(x,u,u_{(1)},u_{(2)},\ldots), \quad \ldots. \qquad (5.17)$$

The resulting *extended transformations* (5.16)–(5.17) leave invariant the contact conditions of any order:

$$\mathrm{d}u - u_i \mathrm{d}x^i = 0, \quad \mathrm{d}u_i - u_{ij}\mathrm{d}x^j = 0, \ldots \text{ to inf.} \tag{5.18}$$

Hence, they can be regarded as *infinite-order tangent transformations*.

The main emphasis of Bäcklund's work was on the restricted types of transformations (5.16). He extensively discussed the following two types.

The first type comprises transformations (5.16)–(5.17) that are closed and invertible in a finite-dimensional space of variables $(x, u, u_{(1)}, \ldots, u_{(s)})$, $s < \infty$:

$$\overline{x}^i = f^i, \quad \overline{u} = g, \quad \overline{u}_i = h_i, \quad \ldots, \quad \overline{u}_{i_1 \cdots i_s} = h_{i_1 \cdots i_s},$$

where $f_i, g, h_i, \ldots, h_{i_1 \cdots i_s}$ are functions of the variables $x, u, u_{(1)}, \ldots, u_{(s)}$. The point and contact transformations belong to this type with $s = 0$ and $s = 1$, respectively. The major part of the paper [5.23] is devoted to the proof of *Bäcklund's non-existence theorem* asserting that the set of the above transformations is confined to point and contact transformations. However, the limiting case $s \to \infty$, i.e. the case of infinite-order tangent transformations, remains as a candidate for a non-trivial generalization of contact transformations. Bäcklund himself did not investigate the details of this possibility.

The second type consists of infinite-valued transformations widely known in the literature as *Bäcklund transformations*. A simple example of this type is provided by the following irreversible transformation [5.24]:

$$\overline{x} = x, \quad \overline{y} = y, \quad \overline{z} = q. \tag{5.19}$$

It maps any second-order partial differential equation of the form

$$r = F(x, y, q, s, t)$$

into a second-order *quasi-linear* equation. Here x and y are independent variables, z is a dependent variable, and $p = z_x$, $q = z_y$, $r = z_{xx}$, $s = z_{xy}$, $t = z_{yy}$.

Bäcklund was awarded a Swedish state research grant to travel abroad for his 1874 work. In 1874 he spent six months in Germany where he met, in particular, F. Klein. Felix Klein clearly realized the significance of Bäcklund's results and subsequently popularized them in his lectures [5.25]. Naturally, these results were well known to colleagues and students of Klein.

Later, various Bäcklund transformations were found and widely used in problems of geometry and the theory of differential equations. However, a possible path of generalization provided by infinite-order tangent transformations was not elaborated until recently. The main obstacle in this direction of study is the fact that the definition and invertibility tests of transformations (5.16)–(5.17) involving infinitely many variables are, in general, incomprehensible.

5.3. TANGENT TRANSFORMATIONS

5.3.4 Modern theory of Lie-Bäcklund transformation groups

Recently, an attempt was undertaken to tackle the problem by imposing the one-parameter Lie group structure on these transformations and using Lie's infinitesimal technique [5.26]. A Lie group of invertible infinite-order tangent transformations of the form (5.16)–(5.17) was termed in [5.26] a *Lie–Bäcklund transformation group* in recognition of the fundamental contribution of Lie and Bäcklund. The extension of point and contact transformation groups to all derivatives provide examples of Lie–Bäcklund transformation groups.

In modern group analysis, there exist a variety of so-called *generalized symmetries* which generalize Lie's infinitesimal generators of point and contact transformation groups. However, the problem still remains whether these generalized symmetries generate, via the Lie equations, a group.

The problem thus far is solved [5.27] for the generalization furnished by *Lie–Bäcklund operators*. The latter have the form

$$X = \xi^i \frac{\partial}{\partial x^i} + \eta^\alpha \frac{\partial}{\partial u^\alpha} + \cdots, \tag{5.20}$$

where ξ^i ($i = 1, \ldots, n$) and η^α ($\alpha = 1, \ldots, m$) are arbitrary locally analytic functions of the independent and dependent variables, x^i and u^α, and of the derivatives $u_{(1)} = \{u_i^\alpha\}, \ldots, u_{(s)} = \{u_{i_1 i_2 \ldots i_s}^\alpha\}$ up to a finite order s. Functions of this type are termed *differential functions*, and the space of all differential functions of all orders s is denoted by $\mathcal{A}$. Thus, a Lie-Bäcklund operator is defined by (5.20) with arbitrary coefficients $\xi^i, \eta^\alpha \in \mathcal{A}$ and with the dots meaning that the action of X is extended to all derivatives $u_{(1)}, u_{(2)}, \ldots$ (see Chapter 8).

The main theorem states (see [5.27] or [5.28], Section 16) that a Lie-Bäcklund operator (5.20) generates a group of Lie-Bäcklund transformations (5.16) represented by the exponential map (3.35):

$$\begin{aligned}\overline{x}^i &= e^{aX}(x^i) \equiv x^i + a\xi^i + \frac{a^2}{2!}X(\xi^i) + \cdots + \frac{a^s}{s!}X^{s-1}(\xi^i) + \cdots, \\ \overline{u}^\alpha &= e^{aX}(u^\alpha) \equiv u^\alpha + a\eta^\alpha + \frac{a^2}{2!}X(\eta^\alpha) + \cdots + \frac{a^s}{s!}X^{s-1}(\eta^\alpha) + \cdots.\end{aligned} \tag{5.21}$$

Example. Let $n = m = 1$, and let u' denote the differentiation of u with respect to x. The operator $X = u''\partial/\partial u + \cdots$ generates the transformation

$$\overline{x} = x, \quad \overline{u} = u + \sum_{s=1}^{\infty} \frac{a^s}{s!} u^{(2s)}. \tag{5.22}$$

The transformation of derivatives is obtained from (5.22) by differentiation:

$$\overline{u}^{(k)} = u^{(k)} + \sum_{s=1}^{\infty} \frac{a^s}{s!} u^{(k+2s)}. \tag{5.23}$$

5.4 Applied group analysis

The rich store of results of Lie's work remained for a long time the special preserve of a few. Consequently, symmetry analysis of differential equations laid dormant until the 1950s.

In the 1960s, however, the group theoretic approach to differential equations was restored and tested in new situations. The result was widespread interest and a flurry of activity in applied group analysis which was regarded as a tool to search for exact solutions of differential equations as well as conservation laws, in particular first integrals of ordinary differential equations.

5.4.1 Symmetry and conservation laws

Conservation equations express unalterable laws of nature. Consequently, fundamental mathematical models are formulated in the form of conservation laws.

The concept of a conservation law generalizes the notion of first integrals of systems of ordinary differential equations of the first order to arbitrary ordinary and partial differential equations. The term *conservation law* is motivated by the conservation of such physical quantities as energy, momentum, etc. It has been known for a long time that the conservation laws of mechanics are connected with the symmetry properties of the physical system [5.29]. This connection is treated, e.g. in Jacobi's *Lectures on Dynamics* [4.12].

E. Noether [5.30], in her paper written in 1918 under the influence of F. Klein, combined the methods of variational calculus with the theory of Lie groups and offered a general approach for constructing conservation laws for Euler-Lagrange equations when their symmetries are known. The general result exhibited by the Noether theorem states, e.g. in the case of Lagrangians of the first order, $L(x^i, u^\alpha, u^\alpha_i)$, that if the integral $\int L dx$ is invariant under an infinitesimal transformation $\overline{x}^i \approx x^i + a\xi^i$, $\overline{u}^\alpha \approx u^\alpha + a\eta^\alpha$, then the quantities

$$T^i = L\xi^i + \left(\eta^\alpha - \xi^k u^\alpha_k\right) \frac{\partial L}{\partial u^\alpha_i}, \quad i = 1, \ldots, n, \tag{5.24}$$

satisfy the conservation equation

$$D_i\left(T^i\right) = 0 \tag{5.25}$$

whenever the functions $u^\alpha = u^\alpha(x)$ solve the Euler-Lagrange equations:

$$\frac{\delta L}{\delta u^\alpha} \equiv \frac{\partial L}{\partial u^\alpha} - D_i\left(\frac{\partial L}{\partial u^\alpha_i}\right) = 0, \quad \alpha = 1, \ldots, m. \tag{5.26}$$

Here D_i is the total differentiation of differential functions (see Chapter 8):

$$D_i = \frac{\partial}{\partial x^i} + u^\alpha_i \frac{\partial}{\partial u^\alpha} + u^\alpha_{ij} \frac{\partial}{\partial u^\alpha_j} + \cdots. \tag{5.27}$$

The functions ξ^i and η^α may depend upon the higher order derivatives of u^α.

5.4. APPLIED GROUP ANALYSIS

Noether's proof is based on cumbersome calculations involving variations of integrals $\int L dx$. Consequently, it is commonly regarded to be difficult to present in university texts. In 1979, a new differential algebraic identity was invented [5.31] that underlies conservation theorems and considerably simplifies the proof. This approach is used in Section 9.7.2 of the present book.

5.4.2 Restoration of group analysis in the 1960s

The application of Lie's theory of differential equations to mathematical models in natural sciences has been essentially advanced in the 1960s, to a great extent due to the work of L.V. Ovsyannikov [5.32]. His monograph [5.33] has awoken a broad interest in the subject amongst applied mathematicians. In fact, its publication was the origin of modern *applied group analysis*. The emphasis in his book is on infinitesimal calculus of symmetries (determining equations), group classification of systems of linear and nonlinear partial differential equations, and construction of their exact (invariant and partially invariant) solutions.

Furthermore, he supplemented [5.34] Lie's group classification of linear second-order differential equations in two independent variables and discovered new invariants of differential equations. Ovsyannikov's invariants are closely related to the Laplace invariants and are used in modern group analysis, e.g. for the group theoretic treatment of Riemann's method of solution of Cauchy's problem.

It was my good fortune to get interested in Lie group analysis at the very beginning of my university work, and to write my first paper (cited in [9.13]) under the supervision of Professor Ovsyannikov. Subsequently we organized and led with him a joint research seminar on *Group Analysis of Differential Equations* at Novosibirsk University (Russia) during 1967-1980. In these seminars and my later work I saw over and over how effective a tool Lie theory is for tackling complicated problems of differential equations.

5.4.3 Invariant and partially invariant solutions

Special types of exact solutions, widely known today as *invariant solutions*, have long been successfully applied to the analysis of specific problems and were in common use in mechanics and physics prior to the development of group theory, becoming in the process a sort of folklore. Lie [5.18] came to these solutions from the point of view of group invariance and proved the main theorem on invariant solutions of partial differential equations [5.11] (see further Section 9.4).

Ovsyannikov's work made it possible to clarify many intuitive ideas and apply them to numerous equations of mechanics, as a result of which the method of invariant solutions could be included as an important integral part of modern group analysis. It was precisely through the concept of an invariant solution that the theater of action of group analysis shifted in the 1960s from ordinary differential equations to problems of mechanics and mathematical physics. Ovsyannikov introduced a new class of solutions, the so-called *partially invariant solutions*. He also suggested to classify all distinctly different invariant and partially invariant solutions by constructing *optimal systems* of subalgebras of Lie algebras.

Invariant and partially invariant solutions are represented via invariants. Consequently, the search for this type of solutions reduces the number of variables. In particular, ordinary differential equations reduce to algebraic equations.

Example 1. Let us consider the Riccati equation discussed in Section 5.2.2,

$$\frac{dy}{dx} + y^2 - \frac{2}{x^2} = 0, \qquad (5.28)$$

and find its invariant solutions with respect to the infinitesimal symmetry:

$$X = x\frac{\partial}{\partial x} - y\frac{\partial}{\partial y}. \qquad (5.29)$$

The invariant test, $X(J) = 0$, provides one independent invariant, $J = xy$. Therefore $J = \text{const.}$ is the only relation written in terms of invariants. Hence, the general form of invariant solutions is $xy = \lambda$, or $y = \lambda/x$, with an arbitrary constant λ. The substitution into (5.28) reduces the latter to a quadratic equation, $\lambda^2 - \lambda - 2 = 0$, whence $\lambda_1 = 2$ and $\lambda_2 = -1$. Thus, one arrives at the following two invariant solutions of equation (5.28):

$$y_1 = \frac{2}{x} \quad \text{and} \quad y_2 = -\frac{1}{x}. \qquad (5.30)$$

Example 2. Lie [5.18] used the group classification of linear partial differential equations to identify those equations (5.6) for which one can obtain solutions with two arbitrary functions by solving ordinary differential equations. Given an equation (5.6) admitting a group, Lie considers two independent invariant solutions involving a parameter, then takes their linear combination with coefficients depending upon this parameter, and integrates with respect to the parameter. His method is exhibited by the following simple example. Let the coefficients a_{ij}, b_i, and c of equation (5.6) be independent of y. Then the equation

$$a_{11}(x)u_{xx} + 2a_{12}(x)u_{xy} + a_{22}(x)u_{yy} + b_1(x)u_x + b_2(x)u_y + c(x)u = 0 \qquad (5.31)$$

admits the following generator with an arbitrary constant λ:

$$X = \frac{\partial}{\partial y} + \lambda u\frac{\partial}{\partial u}.$$

The equation $X(J) = 0$ provides two independent invariants, x and $ue^{-\lambda y}$. Hence, one can take invariant solutions in the form $u = e^{\lambda y}v(x)$. Then (5.31) reduces to

$$a_{11}v'' + (2\lambda a_{12} + b_1)v' + (\lambda^2 a_{22} + \lambda b_2u_y + c)v = 0. \qquad (5.32)$$

If $a_{11}(x) \neq 0$, the ordinary differential equation (5.32) of the second order has two linearly independent solutions, $v_1 = \phi_1(x, \lambda)$ and $v_2 = \phi_2(x, \lambda)$, involving λ. A solution of (5.31) containing two arbitrary functions, $f_1(\lambda)$ and $f_2(\lambda)$, is given by the integral with any constant limits $\alpha_1, \beta_1, \alpha_2, \beta_2$:

$$u = \int_{\alpha_1}^{\beta_1} f_1(\lambda)e^{\lambda y}\phi_1(x, \lambda)d\lambda + \int_{\alpha_2}^{\beta_2} f_2(\lambda)e^{\lambda y}\phi_2(x, \lambda)d\lambda. \qquad (5.33)$$

5.4.4 Symmetry of fluids

Today, there is a considerable literature on group analysis of various models in hydrodynamics (see [5.35] and the references therein). One of the important results is the group classification of the system of partial differential equations governing adiabatic gas motions:

$$\begin{aligned} \boldsymbol{v}_t + (\boldsymbol{v} \cdot \nabla)\boldsymbol{v} + \rho^{-1}\nabla p &= 0, \\ \rho_t + \boldsymbol{v} \cdot \nabla \rho + \rho \operatorname{div} \boldsymbol{v} &= 0, \\ p_t + \boldsymbol{v} \cdot \nabla p + A(p,\rho) \operatorname{div} \boldsymbol{v} &= 0, \end{aligned} \quad (5.34)$$

where $A(p,\rho)$ is an arbitrary function connected with the entropy $S(p,\rho)$ by $A = -\rho S_\rho/S_p$. The Lie algebra admitted by the system (5.34) with arbitrary $A(p,\rho)$ generates the Galilean group augmented by a scaling transformation. According to the group classification presented in [5.33] (see also [5.35], vol. 1), this Lie algebra extends for the following seven particular functions A:

(1) $A = pf\left(p^{1+2k}\rho\right)$, (2) $A = f\left(\rho e^{-2p}\right)$, (3) $A = f(\rho)$,

(4) $A = f(p)$, (5) $A = \rho^k$, (6) $A = \gamma p$, (7) $A = [(n+2)/n]p$. (5.35)

Here f denotes any function of the indicated arguments, γ is a constant known as an adiabatic exponent, k is an arbitrary constant, and n assumes the values 1, 2 and 3 for one-dimensional, planar and spatial flows, respectively. In cases (1) to (4) the admitted Lie algebra extends by one, in cases (5) and (6) by two, and in case (7) by three infinitesimal generators.

Case (7) occurs for the monatomic gas (when $n = 3$) and shallow-water equations (when $n = 2$). In this case the Galilean group is extended by the projective transformation. The physical significance of this extension is that the monatomic gas and shallow-water equations have, along with classical energy, momentum and angular momentum, two additional conservation laws [5.28]:

$$\begin{aligned} \frac{\mathrm{d}}{\mathrm{d}t}\int_\Omega (2t\mathcal{E} - \rho\boldsymbol{x}\cdot\boldsymbol{v})\mathrm{d}\boldsymbol{x} &= -\int_S p(2t\boldsymbol{v} - \boldsymbol{x})\cdot\boldsymbol{\nu}\mathrm{d}S, \\ \frac{\mathrm{d}}{\mathrm{d}t}\int_\Omega [2t^2\mathcal{E} - \rho\boldsymbol{x}\cdot(2t\boldsymbol{v} - \boldsymbol{x})]\mathrm{d}\boldsymbol{x} &= -\int_S 2tp(t\boldsymbol{v} - \boldsymbol{x})\cdot\boldsymbol{\nu}\mathrm{d}S. \end{aligned} \quad (5.36)$$

Here Ω is an arbitrary domain moving with the fluid, S is the boundary of Ω, $\boldsymbol{\nu}$ is a unit external normal to S, and $\mathcal{E} = (\rho|\boldsymbol{v}|^2 + np)/2$ is the energy density.

5.5 New trends

5.5.1 Invariance principle in boundary-value problems

When we pass from ordinary to partial differential equations, it becomes impossible (with rare exceptions) and anyway virtually futile to write out general solutions. But mathematical physics in any case seeks only those solutions that satisfy given boundary conditions.

Group invariant solutions can serve in the solution of *invariant boundary-value problems*. This application is governed by the following rule.

Invariance principle: *If a boundary-value problem is invariant under a group G, then we should seek a solution among functions invariant under G.* Here, invariance of a boundary-value problem means that the differential equation, the manifold where the data are given, and the data themselves are invariant under G.

Invariant solutions are usually regarded to be not particularly useful for solving boundary-value problems because arbitrary data are not invariant. It seems that this argument is irrefutable. However, it was shown recently [5.36] that the invariance principle furnishes a systematic method suited to purposes of solving linear initial-value problems when it is combined with the theory of *distributions* and *fundamental solutions*.

The new approach, however, necessitates an extension of Lie group methods to differential equations in distributions. The method was tested on classical equations of mathematical physics such as the Laplace, heat conduction and wave equations. It was also used in the investigation of Huygens' principle for wave equations in curved space-times as well as for solving an arbitrary initial-value problem for the Black-Scholes equation (1.5) in the mathematics of finance.

An overview of this natural path of development of Lie group analysis venturing into the space of distributions with numerous applications is presented in [5.35] (vol. 3, Chap. 3, and vol. 2, Chap. 3, 4, 7).

5.5.2 "Hidden" symmetries, in particular non-local symmetries

It is known that Kepler's second law (see Section 1.3.4) is equivalent to the conservation of the angular momentum $M = m\boldsymbol{x} \times \boldsymbol{v}$ (see [5.29]). It is also known that the latter is obtained by the Noether theorem (Problem 5.6(ii)) due to the rotational symmetry of Newton's gravitation force (1.33), i.e. of the Lagrangian

$$L = \frac{1}{2}m\sum_{i=1}^{3}(v^i)^2 - \frac{\mu}{r}, \qquad \mu = \text{const.} \tag{5.37}$$

In 1798, Laplace [5.37] discovered the conservation of the vector

$$\boldsymbol{A} = \boldsymbol{v} \times M + \mu\frac{\boldsymbol{x}}{r} \tag{5.38}$$

and showed that Kepler's first law is its direct consequence. The question, however, remained open whether Newton's gravitation field possesses specific symmetries that provide the *Laplace vector* (5.38) via Noether's theorem. This "hidden" symmetry was found recently ([5.28], Section 25.1) by using the theory of Lie-Bäcklund groups. It is the following infinitesimal Lie-Bäcklund symmetry:

$$\delta\boldsymbol{x} = \boldsymbol{x} \times (\boldsymbol{v} \times \boldsymbol{a}) + (\boldsymbol{x} \times \boldsymbol{v}) \times \boldsymbol{a}, \tag{5.39}$$

5.5. NEW TRENDS

where $\boldsymbol{a} = (a_1, a_2, a_3)$ is a vector-parameter (cf. with the infinitesimal rotations written in the vector form $\delta \boldsymbol{x} = \boldsymbol{x} \times \boldsymbol{a}$; see Remark 9.7.3). The Noether theorem is applicable to (5.39) and yields the Laplace vector (5.38).

Other types of "hidden" symmetries are provided by so-called *non-local symmetries*. This approach gives about 35 particular functions $A(p, \rho)$ (including Ovsyannikov's (5.35)), when the one-dimensional equations (5.34),

$$\begin{aligned} v_t + vv_x + \rho^{-1} p_x &= 0, \\ \rho_t + v\rho_x + \rho v_x &= 0, \\ p_t + vp_x + A(p,\rho) v_x &= 0, \end{aligned} \tag{5.40}$$

possess additional symmetries. They can be sorted into 13 basic types [5.38].

Example. In the case (5.35)(3), $A = f(\rho)$, equations (5.40) have point symmetries (an extended Galilean group) generated by

$$X_1 = \frac{\partial}{\partial t},\ X_2 = \frac{\partial}{\partial x},\ X_3 = t\frac{\partial}{\partial x} + \frac{\partial}{\partial v},\ X_4 = t\frac{\partial}{\partial t} + x\frac{\partial}{\partial x},\ X_5 = \frac{\partial}{\partial p}, \tag{5.41}$$

and an infinitesimal non-local symmetry

$$Y = \frac{t^2}{2}\frac{\partial}{\partial x} + t\frac{\partial}{\partial v} - R\frac{\partial}{\partial p}, \tag{5.42}$$

where R is a non-local variable defined by the equations $R_x = \rho,\ R_t = -\rho v$. The physical significance of the group transformations generated by Y:

$$\bar{t} = t,\ \bar{x} = x + \frac{at^2}{2},\ \bar{v} = v + at,\ \bar{\rho} = \rho,\ \bar{p} = p - aR,$$

is that the motion of a compressible fluid with $A(p, \rho) = f(\rho)$ is insensitive with regard to passage to a coordinate frame $\bar{x}$ moving with a uniform acceleration a, provided that the pressure undergoes a non-local transformation $\bar{p} = p - a\int \rho dx$.

5.5.3 Approximate symmetries

A variety of differential equations, recognized as mathematical models in engineering and physical sciences, involve empirical parameters or constitutive laws. Therefore the coefficients of model equations are defined approximately with an inevitable error. Consequently, differential equations depending on a small parameter occur frequently in applications.

For tackling differential equations with a small parameter, a new direction in group analysis, namely a method of approximate symmetries, was initiated in 1987 and then elaborated in our subsequent work [5.38](ii). The core of this approach is the concept of approximate transformation groups. A brief sketch of the theory of approximate transformation groups is given in Section 7.5. For a detailed discussion of the present state of the art and numerous applications, the reader is referred to [5.38](ii) and [5.35], vol. 3, Chap. 2 and 9.

5.5.4 Group theoretic modelling

Knowledge of symmetry groups can be used to enhance our understanding of complex natural phenomena and to solve problems formulated in terms of differential equations. Moreover, experience gained from solving problems of mathematical physics by means of group analysis has convinced me that quite a few phenomena can be modelled directly in group theoretical terms and that differential equations, conservation laws, solutions to initial value problems, and so forth can be obtained as immediate consequences of group theoretic modelling [5.39].

Group classification of differential equations furnishes us an effective method of group theoretic modelling. Indeed, differential equations occurring in engineering and the physical sciences often involve indeterminate parameters or arbitrary functions of certain variables. Usually, these arbitrary elements (parameters or functions) are found experimentally or chosen from a "simplicity criterion". Classification of equations according to their symmetries provides a regular procedure for determining arbitrary elements. In this approach, differential equations admitting more symmetries are considered to be "preferable". Consequently, one often arrives at equations possessing remarkable physical properties.

Example 1. Invariance under the group composed by time translations, rotations, a scaling transformation and the Lie-Bäcklund transformations defined by the increment (5.39), $\delta \boldsymbol{x} = \boldsymbol{x} \times (\boldsymbol{v} \times \boldsymbol{a}) + (\boldsymbol{x} \times \boldsymbol{v}) \times \boldsymbol{a}$, determines uniquely the Lagrangian (5.37) of Newton's gravitation theory, in particular all Kepler's laws.

Example 2. The heat conduction equation $u_t - 4k\Delta u = 0$ is determined by the invariance under the Galilean group extended by a scaling transformation. Here $\Delta = \nabla^2$ is the Laplace operator with respect to the variables $x = (x^1, \ldots, x^n)$. It is assumed that the heat balance law is independent of the choice of an inertial reference frame. This natural assumption leads to the physically surprising conclusion that the temperature in a diffusion process depends on the choice of an inertial reference frame [5.39](i). Namely, *if an observer at rest detects the temperature field* $u = \phi(t, x)$, *then another observer moving with a constant velocity* $V = (V^1, \ldots, V^n)$ *will detect (in his local frame* $\bar{x} = x + tV$*) the temperature*

$$\bar{u}(t, \bar{x}) = e^{k(t|V|^2 - 2\bar{x}\cdot V)} \phi(t, \bar{x} - tV). \tag{5.43}$$

Furthermore, the fundamental solution, and hence the diffusion of any initial temperature $u|_{t=0} = u_0(x)$, is uniquely determined by the invariance principle.

Example 3. Let $ds^2 = -(1+\varepsilon|x|^2)^{-2} dx_\mu dx^\mu$ be the metric of a de Sitter universe with a small constant curvature $K \neq 0$, where $\varepsilon = K/4$. A group analysis of Dirac's theory [5.40] shows that a neutrino described by the Dirac equation in the Minkowski space-time,

$$\gamma^\mu \frac{\partial \psi}{\partial x^\mu} = 0, \tag{5.44}$$

splits into two neutrinos in the de Sitter universe ([5.39](i), (iv)). These two

5.5. NEW TRENDS

neutrinos are described (in the first order of precision in ε) by the equations

$$\gamma^\mu \frac{\partial \phi}{\partial x^\mu} - 3\varepsilon(x\cdot\gamma)\phi = 0 \tag{5.45}$$

and

$$\gamma^\mu \frac{\partial \varphi}{\partial x^\mu} + \varepsilon(x\cdot\gamma)\varphi = 0, \tag{5.46}$$

where $(x\cdot\gamma) = x_\mu \gamma^\mu$. Equations (5.45) and (5.46) can be treated as the Dirac equations $\gamma^\mu \partial\phi/\partial x^\mu + m\phi = 0$ with the variable matrix valued "effective" masses $m_1 = -3\varepsilon(x\cdot\gamma)$ and $m_2 = 3\varepsilon(x\cdot\gamma)$, respectively. Then, in the framework of the usual relativistic theory, neutrinos will have *small but nonzero* mass. These "massive" neutrinos satisfy the *Huygens principle* and have the velocity of light.

We conclude that, in a de Sitter universe with a small curvature $K \neq 0$, a neutrino is a compound particle *neutrino–antineutrino* with the total mass $m = m_1 + m_2 = 0$. It is natural to assume that only the first component of the compound is observable and is perceived as a massive neutrino. The counterpart to the neutrino provides the validity of the zero-mass-neutrino model and has real meaning (i.e. is observable) in the *antiuniverse* with the curvature $(-K)$.

This theoretical conclusion of group theoretic modelling combined with the neutrino-as-dark matter hypothesis may play a significant role in our understanding the nature of the material of dark halos surrounding galaxies.

Example 4. Scaling transformations are of a great importance in mechanics. Group classification of Lagrangians $L = m|\boldsymbol{v}|^2/2 - U(r)$ reveals that

$$U(r) = -\frac{\beta}{r^2}, \quad \beta = \text{const.}, \tag{5.47}$$

is the only central field (discussed by I. Newton and R. Cotes) for which a scaling transformation satisfies the Noether theorem [5.39](i). For the potential (5.47), the projective transformation

$$\bar{t} = \frac{t}{1-at}, \quad \bar{\boldsymbol{x}} = \frac{\boldsymbol{x}}{1-at} \tag{5.48}$$

also satisfies the Noether theorem. This provides two specific integrals of motion of a particle in the Newton-Cotes potential field (5.47):

$$I_1 = 2tE - m\boldsymbol{x}\cdot\boldsymbol{v}, \quad I_2 = 2t^2 E - m\boldsymbol{x}\cdot(2t\boldsymbol{v} - \boldsymbol{x}), \tag{5.49}$$

where $E = m|\boldsymbol{v}|^2/2 + U(r)$ is the energy.

An astonishing similarity between the hydrodynamic conservation laws (5.36) and the mechanical integrals of motion (5.49) is deeper than a transparent likeness. It turns out that the Boltzmann equation of gas kinetics inherits the scaling and projective symmetries of Newton's equation of motion of many bodies in the Newton-Cotes potential [5.41]. Furthermore, these symmetries, as well as the corresponding conservation laws, remain unaltered in the transition to the equations of continuum mechanics, i.e. to the equations of a monatomic gas. Reversing the process, we can assume that a monatomic gas generates the Newton-Cotes potential field (5.47). Since the solar neighborhood contains mainly monatomic gases, this assumption furnishes us a basis for considering a perturbation of the Lagrangian (5.37) due to the interplanetary gas:

$$L = \frac{1}{2}m \sum_{i=1}^{3}(v^i)^2 - \frac{\alpha}{r} + \frac{\beta}{r^2}. \tag{5.50}$$

Here β is a small constant depending on the state of the interplanetary gas. The perturbation will affect mostly Mercury, the planet nearest to the Sun.

5.5.5 Miscellany

The following topics from modern group analysis are not discussed in this book:
- *Group theory in numerical analysis, in particular, symmetry of finite-difference equations:* [5.35], vol. 1 (Chap. 16, 17).
- *Symmetry of integro-differential equations:* [5.35], vol. 1 (Chap. 15), vol. 2 (Chap. 16), vol. 3 (Chap. 5).
- *Nonclassical and conditional symmetries:* [5.35], vol. 3 (Chap. 11).
- *Symbolic software for calculating symmetries:* [5.35], vol. 3 (Chap. 13, 14).

Problems

5.1. Show that the change of variables $t = \ln x, u = xy$ transforms the Riccati equation $dy/dx + y^2 - 2/x^2 = 0$ to the equation $du/dt + u^2 - u - 2 = 0$.

5.2. Show that the Legendre transformation (5.12) obey the contact condition (5.11).

5.3. Check the contact condition for the transformation (5.13).

5.4. The Lie-Bäcklund transformations T_a given by (5.22)–(5.23) satisfy the group property, $T_b T_a = T_{a+b}$. Check this property in the second order of precision in a and b.

5.5. Invert the Lie-Bäcklund transformation given in Example of Section 5.3.4, i.e. solve equations (5.22)–(5.23) with respect to u.

5.6. The motion of a particle in a central potential field is determined by the Lagrangian $L(t, \boldsymbol{x}, \boldsymbol{v}) = m|\boldsymbol{v}|^2/2 - U(r)$, $m = $ const., where t is the time, $r = |\boldsymbol{x}|$ and $|\boldsymbol{v}|$ are the magnitudes of the position vector $\boldsymbol{x} = (x^1, x^2, x^3)$ and velocity $\boldsymbol{v} = (v^1, v^2, v^3) = d\boldsymbol{x}/dt$. Since the Lagrangian does not involve explicitly the time t and depends only upon the magnitudes $|\boldsymbol{x}|$ and $|\boldsymbol{v}|$ of the vectors $\boldsymbol{x}$ and $\boldsymbol{v}$, the action integral $\int L dt$ is invariant under
   (i) time translations, $\bar{t} = t + a$, and
   (ii) rotations with the generators X_{12}, X_{23}, X_{31}, where $X_{ij} = x^j \dfrac{\partial}{\partial x^i} - x^i \dfrac{\partial}{\partial x^j}$.

Find the corresponding conserved quantities (5.24).

5.7. Find Lie's solution (5.33) for the heat conduction equation $u_t - u_{xx} = 0$.

5.8*. Check that equations (5.40) with $A = f(\rho)$ admit the operator Y, (5.42).

Chapter 6

Preliminaries on transformations and groups

This chapter gives a general idea of transformations and exhibits a variety of transformation groups. It also discusses the duality between changes of coordinates and point transformations [6.1].

6.1 Transformations in elementary mathematics

Problems of elementary mathematics can often be solved by the method of transformations. The material in this section is selected and presented in such a way that it is available for senior high schools, and teachers might wish to use it as additional reading for their students inclined to mathematics.

6.1.1 Algebra: Solution of equations

Recall that the roots $x = x_1$ and $x = x_2$ of the general quadratic equation

$$ax^2 + bx + c = 0, \quad a \neq 0, \tag{6.1}$$

are given by

$$x_{1,2} = \frac{-b \pm \sqrt{b^2 - 4ac}}{2a}. \tag{6.2}$$

The vanishing of the expression $b^2 - 4ac$, known as the *discriminant* of the quadratic equation, is the condition for (6.1) to have two equal roots, $x_1 = x_2$.

In accordance with tradition, students learn from school to derive the solution (6.2) by completing the square. Indeed, this method is simple but it is not suitable for tackling the general cubic as well as equations of higher degrees.

The idea of transformation of equations, unlike the *ad hoc* method of completing the square, furnishes a general method appropriate for solution of the quadratic equation as well as for a simplification of equations of higher degrees.

6. PRELIMINARIES ON TRANSFORMATIONS AND GROUPS

The simplest transformation of equations is provided by a linear transformation of the variable x [6.2]:

$$\bar{x} = x + \varepsilon. \tag{6.3}$$

It converts any equation of degree n into an equation of the same degree. In particular, the quadratic equation (6.1) after the substitution $x = \bar{x} - \varepsilon$ becomes $a\bar{x}^2 + (b - 2a\varepsilon)\bar{x} + a\varepsilon^2 - b\varepsilon + c = 0$. Hence, the transformation (6.3) converts (6.1) into a new quadratic equation, $\bar{a}\bar{x}^2 + \bar{b}\bar{x} + \bar{c} = 0$, where

$$\bar{a} = a, \quad \bar{b} = b - 2a\varepsilon, \quad \bar{c} = c + a\varepsilon^2 - b\varepsilon. \tag{6.4}$$

Defining ε from $b - 2a\varepsilon = 0$, one obtains $\bar{b} = 0$ and $\bar{c} = c - b^2/(4a)$. Hence, the transformation $x = \bar{x} - b/(2a)$ converts (6.1) into the equation

$$a\bar{x}^2 - \frac{b^2 - 4ac}{4a} = 0.$$

Substituting its roots, $\bar{x}_{1,2} = \pm\sqrt{b^2 - 4ac}/(2a)$, into $x = \bar{x} - b/(2a)$ one arrives at the roots (6.2) of the original equation (6.1).

Consider now the general cubic equation written with binomial coefficients for convenience of calculations:

$$ax^3 + 3bx^2 + 3cx + d = 0, \quad a \neq 0. \tag{6.5}$$

After the linear transformation [6.3] $y = ax + b$ it takes the form

$$y^3 + 3py + 2q = 0, \tag{6.6}$$

where

$$p = ac - b^2, \quad 2q = a^2 d - 3abc + 2b^3. \tag{6.7}$$

The reduced equation (6.6) is readily solved by setting $y = \sqrt[3]{k} + \sqrt[3]{l}$. Then $y^3 - 3\sqrt[3]{kl}\, y - (k+l) = 0$, and equation (6.6) yields $k + l = -2q$, $\sqrt[3]{kl} = -p$. It follows that k and l are the roots of the quadratic equation $z^2 + 2qz - p^3 = 0$. Hence one of the roots, e.g. k, is given by $k = -q + \sqrt{q^2 + p^3}$. Let

$$u = \sqrt[3]{-q + \sqrt{q^2 + p^3}}$$

be any one of the three values of this cube root. Then all three values of $\sqrt[3]{k}$ are $u, \epsilon u$, and $\epsilon^2 u$, where ϵ is an imaginary cube root of unity, $\epsilon^3 = 1$. The relation $\sqrt[3]{kl} = -p$ shows that the corresponding values of $\sqrt[3]{l}$ are $-p/u, -\epsilon^2 p/u$, and $-\epsilon p/u$. They can be rewritten in the form $v, \epsilon^2 v, \epsilon v$ with

$$v = \sqrt[3]{-q - \sqrt{q^2 + p^3}}.$$

Summing up, one arrives at Cardan's solution mentioned in Section 5.1. Namely, the roots of (6.6) are given by

$$y_1 = u + v, \quad y_2 = \epsilon u + \epsilon^2 v, \quad y_3 = \epsilon^2 u + \epsilon v, \tag{6.8}$$

6.1. TRANSFORMATIONS IN ELEMENTARY MATHEMATICS

where

$$u = \sqrt[3]{-q + \sqrt{q^2 + p^3}}, \quad v = \sqrt[3]{-q - \sqrt{q^2 + p^3}}. \tag{6.9}$$

Here ϵ is an imaginary cube root of unity, $\epsilon = (-1 + i\sqrt{3})/2$. It is apparent that its square, $\epsilon^2 = (-1 - i\sqrt{3})/2$, is the complex conjugate cube root of unity. The expression $q^2 + p^3$ is termed the *discriminant* of the cubic (6.6). It follows from (6.9) that the vanishing of the discriminant is the condition for (6.6) to have two equal roots. The roots of (6.5) are obtained by substituting (6.8) in $x = (y-b)/a$.

Example. The equation $y^3 - 6y + 4 = 0$ has the form (6.6) with $p = -2$ and $q = 2$. Here $q^2 + p^3 = -4$, and hence (6.9) is written

$$u = \sqrt[3]{2(-1+i)}, \quad v = \sqrt[3]{2(-1-i)}.$$

The reckoning shows (Problem 6.5) that $u = 1 + i$, $v = 1 - i$, and formulae (6.8) provide three distinct real roots: $y_1 = 2$, $y_2 = -(1 + \sqrt{3})$, and $y_3 = -1 + \sqrt{3}$.

6.1.2 Geometry: Similarity and calculation of areas

Consider, in the (x, y) plane, a scaling transformation $\bar{x} = ax, \bar{y} = by$ known also as a similarity transformation or dilation. A scaling transformation corresponds geometrically to a *uniform* expansion (contruction) from the origin if $a = b$, and to a *non-uniform* expansion (contruction) otherwise.

Definition. Two geometric figures obtained one from another by a motion on the plane (i.e. by a translation and rotation) and a scaling transformation (uniform or non-uniform) are said to be *similar*.

Example 1. Any rectangle is similar to the unit square $\{0 \leq x \leq 1, 0 \leq y \leq 1\}$. Indeed, given a rectangle with sides a and b, we first move it by a proper translation and rotation to the "standard location" so that it will have the form $\{0 \leq x \leq a, 0 \leq y \leq b\}$. Then the stretching $\bar{x} = x/a$, $\bar{y} = y/b$ converts the rectangle to the unit square $\{0 \leq \bar{x} \leq 1, 0 \leq \bar{y} \leq 1\}$.

Theorem. Let two plain geometric figures, $\mathcal{M}$ and $\overline{\mathcal{M}}$, be similar and let the latter be obtained from the former by a scaling transformation $\bar{x} = ax, \bar{y} = by$. Then the areas S and $\overline{S}$ of $\mathcal{M}$ and $\overline{\mathcal{M}}$, respectively, are related by

$$\overline{S} = abS. \tag{6.10}$$

Proof. Let us first consider a rectangle with sides m and n in the x and y directions, respectively. After the dilation, one obtains a rectangle with sides $\overline{m} = am$ and $\overline{n} = bn$. Hence, the areas of the original and new rectangles, $S = mn$ and $\overline{S} = \overline{m}\,\overline{n}$, are related by (6.10). One can cover an arbitrary figure (provided that it is not too fancy) by a grid of rectangular areas and apply (6.10) to these rectangles. Imagine now the process repeated over and over again with finer and finer grids. Since the area S of the figure in question is the limit of the sums of the areas of the covering rectangles, this complets the proof.

Remark. Translations and rotations do not alter the area of any geometric figure. This is the chief property of motions.

Example 2. Any ellipse (in particular, a circle) is similar to the unit circle $x^2 + y^2 = 1$. We proceed as in Example 1. Namely, we first move a given ellipse to the standard location so that it is given by

$$\frac{x^2}{a^2} + \frac{y^2}{b^2} = 1. \tag{6.11}$$

Then the stretching $\bar{x} = x/a$, $\bar{y} = y/b$ converts it to the unit circle $\bar{x}^2 + \bar{y}^2 = 1$.

This example and the above transformation theory furnish an elementary method for calculating the area of ellipses. Namely, given an ellipse (6.11), it is converted by the stretching $\bar{x} = x/a$, $\bar{y} = y/b$ into the unit circle $\bar{x}^2 + \bar{y}^2 = 1$ with the area $\bar{S} = \pi$. In this case, formula (6.10) is written $\bar{S} = S/(ab)$, where S is the area of the ellipse. Thus, the area of the ellipse (6.11) is given by

$$S = \pi ab. \tag{6.12}$$

6.2 Transformations in $\mathbb{R}^n$

6.2.1 Changes of coordinates and point transformations

The duality between changes of coordinates and point transformations is inherent in group analysis. It is crucial, e.g., in the theory of prolongations of Lie groups.

Example 1. Consider, in the (x, y) plane, a point P having coordinates (x, y) referred to the rectangular axes Ox, Oy. Let $\overline{Ox}, \overline{Oy}$ be another pair of rectangular axes parallel to the former axes, and such that $\overline{O}$ has the coordinates $(-ae_1, -ae_2)$ with respect to the first frame of reference. Here $e = (e_1, e_2)$ is a fixed unit vector, and a is an arbitrary real parameter. Then the coordinates $(\bar{x}, \bar{y})$ of the same point P in the second frame of reference are given by

$$\bar{x} = x + ae_1, \quad \bar{y} = y + ae_2. \tag{6.13}$$

An alternative interpretation of equations (6.13) is as follows. One ignores the new axes $\overline{Ox}, \overline{Oy}$ and regards (x, y) and $(\bar{x}, \bar{y})$ as the coordinates of points P and $\overline{P}$, respectively, each referred to the original frame Ox, Oy. Then (6.13) define a transformation of the point $P(x, y)$ into the new position $\overline{P}(\bar{x}, \bar{y})$ in the (x, y) plane. Specifically, equations (6.13) determine the displacement (translation) of all points P of the plane through the distance a in the direction of the vector e.

Example 2. Given a pair of rectangular axes Ox, Oy, let $O\bar{x}, O\bar{y}$ be the new pair of axes obtained by rotating the former ones round the origin O counter-clockwise through an angle a. Let (x, y) and $(\bar{x}, \bar{y})$ be the coordinates of a point P referred to the axes Ox, Oy and $O\bar{x}, O\bar{y}$, respectively. Then (see Problem 6.6)

$$\bar{x} = x\cos a + y\sin a, \quad \bar{y} = y\cos a - x\sin a. \tag{6.14}$$

6.2. TRANSFORMATIONS IN $\mathbb{R}^n$

An alternative interpretation of these equations is as follows. One regards (x, y) and $(\bar{x}, \bar{y})$ as the coordinates of the points P and $\bar{P}$, respectively, each referred to the same axes Ox, Oy. Then equations (6.14) accomplish the rotation of all points of the plane about O clockwise through the angle a.

Example 3. Everyone travelling by ship, train, etc. encounters the duality between uniform motions of his local frame of reference (a ship, train, etc.) and outside points (another vessel or a whale on the ship's way, people or other objects on a depot, etc.). This remarkable exhibition of the duality, when one cannot determine who is actually moving, is known in mechanics as *Galileo's relativity principle*. It is equivalent to the invariance of mechanical equations of motion under the transformation

$$\bar{t} = t, \quad \bar{x} = x + tV, \tag{6.15}$$

where V is the constant velocity. Differentiation of $\bar{x}$ with respect to $\bar{t} = t$ yields $\bar{v} = v + V$. This transformation law of velocity is a mathematical expression of Galileo's relativity principle. The transformation (6.15), known as the Galilean transformation, lies at the core of the *Galilean group* which is one of the most important groups in physics.

6.2.2 Definition of a group

Consider invertible transformations in an n-dimensional Euclidean space $\mathbb{R}^n$ defined by equations of the form

$$\bar{x}^i = f^i(x), \quad i = 1, \ldots, n, \tag{6.16}$$

where $x = (x^1, \ldots, x^n) \in \mathbb{R}^n$ and $\bar{x} = (\bar{x}^1, \ldots, \bar{x}^n) \in \mathbb{R}^n$. It is assumed that the vector-function $f = (f^1, \ldots, f^n)$ is continuous, together with its derivatives involved in further discussions. It is also assumed that the coordinates x^i and $\bar{x}^i$ of points x and $\bar{x}$ are referred to the same coordinate system.

Since the transformation (6.16) is invertible, there exists the inverse transformation:

$$x^i = (f^{-1})^i(\bar{x}), \quad i = 1, \ldots, n.$$

Let us denote by T the transformation (6.16) and by T^{-1} the inverse transformation. Accordingly, T carries any point $x \in \mathbb{R}^n$ into a new position $\bar{x} \in \mathbb{R}^n$, and T^{-1} returns $\bar{x}$ into the original position x. The identical transformation,

$$\bar{x}^i = x^i, \quad i = 1, \ldots, n,$$

will be denoted by I.

Let T_1 and T_2 be two transformations of the form (6.16) with functions f_1^i and f_2^i. Their *product* (or *composition*) $T_2 T_1$ is defined as the consecutive application of these transformations and is given by

$$\bar{\bar{x}}^i = f_2^i(\bar{x}) = f_2^i(f_1(x)), \quad i = 1 \ldots, n. \tag{6.17}$$

The geometric interpretation of the product is as follows. Since T_1 carries the point x to the point $\bar{x} = T_1(x)$, which T_2 carries to the new position $\bar{\bar{x}} = T_2(\bar{x})$, the effect of the product $T_2 T_1$ is to carry x directly to its final location $\bar{\bar{x}}$, without a stopover at $\bar{x}$. Thus, (6.17) means that

$$\bar{\bar{x}} \stackrel{\text{def}}{=} T_2(\bar{x}) = T_2 T_1(x). \tag{6.18}$$

In this notation, the definition of the inverse transformation is written

$$T T^{-1} = T^{-1} T = I. \tag{6.19}$$

Definition. A group of transformations in $\mathbb{R}^n$ is a set G of transformations (6.16) such that it contains the identity I, as well as the inverse and the product of all transformations pertaining to G. Thus, the attributes of the group G are:

$$I \in G, \quad \text{and} \quad T^{-1} \in G, \quad T_1 T_2 \in G \quad \text{whenever} \quad T, T_1, T_2 \in G. \tag{6.20}$$

6.2.3 Subgroups

Definition. Let G be a group of transformations. Its subset $H \subset G$ is called a subgroup of G, if H possesses all properties (6.20) of a group, viz. $I \in H$ and $T^{-1} \in H$, $T_1 T_2 \in H$ whenever $T, T_1, T_2 \in H$.

6.3 A display of transformation groups

6.3.1 Transformations of the straight line

Example 1. *The translation group* is the set G of all displacements T_a:

$$\bar{x} = x + a. \tag{6.21}$$

Since $\bar{x} = x$ when $a = 0$, the set G contains the identity $I = T_0$. Furthermore, the combined effect of two translations, T_a and T_b, acting in succession, is to displace x through the distance $a + b$. Hence, $T_b T_a = T_{a+b}$. It follows from the latter equation that $T_a^{-1} = T_{-a}$. Thus, transformations (6.21) define a one-parameter group G (with the parameter a) known as the *translation group*.

Example 2. *The linear fractional transformation*

$$\bar{x} = \frac{a_1 + a_2 x}{a_3 x + a_4}, \quad a_1 a_3 - a_2 a_4 \neq 0, \tag{6.22}$$

forms a group and contains, as its subgroup, the general linear group,

$$\bar{x} = a_1 + a_2 x, \quad a_2 \neq 0, \tag{6.23}$$

This subgroup itself has two subgroups – the translation and dilation groups:

$$\bar{x} = x + a_1 \quad \text{and} \quad \bar{x} = a_2 x. \tag{6.24}$$

6.3. A DISPLAY OF TRANSFORMATION GROUPS

The general linear fractional transformation (6.22) involves four arbitrary constants a_i subject to the invertibility condition, $a_1 a_3 - a_2 a_4 \neq 0$. But one of these constants can be eliminated by dividing the right-hand side of (6.22) by any coefficient a_i distinct from zero. We will consider the transformations (6.22) in the vicinity of the identical transformation (i.e. near $a_1 = a_3 = 0, a_2 = a_4 = 1$). Then $a_4 \neq 0$, and after dividing by a_4 one arrives at what is called the *projective group* on the straight line:

$$\bar{x} = \frac{a_1 + a_2 x}{1 + a_3 x}, \quad a_2 \neq a_1 a_3. \tag{6.25}$$

Definition 1. A group containing a finite number of parameters and depending continuously on these parameters was termed by Lie a *finite continuous group* (see also [6.4]). A continuous group which contains $r < \infty$ essential parameters is said to be an r-fold group (Lie's terminology) or an r-parameter group (the modern terminology) and is denoted by G_r.

Thus, the projective group is a three-fold (three-parameter) group, while (6.23) and (6.24) provide its two - and one-parameter subgroups, respectively. Furthermore, Lie proved that the projective group and its subgroups are the only types of finite continuous groups on the straight line. To formulate this result, it is convenient to use the following definition.

Definition 2. Groups obtained one from another by a mere change of variables are called *similar* and are regarded as indistinguishable.

Theorem [6.5]. A finite continuous group G_r on the straight line contains at most three essential parameters, i.e. $r \leq 3$. It is similar to the translation group (6.21) if $r = 1$, to the general linear group (6.23) if $r = 2$, and to the three-parameter projective group (6.25) if $r = 3$.

Exercise. Show that the set G of transformations

$$\bar{x} = x + \alpha_1 - \ln\left(1 + \alpha_2 e^x\right),$$

where $|\alpha_2|$ is sufficiently small, defines a two-parameter group similar to (6.23).

Solution. It is convenient to rewrite the above transformation in the form

$$\bar{x} = -\ln\left[e^{-\alpha_1}\left(e^{-x} + \alpha_2\right)\right]. \tag{6.26}$$

Let $T_{\boldsymbol{\alpha}}$ and $T_{\boldsymbol{\beta}}$ be two transformations (6.26) with $\boldsymbol{\alpha} = (\alpha_1, \alpha_2)$ and $\boldsymbol{\beta} = (\beta_1, \beta_2)$, viz.

$$T_{\boldsymbol{\alpha}} : \bar{x} = -\ln\left[e^{-\alpha_1}\left(e^{-x} + \alpha_2\right)\right], \quad T_{\boldsymbol{\beta}} : \bar{x} = -\ln\left[e^{-\beta_1}\left(e^{-x} + \beta_2\right)\right].$$

We have

$$T_{\boldsymbol{\beta}} T_{\boldsymbol{\alpha}} : \bar{\bar{x}} = -\ln\left[e^{-\beta_1}\left(e^{-\bar{x}} + \beta_2\right)\right] = -\ln\left[e^{-\beta_1}\left(e^{\ln[e^{-\alpha_1}(e^{-x}+\alpha_2)]} + \beta_2\right)\right].$$

The result can be written

$$\bar{\bar{x}} = -\ln\left[e^{-(\alpha_1+\beta_1)}\left(e^{-x} + \alpha_2 + \beta_2 e_1^\alpha\right)\right] = -\ln\left[e^{-\gamma_1}\left(e^{-x} + \gamma_2\right)\right],$$

where
$$\gamma_1 = \alpha_1 + \beta_1, \quad \gamma_2 = \alpha_2 + \beta_2 e^{\alpha_1}. \qquad (6.27)$$

Hence $T_\beta T_\alpha = T_\gamma \in G$, where $\gamma = (\gamma_1, \gamma_2)$ is defined by (6.27). It is clear that G contains the identical transformation obtained from (6.26) when $\alpha_1 = \alpha_2 = 0$. Consequently, the inverse transformation to T_α is obtained from (6.27) by letting $\gamma_1 = \gamma_2 = 0$. Hence,

$$T_\alpha^{-1} = T_{\alpha^{-1}} \in G, \quad \text{where} \quad \alpha^{-1} = (-\alpha_1, -\alpha_2 e^{-\alpha_1}). \qquad (6.28)$$

Thus, G is a group and contains two essential parameters, α_1 and α_2. It is mapped into the general linear group by the substitution $y = e^{-x}$. Indeed, the transformation (6.26) is written then in the form (6.23), $\overline{y} = a_1 + a_2 y$, where $a_1 = \alpha_2 e^{-\alpha_1}$ and $a_2 = e^{-\alpha_1}$.

6.3.2 Groups in the plane

Example 1. *The translation group* is a two-fold group of transformations

$$\overline{x} = x + a_1, \quad \overline{y} = y + a_2. \qquad (6.29)$$

Two independent translations in directions parallel to the x and y axes, viz.

$$\overline{x} = x + a_1, \overline{y} = y \quad \text{and} \quad \overline{x} = x, \overline{y} = y + a_2,$$

provide one-parameter subgroups and compose the whole group (6.29). Furthermore, all translations (6.13) in the direction of any vector $e = (e_1, e_2)$ form one-parameter subgroups of the group (6.29).

Example 2. *The rotation group* is given by (cf. Example 2 in Section 6.2.1)

$$\overline{x} = x \cos a + y \sin a, \quad \overline{y} = y \cos a - x \sin a. \qquad (6.30)$$

One can easily prove that the rotations (6.30), where a is an arbitrary parameter, form a one-parameter group (Problem 6.11). The group property becomes, however, manifest in the polar coordinates (r, θ). Indeed, then (6.30) reduces to a one-parameter translation group (see Problem 6.6), $\overline{r} = r$, $\overline{\theta} = \theta - a$.

Example 3. *The group of motions* comprises the rotation and translations:

$$\overline{x} = x \cos a + y \sin a + a_1, \quad \overline{y} = y \cos a - x \sin a + a_2. \qquad (6.31)$$

This is the largest group that changes only the location and orientation of any geometric figure, but not its magnitude and shape. In other words, the motions (6.31) transpose geometric figures (triangles, circles, etc.) as rigid bodies. Consequently, the transformation (6.31) (and its generalization to three dimensions) is a mainstay of Euclidean geometry. Therefore, (6.31) is sometimes called the group of Euclidean motions, or briefly the *Euclidean group*.

Example 4. *The projective group* is given by invertible linear fractional transformations containing eight essential parameters:

$$\overline{x} = \frac{a_{11}x + a_{12}y + a_1}{b_1 x + b_2 y + b_3}, \quad \overline{y} = \frac{a_{21}x + a_{22}y + a_2}{b_1 x + b_2 y + b_3}. \qquad (6.32)$$

6.3. A DISPLAY OF TRANSFORMATION GROUPS

Its well-known subgroups are the *general linear group*:

$$\bar{x} = a_{11}x + a_{12}y + a_1, \quad \bar{y} = a_{21}x + a_{22}y + a_2,$$

the *linear homogeneous group*:

$$\bar{x} = a_{11}x + a_{12}y, \quad \bar{y} = a_{21}x + a_{22}y,$$

and a proper projective transformation known also as the *special projective group*:

$$\bar{x} = \frac{x}{1-ax}, \quad \bar{y} = \frac{y}{1-ax}. \tag{6.33}$$

Note. The nomenclature *projective group* is due to the chief property of the transformation (6.32) to convert any straight line on the plane again into a straight line. The general linear fractional transformation, unlike (6.32), does not possess this property, i.e. it is not a projective transformation.

Exercise. (i) Demonstrate the chief property of the projective transformation (6.32). (ii) Show that the following one-parameter group of linear fractional transformations:

$$\bar{x} = \frac{x}{1-ax}, \quad \bar{y} = \frac{y}{1-ay},$$

does not convert an arbitrary straight line into a straight line.

Solution. (i) Since the transformation (6.32) is invertible, it suffices to show that if the transformed points $(\bar{x}, \bar{y})$ are located on a straight line, the original points (x, y) also lie on a straight line. Thus, let us assume $A\bar{x} + B\bar{y} + C = 0$. The substitution of (6.32) yields $kx + ly + m = 0$, where $k = Aa_{11} + Ba_{21} + Cb_1$, $l = Aa_{12} + Ba_{22} + Cb_2$, $m = Aa_1 + Ba_2 + Cb_3$. Hence, the image of any straight line is again a straight line.

(ii) Substitute the inverse transormation, $x = \bar{x}/(1 + a\bar{x}), y = \bar{y}/(1 + a\bar{y})$, into the equation of a straight line, $Ax + By + C = 0$, to obtain $A\bar{x} + B\bar{y} + C - (A + B + aC)\bar{x}\bar{y} = 0$. The latter equation represents a straight line (i.e. it is linear in $\bar{x}$ and $\bar{y}$) only in the case $A + B + aC = 0$. Since a is arbitrary, one can first assume $a = 0$ to get $A + B = 0$, then take $a \neq 0$ to get $C = 0$. Hence, the transformation in question converts into a straight line only the line $Ax + By + C = 0$ with $B = -A, C = 0$, i.e. the line $y = x$.

Example 5. *Inversion* with respect to a circle of radius R centered at O is:

$$\bar{x} = R^2 \frac{x}{r^2}, \quad \bar{y} = R^2 \frac{y}{r^2}, \tag{6.34}$$

where $r^2 = x^2 + y^2$. The point $O = (0,0)$ and the positive number R are called the center and the radius of inversion (6.34). It follows from (6.34) that $r\bar{r} = R^2$, where $\bar{r}^2 = \bar{x}^2 + \bar{y}^2$. Therefore the inversion is also known as the *transformation of reciprocal radii*. Inversion has the following important geometric properties.

Theorem. Inversion (6.34) (i) leaves unaltered any straight line passing through the center O of inversion, (ii) maps any circle passing through O into a straight line which does not pass through O, (iii) maps any straight line that does not pass through O into a circle passing through O, and (iv) maps any circle not passing through O again into a circle that does not pass through O [6.6].

Denote the inversion (6.34) by S. One can verify that the repeated action of the inversion is the identical transformation, $S^2 = I$ (Problem 6.13), so that $S^{-1} = S$. Hence the following is a group containing two elements:

$$G = \{I, S\}. \tag{6.35}$$

Note that one can transpose the center of inversion to any point as well as to choose any positive number (or $R = \infty$) as a radius of inversion.

Example 6. *Conformal transformation* $C_a = S_1 T_a S_1$ [6.7]. Here S_1 is the inversion with respect to the circle of unit radius, and T_a the translation group $\bar{x} = x + a, \bar{y} = y$ along the x-axis. The combined transformation $S_1 T_a S_1$ acts as follows. The inversion S_1 carries (x, y) to the point (x_1, y_1),

$$x_1 = \frac{x}{r^2}, \quad y_1 = \frac{y}{r^2},$$

which T_a carries to the new position

$$x_2 = x_1 + a = \frac{x + ar^2}{r^2}, \quad y_2 = y_1 = \frac{y}{r^2},$$

and lastly, S_1 brings (x_2, y_2) to the final location $(\bar{x}, \bar{y})$ of the initial point (x, y):

$$\bar{x} = \frac{x_2}{r_2^2}, \quad \bar{y} = \frac{y_2}{r_2^2}.$$

Substituting the above values of x_2, y_2 into $r_2^2 = x_2^2 + y_2^2$, one obtains

$$r_2^2 = \frac{(x + ar^2)^2 + y^2}{r^4} = \frac{1 + 2ax + a^2 r^2}{r^2}.$$

Hence, the transformation $C_a = S_1 T_a S_1$ has the form:

$$\bar{x} = \frac{x + ar^2}{1 + 2ax + a^2 r^2}, \quad \bar{y} = \frac{y}{1 + 2ax + a^2 r^2}, \quad \text{where} \quad r^2 = x^2 + y^2. \tag{6.36}$$

Since $S^{-1} = S$, the transformation (6.36) can be written $C_a = S_1^{-1} T_a S_1$. Here, the right-hand side is the formula for the change of coordinates by inversion in the translation group. Invoking the duality between changes of coordinates and point transformations (Section 6.2.1) and the fact that a change of coordinates does not disturb the group properties (6.20), we conclude that the transformation (6.36) defines a one-parameter group. Problem 6.15 demands a direct proof without using the formula $S_1^{-1} T_a S_1$ of change of variables.

Example 7. *The Poincaré group* (known also as the *non-homogeneous Lorentz group*) of the special theory of relativity comprises (in the case of one spatial variable x) translations of t and x and the Lorentz transformation:

$$\bar{t} = t \cosh(a/c) + (x/c) \sinh(a/c), \quad \bar{x} = x \cosh(a/c) + ct \sinh(a/c), \tag{6.37}$$

6.3. A DISPLAY OF TRANSFORMATION GROUPS

where a is a group parameter and $c = 2.99793 \times 10^{10}$ cm/s is the velocity of light in a vacuum. By introducing the new parameter $V = c \tanh(a/c)$, one can rewrite the transformation (6.37) in the form similar to the Galilean transformation (6.15):

$$\bar{t} = \frac{t + xV/c^2}{\sqrt{1 - V^2/c^2}}, \quad \bar{x} = \frac{x + Vt}{\sqrt{1 - V^2/c^2}}. \tag{6.38}$$

The physical meaning of the transformation (6.38) and of its parameter V can be illustrated by considering the motion of a particle in a direction parallel to the x-axis. It follows from (6.38) that if an observer at rest detects the velocity v of a particle, then an observer moving with the velocity V along the x-axis will detect the velocity

$$\bar{v} = \frac{v + V}{1 + vV/c^2}. \tag{6.39}$$

Consequently, the transformation (6.38) is known in the physical literature as the Lorentz boost. In the limit of classical mechanics, when $V/c \to 0$, the Lorentz boost takes the form of the Galilean transformation, $\bar{t} = t, \bar{x} = x + V$, and formula (6.39) reduces to $\bar{v} = v + V$ (cf. (6.15)).

6.3.3 Motions, conformal mappings and other groups in $\mathbb{R}^n$

The first example in this section is based on the following fundamental concept of geometry.

Definition 1. A motion (alias an *isometric motion*) is a transformation of points $x = (x^1, \ldots, x^n) \in \mathbb{R}^n$ that preserves the metric properties. A motion transposes any figure as a rigid body and may change only its location and orientation. For example, a motion maps any sphere into a sphere, the two spheres having the same radius but not necessarily the same center. The set of all motions in $\mathbb{R}^n$ is a group called the *group of motions* in the Euclidean space $\mathbb{R}^n$, or the group of Euclidean motions (briefly the *Euclidean group*). In geometry, two figures are said to be equal if one of them is obtained from another by a motion.

Example 1. *The Euclidean group* in $\mathbb{R}^n$ is compounded of n translations:

$$\bar{x}^i = x^i + a^i, \quad i = 1, \ldots, n, \tag{6.40}$$

and $n(n-1)/2$ rotations in all coordinate planes (x^i, x^j) :

$$\bar{x}^i = x^i \cos a + x^j \sin a, \quad \bar{x}^j = x^j \cos a - x^i \sin a. \tag{6.41}$$

Hence, the group of Euclidean motions contains $r = n(n+1)/2$ essential parameters. For example, solid geometry is based on the six-parameter group of motions containing three independent rotations and three translations [6.8].

Example 2. *The general linear group* is given by the invertible transformations

$$\bar{x}^i = a^i_k x^k + a^i, \quad i = 1, \ldots, n,$$

where k is the summation index (see footnote in Section 5.3.1). A compact form of the above transformation is

$$\bar{x} = Ax + a, \qquad (6.42)$$

where $x = (x^1, \ldots, x^n)$, $\bar{x} = (\bar{x}^1, \ldots, \bar{x}^n)$, $a = (a^1, \ldots, a^n)$, and $A = \| a_k^i \|$ is an $n \times n$ non-singular matrix (i.e. $\det A \neq 0$). The general linear group depends on $n(n+1)$ essential parameters and contains the $n(n+1)/2$-parameter Euclidean group as a subgroup.

Example 3. *The linear homogeneous group* is a subgroup of the general linear group. It contains n^2 parameters and has the form:

$$\bar{x} = Ax, \quad \det A \neq 0. \qquad (6.43)$$

The composition of $\bar{x} = Ax$ and $\bar{\bar{x}} = B\bar{x}$ yields $\bar{\bar{x}} = B\bar{x} = BAx$, where BA is the product of the matrices A and B. Hence, the composition of transformations (6.43) is represented by the usual multiplication of matrices. Furthermore, the inverse to the transformation (6.43) is $x = A^{-1}\bar{x}$, where A^{-1} is the inverse matrix.

Thus, the linear homogeneous group (6.43) can be regarded as the *matrix group* consisting of all non-singular matrices A, where the group composition law is the usual multiplication of matrices. The identity element of the group is the unit matrix $I = \| \delta_k^i \|$.

Notation. *The Kronecker symbols* (sometimes termed *Kronecker deltas*) are defined as follows:

$$\delta^{ik} \equiv \delta_k^i \equiv \delta_{ik} = \begin{cases} 1 & \text{if } i = k, \\ 0 & \text{if } i \neq k. \end{cases} \qquad (6.44)$$

Example 4. *The special linear homogeneous group* is a subgroup of the linear homogeneous group. It contains $n^2 - 1$ parameters and is defined by (6.43), where the matrices A are subject to the single condition $\det A = 1$.

Definition 2. A conformal mapping is a transformation in $\mathbb{R}^n$ preserving the angles, but not necessarily the magnitude. For example, a conformal mapping transforms a sphere into a sphere, but the latter may have a radius and a center different from those of the original sphere. The set of all conformal mappings is a group called the *conformal group* in $\mathbb{R}^n$. The group of motions is manifestly a subgroup of the conformal group.

Example 5. *Inversion* in $\mathbb{R}^3$ with respect to a sphere of radius R centered at $O = (0, 0, 0)$ is given by

$$\bar{x} = R^2 \frac{x}{r^2}, \quad \bar{y} = R^2 \frac{y}{r^2}, \quad \bar{z} = R^2 \frac{z}{r^2}, \qquad (6.45)$$

where $r^2 = x^2 + y^2 + z^2$. The radii r and $\bar{r}$ of a point (x, y, z) and of its image $(\bar{x}, \bar{y}, \bar{z})$ are in the relation of reciprocal radii (cf. Example 5 in Section 6.3.2), $r\bar{r} = R^2$. Inversion provides an example of a non-trivial (i.e. different from an

6.3. A DISPLAY OF TRANSFORMATION GROUPS

isometric motion) conformal mapping. Moreover, Liouville [6.9] proved that any non-trivial conformal mapping in $\mathbb{R}^3$ is based on inversion [6.10]. Before the precise formulation of this result (known as Liouville's theorem), let us consider the following example.

Example 6. *The continuous conformal group* in $\mathbb{R}^3$ is a finite group. It can be obtained as follows. Proceeding as in Example 6 of Section 6.3.2, one arrives at the one-parameter group of conformal mappings:

$$\overline{x} = \frac{x + a_1 r^2}{1 + 2a_1 x + a_1^2 r^2}, \quad \overline{y} = \frac{y}{1 + 2a_1 x + a_1^2 r^2}, \quad \overline{z} = \frac{z}{1 + 2a_1 x + a_1^2 r^2}, \quad (6.46)$$

and its counterparts in the *y*- and *z*-directions:

$$\overline{x} = \frac{x}{1 + 2a_2 y + a_2^2 r^2}, \quad \overline{y} = \frac{y + a_2 r^2}{1 + 2a_2 y + a_2^2 r^2}, \quad \overline{z} = \frac{z}{1 + 2a_2 y + a_2^2 r^2}, \quad (6.47)$$

and

$$\overline{x} = \frac{x}{1 + 2a_3 z + a_3^2 r^2}, \quad \overline{y} = \frac{y}{1 + 2a_3 z + a_3^2 r^2}, \quad \overline{z} = \frac{z + a_3 r^2}{1 + 2a_3 z + a_3^2 r^2}. \quad (6.48)$$

Now Liouville's theorem in regard to the continuous conformal group (including isometric motions as its subgroup) can be formulated as follows.

Theorem. *The maximal continuous conformal group in $\mathbb{R}^3$ contains 10 essential parameters. It is compounded of the six-parameter group of Euclidean motions* (i.e. translations and rotations of x, y, z), *three one-parameter groups* (6.46)–(6.48), *and the one-parameter dilation group* $\overline{x} = ax, \overline{y} = ay, \overline{z} = az$.

Example 7. *The continuous conformal group* in a Euclidean space $\mathbb{R}^n$, $n \geq 3$, is an $(n+1)(n+2)/2$-parameter group [6.11]. It comprises the $n(n+1)/2$-fold group of Euclidean motions (6.40)–(6.41), the uniform dilation $\overline{x}^i = ax^i$, and the n-dimensional version of conformal mappings (6.46)–(6.48).

Example 8. *The Galilean group* is an $(n+1)(n+2)/2$-parameter subgroup of the general linear group in the space of spatial variables $x = (x^1, \ldots, x^n)$ and time t. It comprises the Euclidean motions (6.40)–(6.41), the time translation $\overline{t} = t + b$, and the proper Galilean transformation (cf. (6.15) with $\boldsymbol{V} = (b^1, \ldots, b^n)$):

$$\overline{t} = t, \quad \overline{x}^i = x^i + tb^i, \; i = 1, \ldots, n.$$

Example 9. *The Poincaré group* (non-homogeneous Lorentz group) signifies the group of isometric motions in the Minkowski space-time of special relativity (cf. Example 7 of Section 6.3.2). It comprises the Euclidean motions (6.40)–(6.41), the time translation and proper Lorentz transformations (6.37) in n different planes (x^i, t), $i = 1, \ldots, n$, viz.

$$\overline{t} = t \cosh(a/c) + (x^i/c) \sinh(a/c), \quad \overline{x}^i = x^i \cosh(a/c) + ct \sinh(a/c). \quad (6.49)$$

Note that each individual one-parameter group (6.49) is endowed with its own group parameter a. Hence, the Lorentz transformations form a group with n essential parameters.

6.3.4 Different types of groups: Continuous, discontinuous and mixed. Path curves. Global and local groups

A group G of transformations in $\mathbb{R}^n$ is said to be *continuous*, if any two transformations $T_1, T_2 \in G$ can be connected via a continuous set of elements within the group. In other words, T_1 can be continuously deformed into T_2 within the group G. Consider, for example, one-parameter groups defined as follows.

Definition. A set G_1 of transformations T_a in $\mathbb{R}^n$, where a ranges over all real numbers from a given interval $U \subset R$, is called a one-parameter group if there is a unique value $a = a_0$ in U providing the identical transformation, $T_{a_0} = I$, and the following conditions hold for all $a, b \in U$ (cf. Definition 6.2.2):

$$T_a^{-1} = T_{a^{-1}} \in G, \quad T_b T_a = T_c \in G, \tag{6.50}$$

where $a^{-1}, c \in U$ and $c = \phi(a, b)$ is a continuous function.

A one-parameter group G_1 is manifestly continuous in the above sense. It means geometrically that any point $x \in \mathbb{R}^n$ is carried by the group transformations into the points $\bar{x} = T_a(x)$ whose locus is a continuous curve (passing through x) and is called a *path curve* of the group G_1. The group property means that *any point of a path curve is carried by G_1 into points of the same curve*. The locus of the images $T_a(x)$ is also termed the G_1-*orbit* of the point x and denoted $G_1(x)$. In the case of continuous r-parameter groups G_r (Section 6.3.1), orbits $G_r(x)$ are continuous r-dimensional manifolds.

The majority of groups exposed in Sections 6.3.1 to 6.3.3 are continuous. The group (6.35) of two elements is an exception. It furnishes an example of a discontinuous group. Namely, a group of transformations in $\mathbb{R}^n$ is said to be *discontinuous* if it contains no transformations whose effects on $x \in \mathbb{R}^n$ differ infinitesimally. Another example of such a group is given in Problem 6.9.

A continuous group of a different type is provided by invertible transformations of a straight line, $\bar{x} = f(x)$. The set of all these transformations, where $f(x)$ ranges over all continuously differentiable functions, subject to the invertibility condition $f'(x) \neq 0$, is a group. Continuous groups involving arbitrary functions are called *infinite continuous groups*. Thus, $\bar{x} = f(x)$ is an example of an infinite continuous group involving one arbitrary function $f(x)$.

Continuous and discontinuous groups do not exhaust all possible groups of transformations. An example of a *mixed group* is offered by symmetry transformations of the rectangular coordinate axes on the plane, i.e. by the following transformations leaving invariant the equation $xy = 0$:

$$T: \bar{x} = f(x), \quad \bar{y} = g(y); \qquad R: \bar{x} = p(y), \quad \bar{y} = q(x), \tag{6.51}$$

where $f(x)$, $g(y)$, $p(y)$, and $q(x)$ are arbitrary functions such that

$$f(0) = g(0) = p(0) = q(0) = 0, \quad f'(0) = g'(0) = p'(0) = q'(0) = 1. \tag{6.52}$$

The set G of all transformations (6.51) is a group. The group G is infinite since it involves arbitrary functions. But it is not continuous because transformations

6.3. A DISPLAY OF TRANSFORMATION GROUPS

of type T cannot be continuously deformed into transformations of type R, and vice versa. However, G contains an infinite continuous subgroup composed of the transformations (6.51) of type T.

Consider the one-parameter family G of projective transformations T_a :

$$\bar{x} = \frac{x}{1 - ax}. \tag{6.53}$$

We have $I \in G$ when $a = 0$. Furthermore, the consecutive application of two transformations (6.53) T_a, T_b yields (Problem 6.12):

$$\bar{\bar{x}} = \frac{x}{1 - (a+b)x}. \tag{6.54}$$

It follows from (6.54) that $T_a^{-1} = T_{-a} \in G$, and $T_b T_a = T_{a+b} \in G$. One naturally concludes that G is a group. However, this conclusion presumes that the three expressions, $1-ax$, $1-bx$, and $1-(a+b)x$, do not vanish. This simple observation is, however, of fundamental significance for the theory of Lie group analysis.

To discern the true nature of the situation, let us consider a fixed point $x_0 > 0$ and move along the path curve through x_0 when a ranges over all real numbers from an interval $0 \leq a \leq \varepsilon$. Two transformations (6.53), T_a and T_b, are well defined in the interval $0 \leq a \leq 1/x_0$, (i.e. $1 - ax_0$ and $1 - bx_0$ do not vanish in this interval), but their consecutive application may give the prohibited value $a+b = 1/x_0$ when the product (6.54) is not determined. This is the case, e.g. for $a = x_0/3$ and $b = 2x_0/3$. Therefore, let us assume that a and b are taken from a closer vicinity of $a = 0$, e.g. from the interval $0 \leq a \leq 1/(2x_0)$. Then $a+b < 1/x_0$, and the product (6.54) is well defined. However, a further multiplication may again result in an unacceptable value of the parameter. For example, when $a = b = 1/(3x_0)$ one has $a + b = 2/(3x_0)$. Hence $T_a T_b = T_{a+b}$ is determined, but $T_a T_{a+b} = T_{2a+b}$ is not. It can be readily seen that iterated multiplication of the projective transformations (6.53) inevitably leads to the prohibited value of the group parameter. One cannot solve the problem by merely fixing a "tiny" interval $0 \leq a \leq \varepsilon$, the true nature of the problem being that the orbit of any point x (excluding the isolated point $x = 0$) has a singularity when $a = 1/x$.

The projective group (6.53) provides an example of what is called a *local group*, meaning that the composition is defined only for those transformations sufficiently close to the identity. The vicinity of the identical transformation, where the composition is determined, may depend upon a transformed point x. An alternative definition of a local group G is that path curves of G have singularities. The rotation (6.30) provides another example of a local group.

A transformation group is called a *global group* if the composition of any transformations is defined simultaneously at all generic points x. A common representative of a global group is the translation group (6.21). Other global groups that occur frequently are provided by the Lorentz transformation (6.37) and the Galilean transformation (6.15).

Problems

6.1. Show that the discriminant $b^2 - 4ac$ of equation (6.1) is invariant under the linear transformation (6.3).

6.2. Verify that the transformation $\bar{x} = ax + b$ converts (6.5) into (6.6)–(6.7).

6.3*. Find the condition for (6.6) to have three equal roots.

4.4*. Find the conditions for the general cubic (6.5) to have two and three equal roots.

6.5. Calculate $\sqrt[3]{2(-1+i)}$ and $\sqrt[3]{2(-1-i)}$ used in Section 6.1.1.

6.6. In the polar coordinates (r, θ), the rotation by the angle a about the origin clockwise is naturally written $\bar{r} = r$, $\bar{\theta} = \theta - a$. Show that this transformation is represented in rectangular Cartesian coordinates $x = r\cos\theta$, $y = r\sin\theta$ by equations (6.14).

6.7. Prove that the composition of transformations defined by (6.18) is associative, i.e. $T_1(T_2 T_3) = (T_1 T_2) T_3$.

6.8. Deduce the group property $T_b T_a = T_{a+b}$ of translations from the transformation formula (6.21).

6.9. Prove that the following six transformations of the straight line form a group [6.4]:

$$I: \bar{x} = x, \quad T_1: \bar{x} = 1 - x, \quad T_2: \bar{x} = \frac{1}{x}, \quad T_3: \bar{x} = \frac{1}{1-x}, \quad T_4: \bar{x} = \frac{x}{x-1}, \quad T_5: \bar{x} = \frac{x-1}{x}.$$

Show also that the subset $H = \{I, T_1\}$ is a subgroup of the group $G = \{I, T_1, \ldots, T_5\}$.

6.10. Prove that the projective transformations (6.25) form a group.

6.11. Check the group property of rotations (6.30).

6.12. Check that the linear fractional transformation (cf. Exercise 6.3.2)

$$\bar{x} = \frac{x}{1 - ax}, \quad \bar{y} = \frac{y}{1 - ay},$$

defines a one-parameter group.

6.13. Show that $S^2 \equiv SS = I$, where S is the inversion (6.34).

6.14. Make a model of the mechanism described in Note [6.6] and check that it works.

6.15. Prove the group properties (6.50) for the conformal mapping (6.36) without appealing to a change of coordinates.

6.16. Test for the group properties (6.50) the following transformation:

$$\bar{x} = x + a, \quad \bar{y} = y + a + a^3.$$

6.17. Check the group property of the set of all transformations of the form

$$\bar{x} = f(x),$$

where $f(x)$ is an arbitrary continuously differentiable function, $f'(x) \neq 0$.

6.18. Show that the set of transformations (6.51) is a group and that the transformations of type T form its infinite continuous subgroup.

6.19. Justify the statement of Section 6.3.3 that the rotation group (6.30) is local whereas the Lorentz group (6.37) is global.

Chapter 7

Infinitesimal transformations and local groups

In applied group analysis, continuous groups are determined by infinitesimal transformations, or *infinitesimal generators*. Given an infinitesimal generator, the group transformation is found by solving the Lie equations. The existence of solutions of the Lie equations is guaranteed only in a small neighborhood of regular points (see Chapter 3). Consequently, one arrives, in general, at *local groups*.

This chapter provides a detailed presentation of the infinitesimal theory of local one-parameter groups, their invariants and invariant equations.

Infinitesimal transformations of multi-parameter local groups form a Lie algebra. This chapter contains an outline of the basic notions from the theory of Lie algebras, multi-parameter groups and their invariants and invariant equations.

This chapter also provides an introduction to the recent theory of approximate transformation groups.

7.1 One-parameter groups

7.1.1 Notation and assumptions

In what follows we adopt the notation and assumptions of Section 6.2.2. Let us consider transformations T_a:

$$\bar{x} = f(x, a), \tag{7.1}$$

depending on a real parameter a, where $x \in \mathbb{R}^n$. We impose the condition that (7.1) is the identical transformation for a certain value a_0 of the parameter, i.e.

$$f(x, a_0) = x, \tag{7.2}$$

and that there are no other values of a in a vicinity of a_0 reducing T_a to the identity. The functions $f(x, a)$ are at least three times continuously differentiable.

Note. One can let $a_0 = 0$, e.g. by setting $a = a' + a_0$ and denoting the new parameter a' again by a. Then equation (7.2) is written $T_0 = I$. The condition $a_0 = 0$ is achieved, however, in a regular way by introducing a canonical parameter (Section 7.1.6).

In coordinates x^i, equations (7.1) and (7.2) are written:

$$\bar{x}^i = f^i(x, a), \quad f^i(x, a_0) = x^i, \quad i = 1, \ldots, n. \tag{7.3}$$

It is assumed that, given a generic point x, there exists an open interval $U \subset \mathbb{R}$ containing a_0 such that the functions $f^i(x, a)$ satisfy regularity conditions (discussed in Section 6.2.2) in a vicinity of x, whenever $a \in U$. Furthermore, it is supposed that a_0 is the only value of $a \in U$ for which (7.1) reduces to the identical transformation. In short, (7.1) defines single-valued transformations $T_a : \mathbb{R}^n \to \mathbb{R}^n$ for all a from the interval U.

7.1.2 Definition of a local group

Transformations occurring in practical applications often obey the group properties (6.50) only when a and b are restricted to values sufficiently small numerically. Then one arrives at what is called a *local one-parameter group* G. The precise definition is formulated, using the preceding notation and assumptions, as follows.

Definition. A set G of transformations T_a in $\mathbb{R}^n$ given by (7.3) is called a *one-parameter local group* if there exists a subinterval $U' \subset U$ containing a_0 such that the functions $f^i(x, a)$ satisfy the *composition rule*

$$f^i(f(x, a), b) = f^i(x, c), \quad i, \ldots, n, \tag{7.4}$$

for all values $a, b \in U'$. Here $c \in U$ is a *thrice continuously differentiable* function, $c = \phi(a, b)$, of two variables $a, b \in U'$ such that the equation

$$\phi(a, b) = a_0 \tag{7.5}$$

has a unique solution b for any $a \in U'$. Given a, the solution b of equation (7.5) is denoted by a^{-1}. Hence, the inverse transformation T_a^{-1} is given by

$$f^i(\bar{x}, a^{-1}) = x^i, \quad i = 1, \ldots, n. \tag{7.6}$$

For brevity, local groups are also termed groups.

The function

$$c = \phi(a, b) \tag{7.7}$$

defined by the composition rule (7.4) is termed a *group composition law*. It follows from the *initial conditions* $f^i(x, a_0) = x^i$ and (7.4) that

$$\phi(a, a_0) = a, \quad \phi(a_0, b) = b. \tag{7.8}$$

7.1. ONE-PARAMETER GROUPS

According to the above definition, a local group G contains the (unique) identity $I = T_{a_0}$. The composition rule (7.4) means that any two transformations $T_a, T_b \in G$ with $a, b \in U'$ carried out one after the other result in a transformation $T_c \in G$, where c belongs to the interval U but may not belong to the subinterval U' (cf. Definition 6.3.4). Consequently, one cannot obtain a global group merely by assigning U' as a basic interval instead of U. To draw a parallel between global and local one-parameter groups, equations (7.4) and (7.6) can be written in the form similar to (6.50):

$$T_a^{-1} = T_{a^{-1}} \in G, \quad T_b T_a = T_c \in G, \quad \text{where} \quad a, b \in U', \ c \in U.$$

Remark. The symbol a^{-1} for the parameter of the inverse transformation (7.6) does not denote, in general, $1/a$ (see Problem 7.2).

7.1.3 Representation of local groups in a canonical parameter

The composition law (7.7) depends on the choice of a group parameter (see, e.g. Problem 7.2). It will be shown in Section 7.1.6 that all composition laws in local one-parameter groups can be transformed one into another by a suitable change of the parameter a. For example, any composition law (7.7) can be transformed to the simplest form, $\phi(a, b) = a + b$. Then the group transformation (7.1),

$$\bar{x}^i = f^i(x, a), \quad i = 1, \ldots, n, \tag{7.9}$$

is said to be represented in a *canonical* parameter a. It is assumed that the functions $f^i(x, a)$ are defined in a neighborhood of $a = 0$ and that

$$f^i(x, 0) = x^i, \quad i = 1, \ldots, n. \tag{7.10}$$

Thus, the representation of a local one-parameter group (7.9) in a canonical parameter is specified by the following rule of composition (7.4):

$$f^i(f(x, a), b) = f^i(x, a + b), \quad i = 1, \ldots, n. \tag{7.11}$$

Here, a and b are any numerical values of the group parameter taken from a neighborhood of $a = 0$.

It follows from (7.11) and (7.10) that $a^{-1} = -a$. In other words, the inverse transformation (7.6) is obtained merely by changing the sign of a:

$$x^i = f^i(\bar{x}, -a), \quad i = 1, \ldots, n. \tag{7.12}$$

Hence, in a canonical parameter, local groups are defined solely by the *initial conditions* (7.10) and the composition rule (7.11).

7.1.4 Infinitesimal transformations

Given a one-parameter group G of transformations (7.9), let us expand the functions $f^i(x,a)$ into the Taylor series in the parameter a in a neighborhood of $a = 0$. Then, invoking the initial condition (7.10), we arrive at what is called the *infinitesimal transformation* of the group G (cf. Section 5.2.2):

$$\bar{x}^i \approx x^i + a\xi^i(x), \quad i = 1, \ldots, n, \tag{7.13}$$

where

$$\xi^i(x) = \frac{\partial}{\partial a}\left[f^i(x,a)\right]_{a=0}. \tag{7.14}$$

Geometrically, the infinitesimal transformation (7.13)–(7.14) defines the tangent vector $\xi(x) = (\xi^1(x), \ldots, \xi^n(x))$, at the point x, to the G-orbit of x (defined in Section 6.3.4). Therefore, ξ is called the *tangent vector field* of the group G. The tangent vector field is often denoted by ξ^i and is written as a first-order linear differential operator (4.15):

$$X = \xi^i(x)\frac{\partial}{\partial x^i}. \tag{7.15}$$

The operator X behaves, unlike the *vector* ξ^i, as a *scalar* under any change of variables (Section 4.2.2, Lemma 2). As mentioned in Section 5.2.2, Lie called the operator (7.15) the *symbol* of the infinitesimal transformation (7.13), or of the corresponding group (7.11). In this book, the terms *symbol*, *infinitesimal generator* (or *operator*) of a one-parameter group are used interchangeably.

Example. The infinitesimal transformation of the rotation group (6.30),

$$\bar{x} = x\cos a + y\sin a, \quad \bar{y} = y\cos a - x\sin a,$$

has the form $\bar{x} \approx x + ya$, $\bar{y} \approx y - xa$, and its symbol is

$$X = y\frac{\partial}{\partial x} - x\frac{\partial}{\partial y}.$$

7.1.5 The Lie equations. The exponential map

The following theorem, due to Lie, asserts that one-parameter local groups are determined by their infinitesimal transformations.

Theorem. Let G be a local group defined by (7.9) with functions f^i obeying (7.10) and the composition rule (7.11). Let (7.13) be the infinitesimal transformation of the group G. Then the functions $\bar{x}^i = f^i(x,a)$ solve the system of first-order ordinary differential equations (known as the *Lie equations*)

$$\frac{d\bar{x}^i}{da} = \xi^i(\bar{x}), \quad i = 1, \ldots, n, \tag{7.16}$$

7.1. ONE-PARAMETER GROUPS

with the initial conditions $\bar{x}^i|_{a=0} = x^i$, or in the compact form (cf. (3.12)):

$$\frac{d\bar{x}}{da} = \xi(\bar{x}), \quad \bar{x}|_{a=0} = x. \tag{7.17}$$

In other words, the G-orbit of a point $x \in \mathbb{R}^n$ is an integral curve of the Lie equations (7.16) passing through the point x.

Conversely, given an infinitesimal transformation (7.13), or its symbol (7.15), where $\xi^i(x)$ are continuously differentiable, the initial value problem (7.17) has a unique solution $\bar{x} = f(x, a)$ in a neighborhood of $a = 0$. This solution satisfies the group property (7.11). In other words, the solution of Lie's equations provides a one-parameter local group with a given infinitesimal transformation (7.13).

Proof. To prove the direct statement, set $b = \Delta a$ in (7.11):

$$f^i(x, a + \Delta a) = f^i(f(x, a), \Delta a), \quad i = 1, \ldots, n, \tag{7.18}$$

then expand both sides of (7.18) in powers of Δa and single out the leading terms:

$$f^i(x, a + \Delta a) \approx f^i(x, a) + \frac{\partial}{\partial a}\left[f^i(x, a)\right] \cdot \Delta a,$$

$$f^i(f(x, a), \Delta a) \approx f^i(x, a) + \frac{\partial}{\partial \Delta a}\left[f^i\left(f(x, a), \Delta a\right)\right]_{\Delta a = 0} \cdot \Delta a.$$

Invoking (7.14), one obtains

$$\frac{\partial}{\partial \Delta a}\left[f^i\left(f(x, a), \Delta a\right)\right]_{\Delta a = 0} = \xi^i(f(x, a)).$$

Therefore equations (7.18) mean that $f^i(x, a)$ solve the Lie equations (7.16):

$$\frac{\partial}{\partial a}\left[f^i(x, a)\right] = \xi^i\left(f(x, a)\right), \quad i = 1, \ldots, n.$$

To prove the second part of the theorem, it is necessary to verify the group property (7.11) alone. Indeed, existence and uniqueness of the local solution to the problem (7.17) is guaranteed by Theorem 3.1.3. Thus, let us prove that the solution $\bar{x} = f(x, a)$ of (7.17) satisfies (7.11). Consider, for any fixed value a, the functions $u = (u^1, \ldots, u^n)$ and $v = (v^1, \ldots, v^n)$ of a variable b defined as follows:

$$u(b) = f(\bar{x}, b) \equiv f(f(x, a), b), \quad v(b) = f(x, a + b).$$

To prove the group property (7.11), it suffices to show that $u(b) = v(b)$ in a neighborhood of $b = 0$. Since $f(x, a)$ solves equations (7.17), one has:

$$\frac{du}{db} \equiv \frac{df(\bar{x}, b)}{db} = \xi(u), \quad u|_{b=0} = f(x, a);$$

and

$$\frac{dv}{db} \equiv \frac{df(x, a + b)}{db} = \xi(v), \quad v|_{b=0} = f(x, a).$$

Hence, both $z = u(b)$ and $z = v(b)$ solve the initial value problem

$$\frac{dz}{db} = \xi(z), \quad z|_{b=0} = f(x, a).$$

One concludes from the uniqueness theorem that $u(b) = v(b)$ in a neighborhood of $b = 0$, thus completing the proof.

The Lie equations (7.16) are precisely of the form (3.30) with $y^i = \bar{x}^i$. Hence, given a generator (7.15) with analytic $\xi^i(x)$, one can obtain the group transformation (7.9) in the form of an infinite series by using the exponential map (3.35):

$$\bar{x}^i = e^{aX}(x^i), \quad i = 1, \ldots, n, \tag{7.19}$$

where

$$e^{aX} = 1 + \frac{a}{1!}X + \frac{a^2}{2!}X^2 + \cdots + \frac{a^s}{s!}X^s + \cdots. \tag{7.20}$$

7.1.6 Determination of a canonical parameter

Recall the definition of a canonical parameter introduced in Section 7.1.3.

Definition. The group parameter is said to be *canonical* if the composition law (7.7) reduces to the addition, $\phi(a, b) = a + b$, i.e. if the group property (7.4) is written in the form (7.11), $T_b T_a = T_{a+b}$. Furthermore, (7.10) holds, i.e. $I = T_0$.

Theorem. Given an arbitrary composition law (7.7), $c = \phi(a, b)$, subject to the differentiability conditions assumed in Section 7.1.2, there exists a canonical parameter $\tilde{a}$. It is defined by [7.1]

$$\tilde{a} = \int_{a_0}^{a} \frac{ds}{w(s)}, \quad \text{where} \quad w(s) = \left.\frac{\partial \phi(s, b)}{\partial b}\right|_{b=a_0}. \tag{7.21}$$

Proof. Let G be a local one-parameter group defined in Section 7.1.2, with an arbitrary composition law (7.7). Let us rewrite equations (7.4) in the form $f(\bar{x}, b) = f(x, c)$, $c = \phi(a, b)$, and differentiate with respect to b to obtain:

$$\frac{\partial f(\bar{x}, b)}{\partial b} = \frac{\partial f(x, c)}{\partial c} \frac{\partial \phi(a, b)}{\partial b}.$$

Letting here $b = a_0$ and invoking (7.8), we get

$$\left.\frac{\partial f(\bar{x}, b)}{\partial b}\right|_{b=a_0} = \frac{\partial f(x, a)}{\partial a} \left[\frac{\partial \phi(a, b)}{\partial b}\right]_{b=a_0}. \tag{7.22}$$

Let us denote the second factor in the right-hand side of (7.22) by $w(a)$. It follows from the second equation (7.8) that $w(a_0) = 1$ and, by continuity, $w(a) \neq 0$ in a neighborhood of $a = a_0$. Furthermore, we use the notation similar to (7.14):

$$\xi(x) = \left.\frac{\partial f(x, a)}{\partial a}\right|_{a=a_0}.$$

7.1. ONE-PARAMETER GROUPS

Now equation (7.22) is written $\xi(\overline{x}) = w(a)\,\mathrm{d}\overline{x}/\mathrm{d}a$, or in coordinates:

$$\frac{\mathrm{d}\overline{x}^i}{\mathrm{d}\tilde{a}} = \xi^i(\overline{x}), \quad i = 1,\ldots,n, \tag{7.23}$$

where the new parameter $\tilde{a}$ is defined by (7.21). Equations (7.23) are identical to the Lie equations (7.16) and, by Theorem 7.1.5, define a one-parameter group with the composition rule (7.11) with respect to the parameter $\tilde{a}$. Hence, $\phi(\tilde{a},\tilde{b}) = \tilde{a} + \tilde{b}$. Furthermore, it follows from (7.21) that $\tilde{a} = 0$ when $a = a_0$, i.e. the identical transformation corresponds to $\tilde{a} = 0$. Thus, $\tilde{a}$ defined by (7.21) is a canonical parameter.

Example. Let $\overline{x} = x + ax$, $|a| < 1$. According to Problem 7.2(ii), this transformation defines a one-parameter group with the composition law $\phi(a,b) = a+b+ab$ and with $a_0 = 0$. Therefore, formula (7.21) provides the function $w(s) = 1 + s$, and hence the canonical parameter $\tilde{a} = \int_0^a (1+s)^{-1}\mathrm{d}s = \ln(1+a)$.

Note. In what follows we shall adopt the canonical parameter when referring to one-parameter groups unless otherwise stipulated.

7.1.7 Invariants

Definition. A function $F(x)$ is called an invariant of a group G of transformations (7.9) if F remains unaltered when one moves along any path curve of the group G. In other words, F is an invariant if $F(f(x,a)) = F(x)$ identically in x and a in a neighborhood of $a = 0$.

Remark. Given an invariant $F(x)$, any function $\Phi(F(x))$ is also an invariant.

Theorem 1. A function $F(x)$ is an invariant of the group G with the generator X (7.15) if and only if it solves the homogeneous linear partial differential equation (cf. (4.13)):

$$X(F) \equiv \xi^i(x)\frac{\partial F(x)}{\partial x^i} = 0. \tag{7.24}$$

Proof. Let $F(x)$ be an invariant. Since

$$F(f(x,a)) \approx F(x + a\xi(x)) \approx F(x) + a\,\xi^i(x)\frac{\partial F(x)}{\partial x^i},$$

the invariance condition $F(f(x,a)) = F(x)$ yields (7.24).

Conversely, let $F(x)$ be any solution of the differential equation (7.24). Since equation (7.24) is valid at any point, one can consider it at $\overline{x} = f(x,a)$:

$$\xi^i(\overline{x})\frac{\partial F(\overline{x})}{\partial \overline{x}^i} = 0.$$

Hence, invoking the Lie equations (7.16):

$$\frac{\mathrm{d}F(f(x,a))}{\mathrm{d}a} = \frac{\partial F(\overline{x})}{\partial \overline{x}^i}\frac{\mathrm{d}f^i(x,a)}{\mathrm{d}a} = \xi^i(\overline{x})\frac{\partial F(\overline{x})}{\partial \overline{x}^i} = 0.$$

Since $F(f(x,0)) = F(x)$ (see (7.10)), we conclude that $u(a) = F(f(x,a))$ solves the initial value problem

$$\frac{du}{da} = 0, \quad u|_{a=0} = F(x).$$

The solution to the latter problem is unique (Theorem 3.1.2) and is obviously given by $u = F(x)$. Hence, the two solutions, $u = F(f(x,a))$ and $u = F(x)$, are identical for any x. This is precisely the invariance condition, $F(f(x,a)) = F(x)$.

Alternative proof of the second part of Theorem 1 is based on the exponential map (7.19), namely on its extension to analytic functions $F(x)$ (Problem 7.6*):

$$F(\bar{x}) = e^{aX} F(x) \stackrel{\text{def}}{=} \left(1 + \frac{a}{1!}X + \frac{a^2}{2!}X^2 + \cdots + \frac{a^s}{s!}X^s + \cdots\right) F(x). \quad (7.25)$$

Indeed, if $XF = 0$, then $X^2F = X(XF) = 0, \ldots, X^sF = 0$. Consequently, (7.25) yields $F(\bar{x}) = F(x)$, provided that $F(x)$ is an analytic function.

Theorem 2. A one-parameter group G of transformations in $\mathbb{R}^n$ has precisely $n - 1$ functionally independent invariants. Any set of independent invariants, $\psi_1(x), \ldots, \psi_{n-1}(x)$, is termed a *basis of invariants* of G. Basis is not unique. One can take, as basic invariants, the left-hand sides of $n - 1$ first integrals

$$\psi_1(x) = C_1, \ldots, \psi_{n-1}(x) = C_{n-1} \quad (7.26)$$

of the *characteristic system* for equation (7.24):

$$\frac{dx^1}{\xi^1(x)} = \cdots = \frac{dx^n}{\xi^n(x)}. \quad (7.27)$$

An arbitrary invariant $F(x)$ of G is given by the formula

$$F = \Phi\left(\psi_1(x), \ldots, \psi_{n-1}(x)\right). \quad (7.28)$$

Proof. This is the direct consequence of the preceding theorem. Equations (7.26) to (7.28) are given by Theorem 4.2.2 applied to equation (7.24).

Example. Consider the group of non-uniform dilations in $\mathbb{R}^3$:

$$\bar{x} = xe^a, \quad \bar{y} = ye^{2a}, \quad \bar{z} = ze^{-2a},$$

with the generator

$$X = x\frac{\partial}{\partial x} + 2y\frac{\partial}{\partial y} - 2z\frac{\partial}{\partial z}.$$

The system (7.27) has the form $dx/x = dy/(2y) = -dz/(2z)$. Integrating, e.g. the equations $dx/x = dy/(2y)$ and $dx/x = -dz/(2z)$, one obtains two first integrals, $yx^{-2} = C_1$ and $zx^2 = C_2$. Hence, $\psi_1 = yx^{-2}$ and $\psi_2 = zx^2$ constitute a basis of invariants, and an arbitrary invariant is $F = \Phi(yx^{-2}, zx^2)$. Likewise, integrating the equations $dy/y = -dz/z$ and $dx/x = -dz/(2z)$, one obtains another basis of invariants, $\psi_1' = yz$ and $\psi_2' = zx^2$. Then an arbitrary invariant is represented in the form $F = \Phi(yz, zx^2)$.

7.1. ONE-PARAMETER GROUPS

7.1.8 Canonical and semi-canonical variables

Let us reconsider, from a new viewpoint, the material of Sections 4.2.2 and 4.3.2 concerning the behavior of linear operators under changes of variables. We know (Lemma 2 in Section 4.2.2) that in new variables x'^i given by (4.16),

$$x'^i = \varphi^i(x), \quad i = 1, \ldots, n, \tag{7.29}$$

an infinitesimal operator X (7.15) is written in the form (4.17):

$$\overline{X} = X(\varphi^i) \frac{\partial}{\partial x'^i}, \tag{7.30}$$

where the coefficients $X(\varphi^i) = \xi^j \partial \varphi^i / \partial x^j$ of $\overline{X}$ should be expressed in terms of the new variables $x' = (x'^1, \ldots, x'^n)$ by solving (7.29) with respect to x^i.

Definition 1. Given a group G (7.9), *canonical variables* x'^i are defined by the condition that G reduces to translations, e.g. in the direction of the x'^n axis:

$$\bar{x}'^1 = x'^1, \ldots, \bar{x}'^{n-1} = x'^{n-1}, \bar{x}'^n = x'^n + a. \tag{7.31}$$

Theorem 1. *Canonical variables exist for any one-parameter group.*

Proof. Given a group G with the generator X (7.15), canonical variables x'^i for G are furnished by the change of variables:

$$x'^1 = \psi_1(x), \ldots, x'^{n-1} = \psi_{n-1}(x), x'^n = \varphi(x), \tag{7.32}$$

where $\psi_1(x), \ldots, \psi_{n-1}(x)$ is a basis of invariants of the group G, and the function $\varphi(x)$ is any solution to the following non-homogeneous linear equation:

$$X(\varphi) \equiv \xi^i(x) \frac{\partial \varphi(x)}{\partial x^i} = 1. \tag{7.33}$$

Indeed, by the definition of basic invariants, the functions $\psi_s(x)$, $s = 1, \ldots, n-1$, are functionally independent and solve the equations $X(\psi_1) = 0$. Furthermore, $\varphi(x)$ is functionally independent of $\psi_s(x)$, otherwise it would solve the equation $X(\varphi) = 0$ instead of (7.33). Hence, the change of variables (7.32) is well defined and, according to (7.30), reduces X to $\overline{X} = \partial/\partial x'^n$. Consequently, the Lie equations (7.16) become $d\bar{x}'^s/da = 0$ ($s = 1, \ldots, n-1$) and $d\bar{x}'^n/da = 1$. Invoking the initial conditions $\bar{x}'^i|_{a=0} = x'^i$, one arrives at the translation group (7.31).

Example 1. Consider the group of dilations with the generator

$$X = x \frac{\partial}{\partial x} + 2y \frac{\partial}{\partial y} - 2z \frac{\partial}{\partial z}.$$

Let us take a basis of invariants given in Example 7.1.7, $\psi_1 = yx^{-2}, \psi_2 = zx^2$. To solve the equation (7.33) one can follow the procedure given in Exercise 4.2.3.

By letting $\varphi = \varphi(x)$, one reduces $X(\varphi) = 1$ to $x\mathrm{d}\varphi/\mathrm{d}x = 1$, whence $\varphi = \ln|x|$. Hence, the canonical variables (7.32) are given by

$$u = yx^{-2}, \quad v = zx^2, \quad w = \ln|x|.$$

Canonical variables were of a considerable use in Lie's work (see, e.g. [5.14], Chapter 6, §5). Indeed, the introduction of canonical variables is one of the basic methods in the integration of ordinary differential equations with known symmetries (Part III).

Semi-canonical variables were introduced recently [7.2]. This generalization of Lie's canonical variables, though a simple one, provides an approach for tackling the problems of integration when the method of canonical variables fails, e.g. when symmetries become *non-local*. Bearing in mind these applications, let us modify Definition 4.3.2 of semi-canonical variables as follows.

Definition 2. Consider linear differential operators of the form

$$L = \lambda^i(x,q)\frac{\partial}{\partial x^i}, \tag{7.34}$$

where the coefficients λ^i may involve, along with $x = (x^1, \ldots, x^n)$, additional variables $q = (q_1, \ldots, q_\nu)$ (e.g. non-local variables). Variables $x' = (x'^1, \ldots, x'^n)$ are called semi-canonical variables for (7.34) if, in these variables, the *partial differential* operator L reduces to an *ordinary differential* operator, i.e. y is replaced by x')

$$\overline{L} = l(x',q)\frac{\partial}{\partial x'^n}. \tag{7.35}$$

Theorem 2. Semi-canonical variables x'^i can be obtained by a transformation of the form (7.29) provided that the coefficients of the operator (7.34) have the special form $\lambda^i(x,q) = \lambda(x,q)\xi^i(x)$, i.e.

$$L = \lambda(x,q)X, \quad \text{where} \quad X = \xi^i(x)\frac{\partial}{\partial x^i}. \tag{7.36}$$

Proof. According to (7.29)–(7.30), semi-canonical variables are given by

$$x'^1 = \psi_1(x), \ldots, x'^{n-1} = \psi_{n-1}(x), x'^n = \varphi(x), \tag{7.37}$$

where $\psi_1(x), \ldots, \psi_{n-1}(x)$ is a basis of invariants of the group with the generator X and $\varphi(x)$ is any function such that $X(\varphi) \neq 0$ (cf. Theorem 1).

Example 2. It will be shown in Section 9.6 that the differential equation

$$\frac{\mathrm{d}y}{\mathrm{d}x} = 3\frac{y}{x} + \frac{x^2}{2y} + x + 1 \tag{7.38}$$

admits the following non-local infinitesimal symmetry:

$$L = e^q \left(x^2 \frac{\partial}{\partial x} + (3xy + x^2)\frac{\partial}{\partial y} \right), \tag{7.39}$$

7.1. ONE-PARAMETER GROUPS

where q is a non-local variable defined by

$$q = \int \frac{\mathrm{d}x}{y}, \quad \text{or} \quad \frac{\mathrm{d}q}{\mathrm{d}x} = \frac{1}{y}. \tag{7.40}$$

The operator (7.39) has the form (7.36), $L = \lambda(q)X$, with $\lambda = e^q$ and

$$X = x^2 \frac{\partial}{\partial x} + (3xy + x^2)\frac{\partial}{\partial y}.$$

For this operator X, the system (7.27) is $\mathrm{d}x/x^2 = \mathrm{d}y/(3xy + x^2)$, or

$$\frac{\mathrm{d}y}{\mathrm{d}x} = 3\frac{y}{x} + 1.$$

Whence, upon integrating, $y = Cx^3 - x/2$. After solving with respect to the constant of integration, one obtains the first integral $C = (2y+x)/(2x^3)$. Hence, an invariant of the group generated by X is $\psi = (2y+x)/(2x^3)$. Consequently, one can take semi-canonical variables (7.37), e.g. in the form

$$u = \frac{2y + x}{2x^3}, \quad v = x. \tag{7.41}$$

By formula (7.30), the operator (7.39) is written in the variables u and v in the form (7.35), $L = e^q v^2 \partial/\partial v$. Let us rewrite the differential equation (7.38) in the new variables. Let us rewrite (7.41) in the form $y = v^3 u - v/2$ and take, e.g. u to be a new dependent variable. Then (7.38) yields

$$\frac{\mathrm{d}u}{\mathrm{d}v} = \frac{2u}{2v^2 u - 1}.$$

The latter is manifestly a Riccati equation for $v = v(u)$:

$$\frac{\mathrm{d}v}{\mathrm{d}u} = v^2 - \frac{1}{2u}. \tag{7.42}$$

Thus, *the transition to semi-canonical variables converts equation (7.38) with a known non-local symmetry to a Riccati equation.* The true nature of this remarkable property of semi-canonical variables will be disclosed in Section 13.1.2.

7.1.9 Three methods of construction of one-parameter groups with known infinitesimal generators

Given an infinitesimal generator (7.15), the corresponding group transformations (7.9) are usually obtained by solving the Lie equations (7.17) or by using the exponential map (7.19), in accordance with preferences of the reader. To simplify calculation, I add here the third method based on canonical variables (7.32).

Thus, the methods mentioned in the title are: *Solution of Lie equations, Use of the exponential map,* and *Introduction of canonical variables.*

Example. This example illustrates all three approaches. Let us find the group with the infinitesimal generator

$$X = x^2 \frac{\partial}{\partial x} + xy \frac{\partial}{\partial y}. \tag{7.43}$$

(i) *Solution of Lie equations.* For the operator (7.43), the Lie equations (7.16) are written

$$\frac{d\bar{x}}{da} = \bar{x}^2, \quad \frac{d\bar{y}}{da} = \bar{x}\,\bar{y}. \tag{7.44}$$

Hence, upon integration,

$$\bar{x} = \frac{1}{C_1 - a}, \quad \bar{y} = \frac{C_2}{C_1 - a}.$$

Then the initial conditions $\bar{x}|_{a=0} = x, \bar{y}|_{a=0} = y$ are written $x = 1/C_1, y = C_2/C_1$ and define the constants of integration, $C_1 = 1/x$ and $C_2 = y/x$. Thus, the operator (7.43) generates the one-parameter group of special projective transformations (cf. Example 4 in Section 6.3.2):

$$\bar{x} = \frac{x}{1 - ax}, \quad \bar{y} = \frac{y}{1 - ax}. \tag{7.45}$$

(ii) *Use of the exponential map.* Equations (7.19) are written:

$$\bar{x} = e^{aX}(x), \quad \bar{y} = e^{aX}(x).$$

According to (7.19)–(7.20), one has to find $X^s(x)$ and $X^s(y)$ for $s = 1, 2, \ldots$. For the operator X (7.43), one can readily calculate several terms, e.g.

$$X(x) = x^2, \quad X^2(x) = X(X(x)) = X(x^2) = 2!x^3, \quad X^3(x) = X(2!x^3) = 3!x^4,$$

and then make a guess $X^s(x) = s!x^{s+1}$. The proof is given by induction:

$$X^{s+1}(x) = X(s!x^{s+1}) = (s+1)!x^2 x^s = (s+1)!x^{s+2}.$$

Likewise, one obtains

$$X(y) = xy, \quad X^2(y) = X(xy) = yX(x) + xX(y) = y(x^2) + y(xy) = 2!yx^2,$$

$$X^3(y) = 2![yX(x^2) + x^2 X(y)] = 2![y(2x^3) + x^2(xy)] = 3!yx^3,$$

then makes a guess $X^s(y) = s!yx^s$, and proves it by induction:

$$X^{s+1}(y) = s!X(yx^s) = s![syx^{s+1} + x^s(xy)] = (s+1)!yx^{s+1}.$$

Substitution of the above expressions in the exponential map yields:

$$e^{aX}(x) = x + ax^2 + \cdots + a^s x^{s+1} + \cdots.$$

7.1. ONE-PARAMETER GROUPS

One can rewrite the right-hand side as $x(1 + ax + \cdots + a^s x^s + \cdots)$. The series in brackets is manifestly the Taylor expansion of the function $1/(1-ax)$ provided that $|ax| < 1$. Consequently,

$$e^{aX}(x) = \frac{x}{1-ax}.$$

Likewise, one obtains

$$e^{aX}(y) = y + ayx + a^2 yx^2 + \cdots + a^s yx^s + \cdots = y(1 + ax + \cdots + a^s x^s + \cdots).$$

Hence,

$$e^{aX}(y) = \frac{y}{1-ax}.$$

Thus, one arrives at the projective transformations (7.45).

The above procedure throws new light on the locality of the projective group, discussed in Section 6.3.4. Namely, the exponential map provides a representation of (7.45) by power series. These series converge when the group parameter a is sufficiently small, viz. $|a| < 1/|x|$.

(iii) *Introduction of canonical variables.* For the operator (7.43), the system (7.27) is written $dx/x = dy/y$ and provides the first integral $x/y = C$. Hence, the invariant $u = x/y$. The equation (7.33), $X(t) = 1$, has the special form (4.24) with $x^1 = x$. Therefore one can look for its particular solution in the form $t = t(x)$ to obtain $t = -1/x$. Thus, one arrives at the canonical variables,

$$u = \frac{x}{y}, \quad t = -\frac{1}{x},$$

such that the operator (7.43) becomes $X = \partial/\partial t$. Consequently, the group transformations are written $\bar{t} = t + a, \bar{u} = u$, or

$$-\frac{1}{\bar{x}} = -\frac{1}{x} + a, \quad \frac{\bar{x}}{\bar{y}} = \frac{x}{y}. \quad (7.46)$$

Upon solving (7.46) with respect to $\bar{x}$ and $\bar{y}$, one arrives at transformations (7.45).

Exercise. The shallow-water equations, i.e. equations (5.34)–(5.35)(7) with $n = 2$, admit the following operator (where r and θ are the polar coordinates on the plane, and v_r and v_θ the corresponding components of velocity):

$$X = (1+t^2)\frac{\partial}{\partial t} + tr\frac{\partial}{\partial r} + (r - tv_r)\frac{\partial}{\partial v_r} - tv_\theta \frac{\partial}{\partial v_\theta} - 2t\rho\frac{\partial}{\partial \rho} - 4tp\frac{\partial}{\partial p}.$$

Find the group transformations generated by X.

Solution. (i) *Solution of Lie equations.* One can integrate separately the first of Lie's equations, viz.

$$\frac{d\bar{t}}{da} = 1 + \bar{t}^2, \quad \bar{t}\big|_{a=0} = t,$$

to obtain $\arctan \bar{t} = a + \arctan t$, or $\bar{t} = (t + \tan a)/(1 - t \tan a)$. Hence,

$$\bar{t} = \frac{\sin a + t \cos a}{\cos a - t \sin a}.$$

By virtue of this formula, the second Lie equation,

$$\frac{d\bar{r}}{da} = \bar{t}\bar{r},$$

is written in the form

$$\frac{d\bar{r}}{\bar{r}} = \frac{\sin a + t \cos a}{\cos a - t \sin a} da = -\frac{d(\cos a - t \sin a)}{\cos a - t \sin a}.$$

Upon integration, using the initial condition $\bar{r}\big|_{a=0} = r$, one obtains:

$$\bar{r} = \frac{r}{\cos a - t \sin a}.$$

Likewise, solving the remaining Lie equations,

$$\frac{d\bar{v}_r}{da} = \bar{r} - \bar{t}\bar{v}_r, \quad \frac{d\bar{v}_\theta}{da} = -\bar{t}\bar{v}_\theta, \quad \frac{d\bar{\rho}}{da} = -2\bar{t}\bar{\rho}, \quad \frac{d\bar{p}}{da} = -4\bar{t}\bar{p},$$

$$\bar{v}_r\big|_{a=0} = v_r, \quad \bar{v}_\theta\big|_{a=0} = v_\theta, \quad \bar{\rho}\big|_{a=0} = \rho, \quad \bar{p}\big|_{a=0} = p,$$

one arrives at the following transformations:

$$\bar{v}_r = (\cos a - t \sin a)v_r + r \sin a, \quad \bar{v}_\theta = (\cos a - t \sin a)v_\theta,$$

$$\bar{\rho} = (\cos a - t \sin a)^2 \rho, \quad \bar{p} = (\cos a - t \sin a)^4 p.$$

(ii) *Use of the exponential map* is reserved for the reader who prefers to deal with summation of series rather than with integration of differential equations.

(iii) *Introduction of canonical variables.* Solve the equations

$$X(s) = 1, \; X(\lambda) = 0, \; X(U) = 0, \; X(V) = 0, \; X(R) = 0, \; X(P) = 0.$$

to obtain the following canonical variables:

$$s = \arctan t, \; \lambda = \frac{r}{\sqrt{1+t^2}}, \; U = rv_r - \frac{tr^2}{1+t^2}, \; V = rv_\theta, \; R = (1+t^2)\rho, \; P = (1+t^2)^2 p.$$

In these variables, the operator X becomes $X = \partial/\partial s$. Hence, the transformations:

$$\bar{s} = s + a, \quad \bar{\lambda} = \lambda, \quad \bar{U} = U, \quad \bar{V} = V, \quad \bar{R} = R, \quad \bar{P} = P.$$

After returning to the original variables, one readily arrives at the transformations obtained above by solving Lie's equations:

$$\bar{t} = \frac{\sin a + t \cos a}{\cos a - t \sin a}, \quad \bar{r} = \frac{r}{\cos a - t \sin a}, \quad \bar{v}_r = (\cos a - t \sin a)v_r + r \sin a,$$

$$\bar{v}_\theta = (\cos a - t \sin a)v_\theta, \quad \bar{\rho} = (\cos a - t \sin a)^2 \rho, \quad \bar{p} = (\cos a - t \sin a)^4 p.$$

7.1. ONE-PARAMETER GROUPS

7.1.10 Table of usual one-parameter groups in the plane, their generators, invariants and canonical variables

Transformations	Generators	Invariants $\psi(x,y)$	Canonical t, u so that $X = \partial/\partial t$
Translations along x: $\bar{x} = x + a$, $\bar{y} = y$	$X = \dfrac{\partial}{\partial x}$	$\psi = y$	$t = x$, $u = \psi$
along y: $\bar{x} = x$, $\bar{y} = y + a$	$X = \dfrac{\partial}{\partial y}$	$\psi = x$	$t = y$, $u = \psi$
along $kx + ly = 0$: $\bar{x} = x + la$, $\bar{y} = y - ka$	$X = l\dfrac{\partial}{\partial x} - k\dfrac{\partial}{\partial y}$	$\psi = kx + ly$	$t = x/l$, $u = \psi$
Rotation $\bar{x} = x\cos a + y\sin a$, $\bar{y} = y\cos a - x\sin a$	$X = y\dfrac{\partial}{\partial x} - x\dfrac{\partial}{\partial y}$	$\psi = x^2 + y^2$	$t = \arctan(x/y)$, $u = \psi$
Lorentz transformation $\bar{x} = x\cosh a + y\sinh a$, $\bar{y} = y\cosh a + x\sinh a$	$X = y\dfrac{\partial}{\partial x} + x\dfrac{\partial}{\partial y}$	$\psi = y^2 - x^2$	$t = \ln\|y + x\|$, $u = \psi$
Galilean transformation $\bar{x} = x + ay$, $\bar{y} = y$	$X = y\dfrac{\partial}{\partial x}$	$\psi = y$	$t = x/y$, $u = \psi$
Uniform dilation $\bar{x} = xe^a$, $\bar{y} = ye^a$	$X = x\dfrac{\partial}{\partial x} + y\dfrac{\partial}{\partial y}$	$\psi = x/y$	$t = \ln\|x\|$, $u = \psi$
Non-uniform dilation $\bar{x} = xe^a$, $\bar{y} = ye^{ka}$	$X = x\dfrac{\partial}{\partial x} + ky\dfrac{\partial}{\partial y}$	$\psi = x^k/y$	$t = \ln\|x\|$, $u = \psi$
Projective transformations $\bar{x} = \dfrac{x}{1 - ax}$, $\bar{y} = \dfrac{y}{1 - ax}$	$X = x^2\dfrac{\partial}{\partial x} + xy\dfrac{\partial}{\partial y}$	$\psi = x/y$	$t = -1/x$, $u = \psi$
$\bar{x} = \dfrac{x}{1 - ay}$, $\bar{y} = \dfrac{y}{1 - ay}$	$X = xy\dfrac{\partial}{\partial x} + y^2\dfrac{\partial}{\partial y}$	$\psi = x/y$	$t = -1/y$, $u = \psi$
Linear fractional $\bar{x} = \dfrac{x}{1 - ax}$, $\bar{y} = \dfrac{y}{1 - ay}$	$X = x^2\dfrac{\partial}{\partial x} + y^2\dfrac{\partial}{\partial y}$	$\psi = \dfrac{1}{y} - \dfrac{1}{x}$	$t = -1/x$, $u = \psi$
Conformal transformation $\bar{x} = \dfrac{x + ar^2}{1 + 2ax + a^2 r^2}$, $\bar{y} = \dfrac{y}{1 + 2ax + a^2 r^2}$	$X = (y^2 - x^2)\dfrac{\partial}{\partial x} - 2xy\dfrac{\partial}{\partial y}$	$\psi = y/r^2$	$t = x/r^2$, $u = \psi$

7.2 Invariant equations

7.2.1 The infinitesimal test

Consider a system of s equations in $\mathbb{R}^n$:

$$F_\sigma(x) = 0, \quad \sigma = 1, \ldots, s, \tag{7.47}$$

where $x \in \mathbb{R}^n$ and $s < n$. We impose the condition that the Jacobian matrix $\|\partial F/\partial x\|$ is of rank s:

$$\mathrm{rank}\left\|\frac{\partial F_\sigma(x)}{\partial x^i}\right\| = s \tag{7.48}$$

at all points x satisfying equations (7.47). The locus of solutions x of the system of equations (7.47) is an $(n-s)$-dimensional manifold $M \subset \mathbb{R}^n$.

Definition. The system (7.47) is said to be invariant with respect to a group G of transformations $\bar{x} = f(x, a)$, or that equations (7.47) *admit* G, if

$$F_\sigma(\bar{x}) = 0, \quad \sigma = 1, \ldots, s, \tag{7.49}$$

whenever x solves equations (7.47). Geometrically, it means that transformations of the group G carry any point of the variety M along this variety. In other words, the path curve of the group G passing through any point $x \in M$ lies in M. Consequently, M is termed an *invariant manifold* for G.

Let G be a one-parameter group with a generator X (7.15). The *infinitesimal test* for a system of equations (7.47) to be invariant under the group G is given by the following theorem.

Theorem. The system (7.47) is invariant under the group G with the infinitesimal generator X if and only if

$$XF_\sigma(x)\Big|_{(7.47)} = 0, \quad \sigma = 1, \ldots, s, \tag{7.50}$$

where the symbol $|_{(7.47)}$ means evaluated on the manifold M (7.47).

Note. If one rewrites equations (7.50) in the form $(\xi \cdot \nabla F_\sigma)|_M = 0$, it becomes evident that (7.50) is the condition for the vector field $\xi = (\xi^1, \ldots, \xi^n)$ to be tangent to the manifold M.

Proof of Theorem. Let the system (7.47) be invariant. One readily arrives at (7.50) by substituting in (7.49) the infinitesimals, $F_\sigma(\bar{x}) \approx F_\sigma(x) + aXF_\sigma(x)$.

Conversely, let us suppose that equations (7.50) hold and derive the invariance conditions (7.49). It follows from the geometric significance of (7.50) mentioned in the Note, that equations (7.50) remain unaltered after any change of variables (7.29). Therefore, one can simplify the representation of the manifold M by an appropriate change of variables. Namely, invoking the condition (7.48), one can take s functions $\varphi(x)$ in (7.29) to be the left-hand sides of the system (7.47), e.g.

$$x'^1 = F_1(x), \ldots, x'^s = F_s(x), \quad x'^{s+1} = \varphi_{s+1}(x), \ldots, x'^n = \varphi_n(x). \tag{7.51}$$

7.2. INVARIANT EQUATIONS

Then the manifold M is given by the equations

$$x'^{\sigma} = 0, \quad \sigma = 1, \ldots, s, \tag{7.52}$$

and equations (7.50) are simply written

$$\xi^{\sigma}(0, \ldots, 0, x'^{s+1}, \ldots, x'^{n}) = 0, \quad \sigma = 1, \ldots, s. \tag{7.53}$$

Consequently, we have to prove that equations (7.53) imply the invariance of equations (7.52) under the group G.

Let us take any point $x' \in M$, so that $x' = (0, \ldots, 0, x'^{s+1}, \ldots, x'^{n})$ according to (7.52), and split the Lie equations (7.17) as follows:

$$\frac{d\bar{x}'^{\sigma}}{da} = \xi^{\sigma}(\bar{x}'^{1}, \ldots, \bar{x}'^{s}, \bar{x}'^{s+1}, \ldots, \bar{x}'^{n}), \quad \bar{x}'^{\sigma}|_{a=0} = 0, \quad \sigma = 1, \ldots, s, \tag{7.54}$$

$$\frac{d\bar{x}'^{l}}{da} = \xi^{l}(\bar{x}'^{1}, \ldots, \bar{x}'^{s}, \bar{x}'^{s+1}, \ldots, \bar{x}'^{n}), \quad \bar{x}'^{l}|_{a=0} = x'^{l}, \quad l = s+1, \ldots, n. \tag{7.55}$$

It follows from (7.53) that the functions $\bar{x}'^{\sigma} = 0$, $\sigma = 1, \ldots, s$, satisfy (7.54). The functions $\bar{x}'^{l}$ are obtained from equations (7.55), after substituting there $\bar{x}'^{1} = \cdots = \bar{x}'^{s} = 0$. The uniqueness theorem of Section 3.1.3 guarantees that this is the only solution of the Cauchy problem (7.54)–(7.55). Thus, any point $x' \in M$ is transformed into a point $\bar{x}' = (0, \ldots, 0, \bar{x}'^{s+1}, \ldots, \bar{x}'^{n}) \in M$. In other words, equations (7.52) are invariant under the group G.

Remark. Provided that $F_\sigma(x)$ and $XF_\sigma(x)$ are analytic in a neighborhood of the manifold M (7.47), the invariance test (7.50) can be written in the form

$$XF_\sigma(x) = \lambda_\sigma^\nu(x)F_\nu(x), \quad \sigma = 1, \ldots, s, \tag{7.56}$$

where the coefficients $\lambda_\sigma^\nu(x)$ are bounded in a neighborhood of the variety M.

Equation (7.56), together with (7.25), provide a simplified proof of the second part of the above theorem. Indeed, it follows from (7.56) that

$$X^2 F_\sigma = X(\lambda_\sigma^\nu)F_\nu + \lambda_\sigma^\nu X(F_\nu) = [X(\lambda_\sigma^\mu) + \lambda_\sigma^\nu \lambda_\nu^\mu] F_\mu.$$

Iteration and substitution into (7.25) yields $F_\sigma(\bar{x}) = \Lambda_\sigma^\nu(x) F_\nu(x)$, hence (7.49).

7.2.2 Invariant representation of invariant manifolds

Given a group G, invariant manifolds for G can be equivalently represented by many different systems of equations (7.47). To develop a general procedure for constructing invariant manifolds, it is advantageous to distinguish *singular* and *regular* (non-singular) invariant manifolds.

Definition 1. Let M be an invariant manifold for a group G. Since G transports points x of M along this variety, the action of the group G confined to M is also a local group. It is termed a group induced on M by G, or briefly an *induced group*.

The induced group, its generator (termed an *induced generator*) and invariants will be denoted here by the symbols $\widetilde{G}$, $\widetilde{X}$ and $\widetilde{\psi}$, respectively.

Definition 2. Let G be a one-parameter group, and let $X = \xi^i(x)\partial/\partial x^i$ be its generator (7.15). An invariant manifold M for G is said to be *regular* (with respect to G) if the induced generator $\widetilde{X}$ does not vanish, and *singular* if $\widetilde{X} = 0$. In other words, M is a regular invariant manifold if at least one of the coefficients $\xi^i(x)$ does not vanish on M, and it is singular if all coefficients $\xi^i(x)$ of the generator X vanish identically[1] on M.

Theorem. Let a system of equations (7.47) admit a group G. Let the invariant manifold M defined by the system (7.47) be regular with respect to G. Then M can be represented by a system whose equations are given by invariant functions with respect to G, i.e. in the form (cf. formula (7.28)):

$$\Phi_\sigma(\psi_1(x), \ldots, \psi_{n-1}(x)) = 0, \quad \sigma = 1, \ldots, s, \tag{7.57}$$

where

$$\psi_1(x), \ldots, \psi_{n-1}(x) \tag{7.58}$$

is a basis of invariants of G. Hence, equations (7.57) with arbitrary functions Φ_σ of $n - 1$ variables furnish the general form of regular invariant manifolds for G.

Proof [7.3]. Let M be a regular invariant manifold given by equations (7.47). In the variables x'^i (7.51), the manifold M is given by (7.52), and the induced group $\widetilde{G}$ acts in the $(n - s)$-dimensional space $\mathbb{R}^{n-s}$ of variables $\widetilde{x}' = (x'^{s+1}, \ldots, x'^n)$. Since M is regular, the induced generator $\widetilde{X} \neq 0$. Hence, the induced group has precisely $n - s - 1$ independent invariants. In particular,

$$\widetilde{\psi}_1(\widetilde{x}'), \ldots, \widetilde{\psi}_{n-1}(\widetilde{x}') \tag{7.59}$$

obtained from (7.58) by restricting them to M, are invariants of $\widetilde{G}$. Some of these $n - 1$ invariants will be functionally dependent, since the number of independent invariants is not greater than $n - s - 1$. Let $n - s' - 1$ of the invariants (7.59) be independent. Since $n - s' - 1 \leq n - s - 1$, there exist $s' \geq s$ relations[2]:

$$\Phi_\sigma\left(\widetilde{\psi}_1(\widetilde{x}'), \ldots, \widetilde{\psi}_{n-1}(\widetilde{x}')\right) = 0, \quad \sigma = 1, \ldots, s'. \tag{7.60}$$

Assuming that the relations (7.60) (i.e. the functions Φ_σ therein) are known, define an $(n - s')$-dimensional manifold $M' \subset \mathbb{R}^n$ by the equations:

$$\Phi_\sigma(\psi_1(x), \ldots, \psi_{n-1}(x)) = 0, \quad \sigma = 1, \ldots, s'. \tag{7.61}$$

If $x \in M$, equations (7.61) reduce to (7.60). Hence, $M \subset M'$. Consequently, $\dim M' \geq \dim M$, i.e. $n - s' \geq n - s$. It follows that $s' \leq s$. However, $s' \geq s$ by definition of relations (7.60). Hence, $s' = s$. It follows from $\dim M' = \dim M$ and $M \subset M'$ that, locally, $M' = M$. Thus, equations (7.61) furnish an invariant representation (7.57) of the manifold M.

---

[1] Since we treat manifolds locally near generic points and suppose that all functions under consideration are continuous, these two possibilities cover all cases.

[2] When $n = 2$, the relation (7.60) reduces to $\widetilde{\psi}_1(\widetilde{x}') = C$, $C =$const.

7.2.3 Examples on Theorem 7.2.2

The first example provides a simple illustration of the fact that the theorem on invariant representation is not valid for singular invariant manifolds. The remaining examples deal with regular invariant manifolds given by a single equation (Examples 2 and 3) and by a system of equations (Example 4).

Example 1. Consider the group G of transformations $\bar{x} = xe^a, \bar{y} = y$ with the generator $X = x\partial/\partial x$. The equation $x = 0$ is invariant under G, and hence the y-axis is an invariant manifold for G. It is singular since $X = 0$ when $x = 0$. Accordingly, the y-axis cannot be represented via invariants of the group G. Indeed, the general form of invariants of G is $\psi(y)$, so that the most general equation of the form (7.57) provides $y = $ const., i.e. straight lines parallel to the x-axis.

Example 2. Consider, in the (x, y, z) space, the paraboloid given by

$$F(x, y, z) \equiv x^2 + y^2 - z = 0. \tag{7.62}$$

Let G be the group of dilations $\bar{x} = xe^a, \bar{y} = ye^a, \bar{z} = ze^{2a}$ with the generator

$$X = x\frac{\partial}{\partial x} + y\frac{\partial}{\partial y} + 2z\frac{\partial}{\partial z}.$$

We have $F(\bar{x}, \bar{y}, \bar{z}) = e^{2a} F(x, y, z)$, or in infinitesimals, $XF = 2F$. It follows that equation (7.62), $F = 0$, is invariant under the group G, but the function F is not an invariant of this group.

The generator X vanishes at the point $O = (0, 0, 0)$. Let us exclude this singular point from the paraboloid (7.62), denote by M the resulting regular invariant surface and find its invariant representation (7.57). In our example, a basis of invariants (7.58) is provided, e.g. by $\psi_1 = x^2/z$, $\psi_2 = y^2/z$. Let us take the change of variables (7.51) in the form $z' = z - x^2 - y^2$, $x' = x$, $y' = y$. Then the manifold M is given by $z' = 0$, and the induced invariants are written $\tilde{\psi}_1 = x'^2/r^2$, $\tilde{\psi}_2 = y'^2/r^2$, where $r^2 = x'^2 + y'^2$. It is manifest that $\tilde{\psi}_1 + \tilde{\psi}_2 = 1$. This is the relation (7.59). Hence, the invariant representation (7.57) of the paraboloid (7.62) with the origin O excluded has the form $\psi_1 + \psi_2 = 1$, i.e.

$$\frac{x^2 + y^2}{z} - 1 = 0. \tag{7.63}$$

Example 3. Consider the logarithmic spiral $\ln r = \theta$, or

$$r = e^\theta, \tag{7.64}$$

where (r, θ) are polar coordinates, $r = \sqrt{x^2 + y^2}$ and $\theta = \arctan(y/x)$. In rectangular Cartesian coordinates $x = r\cos\theta$, $y = r\sin\theta$, equation (7.64) is written:

$$x^2 + y^2 = e^{2\arctan(y/x)}. \tag{7.65}$$

The spiral (7.65) is a regular invariant manifold for the group G obtained by simultaneous rotations and dilations. The group G, according to its geometric significance, will be termed the *spiral transformation group*. Its generator

$$X = (x-y)\frac{\partial}{\partial x} + (x+y)\frac{\partial}{\partial y}. \tag{7.66}$$

is the sum of the generators for rotations and dilations, $X_1 = x\,\partial/\partial y - y\,\partial/\partial x$ and $X_2 = x\,\partial/\partial x + y\,\partial/\partial y$. In polar coordinates, the generator (7.66) is written

$$X = \frac{\partial}{\partial \theta} + r\frac{\partial}{\partial r}. \tag{7.67}$$

The corresponding group transformation is $\bar\theta = \theta + a,\ \bar r = re^a$.

Exercise. Find the one-parameter group generated by (7.66).

Solution. The Lie equations have the form:

$$\frac{d\bar x}{da} = \bar x - \bar y,\quad \frac{d\bar y}{da} = \bar x + \bar y;\quad \bar x\big|_{a=0} = x,\ \bar y\big|_{a=0} = y. \tag{7.68}$$

They can be solved, e.g. by Euler's method discussed in Section 2.3. Here the dependent variables are written as the vector $y = (\bar x, \bar y)$. The differential equations (7.68) have the form (2.61), where the independent variable is the group parameter a and A is the 2×2 matrix with $a_{11} = a_{22} = -1, a_{12} = 1, a_{21} = -1$. The characteristic equation (2.64), $P_2(\lambda) \equiv (\lambda-1)^2 + 1 = 0$, has a pair of complex roots, $\lambda = 1+i, \bar\lambda = 1-i$. The solution of the system (2.63) yields $l = p + iq$ with $p = (1,0)$, $q = (0,-1)$. The formula (2.67) provides a fundamental system of solutions $y_{(1)} = (\bar x_{(1)}, \bar y_{(1)})$ and $y_{(2)} = (\bar x_{(2)}, \bar y_{(2)})$, where $\bar x_{(1)} = e^a \cos a,\ \bar y_{(1)} = e^a \sin a$ and $\bar x_{(2)} = e^a \sin a,\ \bar y_{(2)} = e^a \cos a$. Hence, the general solution has the form $\bar x = e^a(C_1 \cos a + C_2 \sin a),\ \bar y = e^a(C_1 \sin a - C_2 \cos a)$. Whence, using the initial conditions, one arrives at the spiral transformation group:

$$\bar x = e^a(x\cos a - y\sin a),\quad \bar y = e^a(y\cos a + x\sin a). \tag{7.69}$$

Example 4. Consider Newton's equations (1.34) in two dimensions:

$$w^i + \alpha\frac{x^i}{r^3} = 0,\quad i = 1, 2, \tag{7.70}$$

where $w^i = d^2 x^i/dt^2$ and $\alpha > 0$. Let G be the group of simultaneous rotations of vectors $\boldsymbol{x} = (x^1, x^2)$ and $\boldsymbol{w} = (w^1, w^2)$, i.e. the group with the generator

$$X = x^2\frac{\partial}{\partial x^1} - x^1\frac{\partial}{\partial x^2} + w^2\frac{\partial}{\partial w^1} - w^1\frac{\partial}{\partial w^2}. \tag{7.71}$$

Geometry provides the following invariants:

$$\psi_1 = |\boldsymbol{x}| \equiv \sqrt{(x^1)^2 + (x^2)^2},\quad \psi_2 = |\boldsymbol{w}| \equiv \sqrt{(w^1)^2 + (w^2)^2},$$
$$\psi_3 = |\boldsymbol{x}\times\boldsymbol{w}| \equiv |x^1 w^2 - x^2 w^1|,\quad \psi_4 = \boldsymbol{x}\cdot\boldsymbol{w} \equiv x^1 w^1 + x^2 w^2.$$

However, only three of them are independent. Indeed, they are connected by the relation $\psi_3^2 + \psi_4^2 = \psi_1^2 \psi_2^2$, the latter being manifest from the definition of scalar

7.3. INTRODUCTION TO LIE ALGEBRAS

and vector products, $x \cdot w = |x||w|\cos\omega$, $|x \times w| = |x||w|\cos\omega$, where ω is the angle between x and w. A basis of invariants is furnished, e.g. by

$$\psi_1 = |x|, \quad \psi_2 = |w|, \quad \psi_3 = |x \times w|. \tag{7.72}$$

The induced invariants are $\tilde{\psi}_1 = |x|, \tilde{\psi}_2 = \alpha|x|^{-2}$, and $\tilde{\psi}_3 = 0$. Hence, two relations of the form (7.60) are $\psi_1^2 \psi_2 = \alpha$ and $\tilde{\psi}_3 = 0$. Therefore, the invariant representation (7.57) of equations (7.70) is provided by $\psi_1^2 \psi_2 = \alpha$, $\psi_3 = 0$, or

$$|x|^2|w| = \alpha, \quad |x \times w| = 0. \tag{7.73}$$

7.3 Introduction to Lie algebras

As we already know, every one-parameter group is determined by one infinitesimal generator X. The theory of multi-parameter local groups [7.4] leads to *Lie algebras*. Specifically, infinitesimal generators of an r-parameter continuous group span a Lie algebra [7.5] of dimension r.

The theory of Lie algebras is one of the well-developed fields of modern mathematics. A rigorous treatment of this subject can be found in a special literature. We shall here be content with outlining the basic nomenclature used in Lie group analysis. Yet, some of the results are proved if the proofs serve integration of ordinary differential equations.

7.3.1 Definition of Lie algebras of operators

We shall consider Lie algebras of operators, i.e. vector spaces of linear differential operators of the form (7.15) endowed with the commutator defined by (4.95).

Definition. A Lie algebra is a vector space L of operators $X = \xi^i(x)\, \partial/\partial x^i$ with the following property. If the operators

$$X_1 = \xi_1^i(x)\frac{\partial}{\partial x^i}, \quad X_2 = \xi_2^i(x)\frac{\partial}{\partial x^i},$$

are elements of L, then their commutator

$$[X_1, X_2] \equiv X_1 X_2 - X_2 X_1 = \left(X_1(\xi_2^i) - X_2(\xi_1^i)\right)\frac{\partial}{\partial x^i} \tag{7.74}$$

is also an element of L. The Lie algebra is denoted by the same letter L, and the dimension $\dim L$ of the Lie algebra is the dimension of the vector space L. We shall use the symbol L_r to denote an r-dimensional Lie algebra.

It follows from (7.74) that the commutator is bilinear:

$$[c_1 X_1 + c_2 X_2, X] = c_1[X_1, X] + c_2[X_2, X],$$

$$[X, c_1 X_1 + c_2 X_2] = c_1[X, X_1] + c_2[X, X_2],$$

skew-symmetric:
$$[X_1, X_2] = -[X_2, X_1],$$
and satisfies the Jacobi identity:
$$[X_1, [X_2, X_3]] + [X_2, [X_3, X_1]] + [X_3, [X_1, X_2]] = 0.$$
Consider a Lie algebra L_r of a finite dimension r. Let
$$X_\alpha = \xi_\alpha^i(x)\frac{\partial}{\partial x^i}, \quad \alpha = 1, \ldots, r, \qquad (7.75)$$
be a basis of the vector space L_r. Then any $X \in L_r$ is their linear combination, $X = c^\alpha X_\alpha$, with constant coefficients c^α. In particular, $[X_\alpha, X_\beta] \in L_r$, hence
$$[X_\alpha, X_\beta] = c_{\alpha\beta}^\gamma X_\gamma, \quad \alpha, \beta = 1, \ldots, r. \qquad (7.76)$$
Conversely, given r linearly independent operators (7.75), their linear span is an r-dimensional Lie algebra if the relations (7.76) hold. This is an immediate consequence of the bilinearity of the commutator.

Thus, the relations (7.76) furnish a simple test for the linear span of operators X_α to be a Lie algebra. The constant coefficients $c_{\alpha\beta}^\gamma$ are called the *structure constants* of the algebra L_r.

7.3.2 Notation. Table of commutators

Given any set of operators $X_1, \ldots, X_s$, their linear span will be denoted by $\langle X_1, \ldots, X_s \rangle$. For example, a Lie algebra L_r with the basis (7.75) is written $L_r = \langle X_1, \ldots, X_r \rangle$.

Let L be a vector space, and let K and N be subspaces of L. The set of all operators $X + Y$ with $X \in K$ and $Y \in N$ is denoted by $K + N$. Likewise, the linear span of all commutators $[X, Y]$ with $X \in K$ and $Y \in N$ will be denoted by $[K, N]$. For example, the condition for L to be closed under commutation, i.e. to be a Lie algebra, is written $[L, L] \subset L$.

In applications, it is convenient to use the relations (7.76) written in the form of a *table of commutators* of the basis (7.75). Consider an example.

Example. Let L_6 be the vector space with the following basis [7.6]:
$$X_1 = \frac{\partial}{\partial x}, \quad X_2 = \frac{\partial}{\partial y}, \quad X_3 = \frac{\partial}{\partial u}, \quad X_4 = y\frac{\partial}{\partial u},$$
$$X_5 = x\frac{\partial}{\partial x} + 3u\frac{\partial}{\partial u}, \quad X_6 = y\frac{\partial}{\partial y} - 2u\frac{\partial}{\partial u}. \qquad (7.77)$$

Thus, L_6 is the linear span of the operators (7.77), $L_6 = \langle X_1, \ldots, X_6 \rangle$. One can readily calculate, by definition (7.74), the commutators $[X_\alpha, X_\beta]$ of the basic operators (7.77) and verify that they satisfy the test (7.76) for a Lie algebra. The result becomes directly visual if the commutators are disposed as in the following table, where the intersection of the X_α row with the X_β column represents the commutator $[X_\alpha, X_\beta]$. For example, the table of commutators shows that $[X_2, X_4] = X_3$.

7.3. INTRODUCTION TO LIE ALGEBRAS

	X_1	X_2	X_3	X_4	X_5	X_6
X_1	0	0	0	0	X_1	0
X_2	0	0	0	X_3	0	X_2
X_3	0	0	0	0	$3X_3$	$-2X_3$
X_4	0	$-X_3$	0	0	$3X_4$	$-3X_4$
X_5	$-X_1$	0	$-3X_3$	$-3X_4$	0	0
X_6	0	$-X_2$	$2X_3$	$3X_4$	0	0

(7.78)

7.3.3 Subalgebra and ideal

Let L be a Lie algebra. A subspace $K \subset L$ of the vector space L is called a *subalgebra* of the Lie algebra L if K is closed under commutation, i.e.

$$[K, K] \subset K.$$

The subalgebra K is called an *ideal* of L if

$$[K, L] \subset K.$$

Any Lie algebra contains one-dimensional subalgebras. Indeed an arbitrary operator $X \in L$ spans a one-dimensional subalgebra with elements cX, $c =$ const. The existence of two-dimensional subalgebras in arbitrary Lie algebras is guaranteed only in the complex domain. The result is as follows.

Theorem. Any Lie algebra L_r, $r > 2$, contains a two-dimensional subalgebra. Specifically, given an arbitrary operator $X \in L_r$, there exists $Y \in L_r$ such that the linear span of X, Y is a two-dimensional subalgebra, i.e.

$$[X, Y] = aX + bY, \qquad (7.79)$$

where the constants a and b and the coefficients of Y may assume complex values.

Outline of the proof. A two-dimensional subalgebra is constructed by solving algebraic equations, and hence it involves, in general, complex numbers. To simplify calculations, I shall imitate the general proof by constructing a two-dimensional subalgebra of the Lie algebra L_3 spanned by

$$X_1 = \left(1 + x^2\right)\frac{\partial}{\partial x} + xy\frac{\partial}{\partial y}, \quad X_2 = xy\frac{\partial}{\partial x} + \left(1 + y^2\right)\frac{\partial}{\partial y}, \quad X_3 = y\frac{\partial}{\partial x} - x\frac{\partial}{\partial y}.$$

Their commutators are $[X_1, X_2] = X_3$, $[X_2, X_3] = X_1$, and $[X_3, X_1] = X_2$. Let, e.g. $X = X_1$. Then the companion operator Y in (7.79) can be taken in the form $Y = cX_2 + dX_3$. Now the subalgebra condition (7.79) is written

$$cX_3 - dX_2 = aX_1 + bcX_2 + bdX_3.$$

Hence, $a = 0$, and the unknown coefficients b, c, and d are to be found from

$$bc + d = 0, \quad -c + bd = 0.$$

These algebraic equations are regarded as a system of linear equations with respect to c, d. It has a non-trivial solution (so that $Y = cX_2 + dX_3 \neq 0$) only if the determinant of the system, $1 + b^2$, vanishes. Hence, $b = \pm\sqrt{-1}$. Let us take, e.g. $b = -\sqrt{-1}$. Then $d = c\sqrt{-1}$. Thus, the operators $X = X_1, Y = X_2 + \sqrt{-1}X_3$ span a subalgebra $L_2 \subset L_3$, the commutator relation (7.79) being $[X, Y] = -\sqrt{-1}Y$.

Remark. We will see in Section 7.3.6 that solvable Lie algebras L_r have subalgebras of any dimension $s < r$ in the real domain.

7.3.4 Quotient algebra

Let K be an ideal of a Lie algebra L. The operators $X_1, X_2 \in L$ are said to be equivalent if $X_1 - X_2 \in K$. The collection of all operators equivalent to $X \in L$ is termed a *coset* and is written $X + K$ (notation of Section 7.3.2). A coset can be identified with any of its representatives, e.g. X. The family of pairwise disjoint cosets is naturally endowed with a Lie-algebraic structure. Namely, a linear combination and the commutator of two cosets, $X_1 + K$ and $X_2 + K$, are defined via their representatives, viz. $c_1(X_1 + K) + c_2(X_2 + K) = (c_1X_1 + c_2X_2) + K$ and $[X_1 + K, X_2 + K] = [X_1, X_2] + K$. The resulting Lie algebra is called the *quotient algebra* of the Lie algebra L by its ideal K. It is denoted by L/K.

A simple way to construct the quotient algebra of L_r by its ideal L_s is as follows. One chooses a basis $X_1, \ldots, X_r$ such that its first s operators $X_1, \ldots, X_s$ span the ideal L_s. Then the quotient algebra L_r/L_s is the linear span of $X_1, \ldots, X_s$. The commutators in L_r/L_s are obtained merely by dropping in $[X_\mu, X_\nu] = c_{\mu\nu}^\gamma X_\gamma$ ($\mu, \nu = s+1, \ldots, r; \gamma = 1, \ldots, r$) the right-hand terms with $\gamma = 1, \ldots, s$. See the Example in Section 9.4.2.

7.3.5 Derived algebras

Given a Lie algebra L, the commutators $[X, Y]$ of all operators $X, Y \in L$ form a Lie algebra termed the *derived algebra* of the Lie algebra L. The derived algebra is denoted by $L^{(1)}$. Thus,

$$L^{(1)} = [L, L].$$

By construction, $L^{(1)}$ is an ideal of L. Higher derivatives are defined by

$$L^{(s+1)} = (L^{(s)})^{(1)} \equiv [L^{(s)}, L^{(s)}], \quad s = 1, 2, \ldots.$$

A Lie algebra L is said to be *Abelian* if $L^{(1)} = 0$, i.e. if all elements of L commute.

Remark. Let L_r be a Lie algebra with a basis (7.75). The derived algebra $L_r^{(1)}$ is the linear span of the commutators $[X_\alpha, X_\beta]$ of the basis. In other words, $L_r^{(1)}$ is spanned by linearly independent operators that appear in the table of commutators for L_r. For example, table (7.78) shows that $L_6^{(1)} = \langle X_1, X_2, X_3, X_4 \rangle$.

7.3. INTRODUCTION TO LIE ALGEBRAS

7.3.6 Solvable Lie algebras

Definition. A Lie algebra L_r, $r < \infty$, is said to be solvable if there is a sequence

$$L_r \supset L_{r-1} \supset \cdots \supset L_1 \qquad (7.80)$$

of subalgebras of respective dimensions $r, r-1, \ldots, 1$ such that L_k is an ideal in L_{k+1}, $k = 1, \ldots, r-1$.

Theorem. A Lie algebra L_r is solvable if and only if its derived algebra of a finite order s vanishes: $L_r^{(s)} = 0$, $0 < s < \infty$.

Proof. Let L_r be a Lie algebra such that $L_r^{(s)} = 0$. Consider the sequence of successive derived algebras:

$$L_r \supset L_r^{(1)} \supset L_r^{(2)} \cdots \supset L_r^{(s)} = 0. \qquad (7.81)$$

By the definition of derived algebras, $L_r^{(k)}$ and subspaces of $L_r^{(k-1)}$ containing $L_r^{(k)}$ are ideals of $L_r^{(k-1)}$. Hence, one can construct a link in the chain (7.80) between $L_r^{(k-1)}$ and $L_r^{(k)}$ merely by adding to $L_r^{(k)}$ one operator from $L_r^{(k-1)}$, then one more, etc. if $\dim L_r^{(k)} < \dim L_r^{(k-1)} - 1$. Proceeding in this manner for all parts of (7.81), one accomplishes the construction of (7.80).

Conversely, let L_r be a solvable Lie algebra. It follows from (7.80) that the derived algebra of order r is zero. Indeed,

$$L_r^{(r)} = \left(L_r^{(1)}\right)^{(r-1)} \subset L_{r-1}^{(r-1)} \subset \cdots \subset L_1^{(1)} = 0.$$

Example. Let us examine the Lie algebra L_6 spanned by (7.77). One finds the derived algebras by inspecting table (7.78) and arrives at the sequence (7.81):

$$L_6 \supset L_6^{(1)} \supset L_6^{(2)} \supset L_6^{(3)} = 0,$$

where $L_6 = \langle X_1, X_2, X_3, X_4, X_5, X_6 \rangle$, $L_6^{(1)} = \langle X_1, X_2, X_3, X_4 \rangle$, $L_6^{(2)} = \langle X_3 \rangle$. Hence, the Lie algebra L_6 is solvable. The chain (7.80), $L_6 \supset L_5 \supset \cdots \supset L_1$, given by the above proof comprises the subalgebras $L_5 = \langle X_1, X_2, X_3, X_4, X_5 \rangle$, $L_4 = \langle X_1, X_2, X_3, X_4 \rangle$, $L_3 = \langle X_1, X_2, X_3 \rangle$, $L_2 = \langle X_2, X_3 \rangle$, and $L_1 = \langle X_3 \rangle$.

7.3.7 Isomorphic and similar Lie algebras

Lie's group classification (see Section 5.2.3) of ordinary differential equations $f(x, y, y', \ldots, y^{(m)}) = 0$ is based on the enumeration of all possible Lie algebras (infinitesimal groups in Lie's terminology) in the (x, y) plane. In this enumeration, the algebras are maximally simplified by a proper choice of bases and by means of a suitable change of variables. Associated with these two types of simplifying transformations are two distinctly different notions – *isomorphic* and *similar* Lie algebras.

Definition 1. Let L and K be two Lie algebras, and let $\dim L = \dim K$. A linear one-to-one map f of L onto K is called an isomorphism (automorphism if $K = L$) if it preserves commutators, i.e.

$$f([X_1, X_2]) = [f(X_1), f(X_2)]$$

for any $X_1, X_2 \in L$. If the Lie algebras L and K can be related by an isomorphism, they are termed *isomorphic Lie algebras*.

Theorem 1. Two finite-dimensional Lie algebras are isomorphic if and only if one can choose bases for the algebras such that the algebras have, in these bases, equal structure constants, i.e. the same table of commutators.

The concept of an isomorphism is independent of a particular realization of an abstract Lie algebra [7.5] as, say, an algebra of differential operators (7.15). In the latter case, however, there is an additional way to simplify Lie algebras by transforming one algebra into another by a change of variables x^i.

Definition 2. The Lie algebras L and $\bar{L}$ are *similar* if one is obtained from the other by a change of variables. It means that the operators $X = \xi^i(x)\partial/\partial x^i$ and $\bar{X} = \xi'^i(x')\partial/\partial x'^i$ of L and $\bar{L}$ are related by (7.29)–(7.30), i.e. by

$$x'^i = x'^i(x), \quad \xi'^i = \xi^k \frac{\partial x'^i}{\partial x^k}, \quad i = 1, \ldots, n. \tag{7.82}$$

Lemma. In order that two r-dimensional Lie algebras L_r and $\bar{L}_r$ with the same number of variables be similar, it is necessary that L_r and $\bar{L}_r$ be isomorphic.

Proof. Let the structure constants $c_{\alpha\beta}^\gamma$ of the algebra L_r be defined by (7.76), $[X_\alpha, X_\beta] = c_{\alpha\beta}^\gamma X_\gamma$. Let bases of L_r and $\bar{L}_r$ be given by the operators

$$X_\alpha = \xi_\alpha^i(x)\frac{\partial}{\partial x^i} \quad \text{and} \quad \bar{X}_\alpha = \xi_\alpha'^i(x')\frac{\partial}{\partial x'^i}, \quad \alpha = 1, \ldots, r, \tag{7.83}$$

respectively, where $\bar{X}_\alpha$ are obtained from X_α by transformation (7.82). Then Lemma 1 of Section 4.5.2 yields (see (4.97)):

$$\left[\bar{X}_\alpha, \bar{X}_\beta\right] = \overline{[X_\alpha, X_\beta]} = \overline{c_{\alpha\beta}^\gamma X_\gamma} = c_{\alpha\beta}^\gamma \bar{X}_\gamma.$$

Thus, L_r and $\bar{L}_r$ have the same structure constants in the bases (7.83), and hence they are isomorphic.

Remark. The converse is not true: two Lie algebras may be isomorphic but not similar. Therefore, there exist, e.g. *four* non-similar two-dimensional Lie algebras L_2 in the plane, but only *two* non-isomorphic L_2 (see the next section).

It is precisely similarity (and not simply isomorphism) that is of use in group analysis as a criterion of reducibility of one differential equation to another by a suitable change of variables. Nonetheless, establishing isomorphism is important as a first (and necessary due to Lemma) step for the determination of similarity.

7.3. INTRODUCTION TO LIE ALGEBRAS

Let L_r and $\overline{L}_r$ be two isomorphic r-dimensional Lie algebras with the bases (7.83), so that they have equal structure constants. We shall consider here the case when $r > n$ and assume that $\text{rank} \|\xi_\alpha^i(x)\| = \text{rank}\|\xi_\alpha^{\prime i}(x')\| = n$. Renumbering the indices, if necessary, one can assume that this is the value of the rank of the square matrices $\|\xi_h^i\|$, $\|\xi_h^{\prime i}\|$ ($i, h = 1, \ldots, n$). Then the following conditions hold:

$$\xi_p^i = \varphi_p^h \xi_h^i, \quad \xi_p^{\prime i} = \varphi_p^{\prime h} \xi_h^{\prime i}, \qquad i = 1, \ldots, n;\ p = n+1, \ldots, r, \qquad (7.84)$$

where φ_p^h and $\varphi_p^{\prime h}$ are functions of x and x', respectively, and where we sum over h from 1 to n. Under the assumptions made, we have the following theorem [7.7].

Theorem 2. A necessary and sufficient condition that the Lie algebras L_r and $\overline{L}_r$ with the same structure constants and the same number of variables x^i and $x^{\prime i}$ be similar is that the functions φ_p^h, $\varphi_p^{\prime h}$ satisfy the equations

$$\varphi_p^{\prime h}(x') = \varphi_p^h(x), \quad h = 1, \ldots, n;\ p = n+1, \ldots, r, \qquad (7.85)$$

and that these equations do not lead to relations between the x's alone or the x''s alone.

Proof. Recall that the similarity of the algebras L_r and $\overline{L}_r$ means that the coordinates of the bases (7.83) are related by (7.82), i.e. there is a change of variables $x^{\prime i} = \varphi^i(x)$, such that

$$\xi_\alpha^{\prime i} = \xi_\alpha^k \frac{\partial x^{\prime i}}{\partial x^k}, \quad i = 1, \ldots, n;\ \alpha = 1, \ldots, r. \qquad (7.86)$$

Necessity. Let the algebras L_r and $\overline{L}_r$ be similar. Then (7.86) and (7.84) yield $n(r - n)$ equations (7.85). By construction, the equations (7.85) are consistent, and it is impossible to eliminate all variables $x^{\prime i}$ to arrive at relations between the x's alone, and vice versa.

Sufficiency. Let equations (7.85) be consistent and not lead to relations between the x's alone or the x''s alone. Let us consider equations (7.86) for $\alpha = 1, \ldots, n$. Under the assumptions on the ranks, these equations can be solved for derivatives and written:

$$\frac{\partial x^{\prime i}}{\partial x^k} = \xi_h^{\prime i}(x') \xi_k^h(x), \quad i, k = 1, \ldots, n, \qquad (7.87)$$

where $\|\xi_k^h\|$ is the square matrix inverse to $\|\xi_h^i\|$, i.e. $\xi_k^h \xi_h^i = \delta_k^i$, $\xi_i^h \xi_l^i = \delta_l^h$ with $i, k, h, l = 1, \ldots, n$. It follows from the conditions (7.85) that the system (7.87) (and hence the system (7.86)) is integrable (a detailed treatment of this non-trivial step in the proof is to be found in [7.4](i), §22). This means that there is a change of variables that establishes the similarity of the algebras L_r and $\overline{L}_r$.

Determination of the change of variables. The similarity transformation $x^{\prime i} = \varphi^i(x)$ is determined by solving, with respect to the x''s, the mixed system of equations comprising the functional equations (7.85) and the differential equations (7.87), the latter being integrable in view of (7.85).

7.3.8 Non-isomorphic structures and non-similar realizations in the plane of 1, 2, and 3-dimensional Lie algebras

Lie gave a classification of continuous groups in the (x,y) plane, using, where necessary, complex changes of variables or bases of algebras. Enumeration of all non-similar finite-dimensional algebras in the plane is to be found in [5.13], vol. 3, Chap. 3, 4. Lie's complex classification of groups and invariant equations is quite sufficient for integration purposes.

Here, we focus attention on one-, two-, and three-dimensional Lie algebras of operators in the plane:

$$X = \xi(x,y)\frac{\partial}{\partial x} + \eta(x,y)\frac{\partial}{\partial y}. \tag{7.88}$$

They will be used in Part III for integration of first-, second-, and third-order differential equations.

ONE- AND TWO-DIMENSIONAL ALGEBRAS

Any one-dimensional algebra L_1 is Abelian. Furthermore, its basis X_1 can be written in canonical variables (Section 7.1.8). Hence, the structure and realization of L_1 can be taken to be

$$[X_1, X_1] = 0, \quad X_1 = \frac{\partial}{\partial x}. \tag{7.89}$$

Consider a two-dimensional Lie algebra L_2 with a basis

$$X_\alpha = \xi_\alpha(x,y)\frac{\partial}{\partial x} + \eta_\alpha(x,y)\frac{\partial}{\partial y}, \quad \alpha = 1,2. \tag{7.90}$$

The structure of L_2 is defined by the commutator (cf. (7.74)):

$$[X_1, X_2] = \left(X_1(\xi_2) - X_2(\xi_1)\right)\frac{\partial}{\partial x} + \left(X_1(\eta_2) - X_2(\eta_1)\right)\frac{\partial}{\partial y}. \tag{7.91}$$

Furthermore, the basic operators (7.90) may be linearly connected or not, in accordance with the value of the rank r_* (4.94). Specifically, $r_* = 1$ if the determinant

$$\det\begin{Vmatrix} \xi_1 & \eta_1 \\ \xi_2 & \eta_2 \end{Vmatrix} = \xi_1\eta_2 - \eta_1\xi_2 \tag{7.92}$$

vanishes. Then the operators X_1 and X_2 (7.90) are linearly connected. If the determinant does not vanish, then $r_* = 2$ and the operators X_1 and X_2 are not linearly connected.

The algebra L_2 has one of the following non-isomorphic structures. It can be either Abelian, $[X_1, X_2] = 0$, or non-Abelian, $[X_1, X_2] \neq 0$, in which case the commutator can be taken to be $[X_1, X_2] = X_1$ by choosing a suitable basis (see Problem 7.18). Each of the non-isomorphic structures has two non-similar realizations according to whether or not the basic operators (7.90) are linearly connected. Hence, there exist four distinctly different *canonical forms* of L_2 presented in Table 7.1 (see Section 12.2.2 and [7.4](i), Theorem 25.9).

7.3. INTRODUCTION TO LIE ALGEBRAS

Table 7.1 Structure and non-similar realizations of L_2

Type	Structure of L_2	Canonical forms of L_2
I	$[X_1, X_2] = 0$, $\xi_1\eta_2 - \eta_1\xi_2 \neq 0$	$X_1 = \dfrac{\partial}{\partial x}$, $X_2 = \dfrac{\partial}{\partial y}$
II	$[X_1, X_2] = 0$, $\xi_1\eta_2 - \eta_1\xi_2 = 0$	$X_1 = \dfrac{\partial}{\partial y}$, $X_2 = x\dfrac{\partial}{\partial y}$
III	$[X_1, X_2] = X_1$, $\xi_1\eta_2 - \eta_1\xi_2 \neq 0$	$X_1 = \dfrac{\partial}{\partial y}$, $X_2 = x\dfrac{\partial}{\partial x} + y\dfrac{\partial}{\partial y}$
IV	$[X_1, X_2] = X_1$, $\xi_1\eta_2 - \eta_1\xi_2 = 0$	$X_1 = \dfrac{\partial}{\partial y}$, $X_2 = y\dfrac{\partial}{\partial y}$

THREE-DIMENSIONAL ALGEBRAS

Two possibilities, $[X_1, X_2] \neq 0$ or $[X_1, X_2] = 0$, for bases of two-dimensional algebras L_2 mean that the first derivative $L_2^{(1)}$ has the dimension 1 or 0, respectively. In the case of three-dimensional algebras, the derived algebra may be of dimension 3, 2, 1, or 0. Lie classified non-isomorphic structures of L_3 according to the dimension of the derived algebra $L_3^{(1)}$ and proved the following.

Theorem 1 ([7.8]). The structure of any three-dimensional algebra L_3 can be transformed, by a complex change of its basis X_1, X_2, X_3, to one and only one of the following seven forms, where the letters A, B, C, and D indicate that the first derived algebra $L_3^{(1)}$ has the dimension 3, 2, 1, and 0, respectively:

A. 1) $[X_1, X_2] = X_1$ $\quad [X_1, X_3] = 2X_2$ $\quad [X_2, X_3] = X_3$,
B. 2) $[X_1, X_2] = 0$ $\quad [X_1, X_3] = X_1$ $\quad [X_2, X_3] = cX_2$ $(c \neq 0, \neq 1)$,
   2') $[X_1, X_2] = 0$ $\quad [X_1, X_3] = X_1$ $\quad [X_2, X_3] = X_2$,
   3) $[X_1, X_2] = 0$ $\quad [X_1, X_3] = X_1$ $\quad [X_2, X_3] = X_1 + X_2$,
C. 4) $[X_1, X_2] = 0$ $\quad [X_1, X_3] = X_1$ $\quad [X_2, X_3] = 0$,
   5) $[X_1, X_2] = 0$ $\quad [X_1, X_3] = 0$ $\quad [X_2, X_3] = X_1$,
D. 6) $[X_1, X_2] = 0$ $\quad [X_1, X_3] = 0$ $\quad [X_2, X_3] = 0$.

Type 2') is distinguished from type 2) due to its autonomous significance.

The classification of non-isomorphic structures of Lie algebras does not depend on a particular realization of algebras, e.g. in Lie algebras of operators, on the number of variables. In the case of Lie algebras of operators in two variables, considered here, each type of L_3 furnished by Theorem 1 gives rise to one or several non-similar realizations. In these calculations, one encounters algebraic equations, and hence one naturally needs complex numbers. Lie's complex classification of all non-similar three-dimensional algebras of operators in two variables, x and y, is given by the following theorem.

Theorem 2 ([7.9]). *Any three-dimensional algebra of operators in two variables can be transformed, by a complex change of variables, to one and only one of the following 13 types called* canonical forms *of L_3.*

A. The first derived algebra has the dimension three :

1) $X_1 = \dfrac{\partial}{\partial x} + \dfrac{\partial}{\partial y}$, $X_2 = x\dfrac{\partial}{\partial x} + y\dfrac{\partial}{\partial y}$, $X_3 = x^2\dfrac{\partial}{\partial x} + y^2\dfrac{\partial}{\partial y}$,

2) $X_1 = \dfrac{\partial}{\partial x}$, $X_2 = 2x\dfrac{\partial}{\partial x} + y\dfrac{\partial}{\partial y}$, $X_3 = x^2\dfrac{\partial}{\partial x} + xy\dfrac{\partial}{\partial y}$,

3) $X_1 = \dfrac{\partial}{\partial y}$, $X_2 = y\dfrac{\partial}{\partial y}$, $X_3 = y^2\dfrac{\partial}{\partial y}$.

B. The first derived algebra has the dimension two :

4) $X_1 = \dfrac{\partial}{\partial x}$, $X_2 = \dfrac{\partial}{\partial y}$, $X_3 = x\dfrac{\partial}{\partial x} + cy\dfrac{\partial}{\partial y}$ $(c \neq 0, \neq 1)$,

5) $X_1 = \dfrac{\partial}{\partial y}$, $X_2 = x\dfrac{\partial}{\partial y}$, $X_3 = (1-c)x\dfrac{\partial}{\partial x} + y\dfrac{\partial}{\partial y}$ $(c \neq 0, \neq 1)$,

6) $X_1 = \dfrac{\partial}{\partial x}$, $X_2 = \dfrac{\partial}{\partial y}$, $X_3 = x\dfrac{\partial}{\partial x} + y\dfrac{\partial}{\partial y}$,

7) $X_1 = \dfrac{\partial}{\partial y}$, $X_2 = x\dfrac{\partial}{\partial y}$, $X_3 = y\dfrac{\partial}{\partial y}$,

8) $X_1 = \dfrac{\partial}{\partial x}$, $X_2 = \dfrac{\partial}{\partial y}$, $X_3 = (x+y)\dfrac{\partial}{\partial x} + y\dfrac{\partial}{\partial y}$,

9) $X_1 = \dfrac{\partial}{\partial y}$, $X_2 = x\dfrac{\partial}{\partial y}$, $X_3 = \dfrac{\partial}{\partial x} + y\dfrac{\partial}{\partial y}$.

C. The first derived algebra has the dimension one :

10) $X_1 = \dfrac{\partial}{\partial x}$, $X_2 = \dfrac{\partial}{\partial y}$, $X_3 = x\dfrac{\partial}{\partial x}$,

11) $X_1 = \dfrac{\partial}{\partial y}$, $X_2 = x\dfrac{\partial}{\partial y}$, $X_3 = x\dfrac{\partial}{\partial x} + y\dfrac{\partial}{\partial y}$,

12) $X_1 = \dfrac{\partial}{\partial x}$, $X_2 = \dfrac{\partial}{\partial y}$, $X_3 = x\dfrac{\partial}{\partial y}$.

D. The first derived algebra has the dimension zero :

13) $X_1 = \dfrac{\partial}{\partial y}$, $X_2 = x\dfrac{\partial}{\partial y}$, $X_3 = p(x)\dfrac{\partial}{\partial y}$.

For one- and two-dimensional algebras, classification over the reals is the same as that over the complex numbers. The following classification (Table 7.2) of three-dimensional Lie algebras in the real domain, due to Bianchi [7.10], shows that there are two more non-isomorphic real algebras than complex algebras.

7.3. INTRODUCTION TO LIE ALGEBRAS

Table 7.2 Non-isomorphic structures of real L_3 (by Bianchi [7.10])

Type	$[X_1, X_2]$	$[X_2, X_3]$	$[X_3, X_1]$	Complex structures
I	0	0	0	Lie's classification in
II	0	0	$-X_2$	the complex domain
III	0	0	$-X_1$	given in Theorem 1 is
IV	0	X_2	$-(X_1 + X_2)$	obtained by excluding
V	0	X_2	$-X_1$	types VII and IX
VI	0	$cX_2, c \neq 0, \neq 1$	$-X_1$	(see the Remark) and
VII	0	$X_1 + sX_2$	$X_2 - sX_1$	interchanging X_1 and
VIII	X_1	X_3	$-2X_2$	X_2 in types II and IV.
IX	X_3	X_1	X_2	

Remark. Type VII reduces to VI with $c = (s+i)/(s-i)$ in the complex basis $X_1' = X_2 + iX_1$, $X_2' = X_2 - iX_1$, $X_3' = X_3/(s-i)$, while type IX becomes VIII in the basis $X_1' = X_1 - iX_3$, $X_2' = -iX_2$, $X_3' = X_1 + iX_3$.

Example 1 ([7.11]). Consider the algebra L_3 spanned by the operators

$$X_1 = \frac{\partial}{\partial x}, \quad X_2 = \sin x \frac{\partial}{\partial x} + \cos x \frac{\partial}{\partial y}, \quad X_3 = \cos x \frac{\partial}{\partial x} - \sin x \frac{\partial}{\partial y}.$$

Their commutators, $[X_1, X_2] = X_3$, $[X_1, X_3] = -X_2$, and $[X_2, X_3] = -X_1$, take the form 1) of Theorem 1 in the basis $X_1' = X_1 + X_3$, $X_2' = X_2$, $X_3' = X_1 - X_3$:

$$X_1' = (1+\cos x)\frac{\partial}{\partial x} - \sin x \frac{\partial}{\partial y}, \quad X_2' = X_2, \quad X_3' = (1-\cos x)\frac{\partial}{\partial x} + \sin x \frac{\partial}{\partial y}.$$

Example 2 ([7.11]). Consider the algebra L_3 spanned by the operators

$$X_1 = \frac{\partial}{\partial x}, \quad X_2 = y\frac{\partial}{\partial x}, \quad X_3 = xy\frac{\partial}{\partial x} + (1+y^2)\frac{\partial}{\partial y}.$$

Here $[X_1, X_2] = 0$, $[X_1, X_3] = X_2$, and $[X_2, X_3] = -X_1$. They take the form 2) of Theorem 1 in the complex basis $X_1' = X_1 - iX_2$, $X_2' = iX_1 - X_2$, $X_3 = -iX_3$:

$$X_1' = (1-iy)\frac{\partial}{\partial x}, \quad X_2' = (i-y)\frac{\partial}{\partial x}, \quad X_3' = -ixy\frac{\partial}{\partial x} - i(1+y^2)\frac{\partial}{\partial y}.$$

The transformation of an algebra L_2 or L_3 to its canonical form, given by Table 7.1 and Theorem 2, is crucial for solving second- and third-order differential equations admitting L_2 and L_3, respectively. The transformations of L_2 to their canonical forms will be used in Section 12.2. Transformations of L_3 are not employed in this book. Yet, it is useful to consider the following instructive exercises (see also Problems 7.20 and 7.21). The reader can find more details in [5.14].

166   7. INFINITESIMAL TRANSFORMATIONS AND LOCAL GROUPS

Exercise 1. Let L_3 be the Lie algebra spanned by [7.12]

$$X_1 = (1+x^2)\frac{\partial}{\partial x} + xy\frac{\partial}{\partial y}, \quad X_2 = x\frac{\partial}{\partial y} - y\frac{\partial}{\partial x}, \quad X_3 = xy\frac{\partial}{\partial x} + (1+y^2)\frac{\partial}{\partial y}. \quad (7.93)$$

Find a complex change of variables transforming the operators (7.93) to Lie's canonical form 1) of Theorem 2.

Solution. L_3 is of type IX, Table 2: $[X_1, X_2] = X_3$, $[X_2, X_3] = X_1$, $[X_3, X_1] = X_2$. Denote by $\overline{L}_3 = \langle \overline{X}_1, \overline{X}_2, \overline{X}_3 \rangle$ the algebra given in 1) of Theorem 2. According to the remark to Table 2, the algebras L_3 and $\overline{L}_3$ are complex isomorphic. Following the remark, we bring them to the same commutator relations, e.g., to IX of Table 2, by the complex change of the basis in $\overline{L}_3$, viz. $X_1' = (\overline{X}_1 + \overline{X}_3)/2$, $X_2' = i\overline{X}_2$, $X_3' = i(\overline{X}_1 - \overline{X}_3)/2$:

$$X_1' = \frac{1}{2}\left[(1+x'^2)\frac{\partial}{\partial x'} + (1+y'^2)\frac{\partial}{\partial y'}\right], \quad X_2' = i\left[x'\frac{\partial}{\partial x'} + y'\frac{\partial}{\partial y'}\right],$$

$$X_3' = \frac{i}{2}\left[(1-x'^2)\frac{\partial}{\partial x'} + (1-y'^2)\frac{\partial}{\partial y'}\right]. \quad (7.94)$$

Thus, the algebras L_3 and $\overline{L}_3$ have the same structure constants in the bases (7.93) and (7.93), respectively. Equations (7.84), written in the form

$$\xi_3^k = \sum_{h=1}^{2} \varphi^h \xi_h^k, \quad \xi_3'^k = \sum_{h=1}^{2} \varphi'^h \xi_h'^k,$$

yield

$$xy = (1+x^2)\varphi^1 - y\varphi^2, \quad i(1-x'^2) = (1+x'^2)\varphi'^1 + 2ix'\varphi'^2,$$
$$1+y^2 = xy\varphi^1 + x\varphi^2, \quad i(1-y'^2) = (1+y'^2)\varphi'^1 + 2iy'\varphi'^2.$$

It follows that

$$\varphi^1 = \frac{y}{x}, \quad \varphi^2 = \frac{1}{x}, \quad \text{and} \quad \varphi'^1 = i\frac{1+x'y'}{1-x'y'}, \quad \varphi'^2 = -\frac{x'+y'}{1-x'y'}.$$

With these functions φ^h and φ'^h, the system (7.85) is written

$$i\frac{1+x'y'}{1-x'y'} = \frac{y}{x}, \quad -\frac{x'+y'}{1-x'y'} = \frac{1}{x},$$

and furnishes the change of variables:

$$x = -\frac{1-x'y'}{x'+y'}, \quad y = -i\frac{1+x'y'}{x'+y'},$$

which establishes the similarity of the algebras L_3 and $\overline{L}_3$ in the complex plane. In this example, in order to find the change of variable, it was enough to solve equations (7.85).

Exercise 2. Test for similarity the algebras L_3 and $\overline{L}_3$ spanned by

$$X_1 = \frac{\partial}{\partial x} + \frac{\partial}{\partial y}, \quad X_2 = x\frac{\partial}{\partial x} + x\frac{\partial}{\partial y}, \quad X_3 = x^2\frac{\partial}{\partial x} + y^2\frac{\partial}{\partial y}$$

and

$$\overline{X}_1 = \frac{\partial}{\partial x'}, \quad \overline{X}_2 = x'\frac{\partial}{\partial x'} + \frac{y'}{2}\frac{\partial}{\partial y'}, \quad \overline{X}_3 = x'^2\frac{\partial}{\partial x'} + x'y'\frac{\partial}{\partial y'}.$$

Solution. L_3 and $\overline{L}_3$ are the first and the second canonical algebras of Theorem 2. They have the same structure constants. Equations (7.84), written as in the previous exercise, have the form
$$x^2 = \varphi^1 + x\varphi^2, \qquad x'^2 = \varphi'^1 + x'\varphi'^2,$$
$$y^2 = \varphi^1 + y\varphi^2, \qquad 2x'y' = y'\varphi'^2.$$
Hence,
$$\varphi^1 = -xy, \quad \varphi^2 = x+y, \quad \varphi'^1 = -x'^2, \quad \varphi'^2 = 2x'.$$
With these functions φ^h and φ'^h, equations (7.85) assume the form
$$x'^2 = xy, \quad 2x' = x+y,$$
whence, after eliminating x', one arrives at the relation $x - y = 0$. By Theorem 2 of Section 7.3.7, it follows that L_3 and $\overline{L}_3$ are not similar (neither over the reals, nor over the complex numbers), though they are isomorphic.

7.4 Multi-parameter groups: Outline of basic notions

7.4.1 Definition

We shall use the notation and assumptions of Section 7.1.1, but the parameter a will be replaced by r-parameters, i.e. by the vector-parameter $a = (a^1, \ldots, a^r)$. We consider invertible transformations $T_a : \mathbb{R}^n \to \mathbb{R}^n$,
$$\overline{x} = f(x, a), \qquad (7.95)$$
defined in a neighborhood of $a = 0 \equiv (0, \ldots, 0)$. Here $f = (f^1, \ldots, f^n)$, where the functions $f^i = f^i(x, a)$ are at least three times continuously differentiable with respect to all variables $x^1, \ldots, x^n, a^1, \ldots, a^r$. We impose the initial condition:
$$f|_{a=0} = x. \qquad (7.96)$$

Definition. The transformations (7.95) are said to form an *r-parameter local group* G_r if
$$f(f(z, a), b) = f(z, c) \qquad (7.97)$$
for all values a and b of the parameter sufficiently close to $a = 0$. Here $c = c(a, b)$ is a vector-function with components
$$c^\alpha = \phi^\alpha(a, b), \quad \alpha = 1, \ldots, r, \qquad (7.98)$$
defined and thrice continuously differentiable for sufficiently small a and b (in the sense of Definition 7.1.2). The functions (7.98) define a composition law in the group G_r. It is assumed that the system of equations
$$\phi^\alpha(a, b) = 0, \quad \alpha = 1, \ldots, r, \qquad (7.99)$$
has a unique solution $b = (b^1, \ldots, b^r)$ for any small a. Given a, the solution b of the system (7.99) is denoted by a^{-1}. Hence, the inverse transformation T_a^{-1} is
$$x = f(\overline{x}, a^{-1}).$$

7.4.2 Composition via one-parameter groups

The following theorem is sufficient to meet the needs of applied group analysis when the practical construction of multi-parameter groups is desired.

Theorem. Let L_r be an r-dimensional vector space spanned by the operators

$$X_\alpha = \xi_\alpha^i(x)\frac{\partial}{\partial x^i}, \quad \alpha = 1,\ldots,r. \tag{7.100}$$

The composition $T_a = T_{a^r}\cdots T_{a^1}$ of r one-parameter groups of transformations T_{a^α} generated individually by each of the base operators X_α via the Lie equations

$$\frac{d\bar{x}^i}{da^\alpha} = \xi_\alpha^i(\bar{x}), \quad \bar{x}^i|_{a^\alpha=0} = x^i, \quad i = 1,\ldots,n, \tag{7.101}$$

is an r-parameter (local) group G_r if and only if L_r is a Lie algebra. By applying the same construction to any s-dimensional subalgebra of L_r, one generates an s-parameter subgroup of the group G_r.

Remark. The above construction of G_r depends upon the choice of a basis in L_r. Therefore, the theory of Lie groups considers more general constructions and different representations of Lie groups with a given Lie algebra. However, all these representations are *similar*, and hence are indistinguishable in applications.

Example. Consider a three-dimensional Lie algebra spanned by the operators

$$X_1 = \frac{\partial}{\partial x}, \quad X_2 = \frac{\partial}{\partial y}, \quad X_3 = y\frac{\partial}{\partial x}.$$

Solution of the Lie equations (7.101) for these operators provides the following three one-parameter groups of transformations in the (x,y) plane, with the respective parameters a^1, a^2, and a^3:

$$T_{a^1}: \bar{x} = x + a^1, \bar{y} = y; \quad T_{a^2}: \bar{x} = x, \bar{y} = y + a^2; \quad T_{a^3}: \bar{x} = x + ya^3, \bar{y} = y.$$

Their composition $T_a = T_{a^3}T_{a^2}T_{a^1}$, where $a = (a^1, a^2, a^3)$, has the form:

$$\bar{x} = x + ya^3 + a^1 + a^2 a^3, \quad \bar{y} = y + a^2.$$

The consecutive application of T_a and T_b, where $b = (b^1, b^2, b^3)$, yields the transformations $T_b T_a$:

$$\bar{\bar{x}} = x + y(a^3 + b^3) + a^2(a^3 + b^3) + b^2 b^3 + a^1 + b^1, \quad \bar{\bar{y}} = y + a^2 + b^2.$$

Now $T_b T_a = T_c$, i.e. the equations $\bar{\bar{x}} = x + yc^3 + c^1 + c^2 c^3$ and $\bar{\bar{y}} = y + c^2$ yield:

$$c^1 = a^1 + b^1 - b^2 a^3, \quad c^2 = a^2 + b^2, \quad c^3 = a^3 + b^3.$$

Thus, the transformations $T_a = T_{a^3}T_{a^2}T_{a^1}$ form a three-parameter group with the composition law (7.98) given by

$$\phi^1(a,b) = a^1 + b^1 - b^2 a^3, \quad \phi^2(a,b) = a^2 + b^2, \quad \phi^3(a,b) = a^3 + b^3.$$

7.4. MULTI-PARAMETER GROUPS

7.4.3 Basis of invariants

The definition of the invariants of multi-parameter groups is the same as that for one-parameter groups (Definition 7.1.7). Thus, $F(x)$ is an invariant of a group G_r if $F(T_a(x)) = F(x)$ identically in $x = (x^1, \ldots, x^n)$ for all $T_a \in G_r$.

Theorem 1. A function $F(x)$ is an invariant of the group G_r with the base operators (7.100) if and only if it solves the following system of homogeneous linear partial differential equations (cf. (4.92)):

$$X_\alpha F \equiv \xi_\alpha^i(x) \frac{\partial F}{\partial x^i} = 0, \quad \alpha = 1, \ldots, r. \tag{7.102}$$

Proof. Let $F(x)$ be an invariant. In particular, it is invariant under the transformations T_{a^α} generated by each operator (7.100). Consequently, $F(x)$ solves the equation (7.24) for each operator X_α, i.e. the system (7.102).

Conversely, let $F(x)$ solve the system (7.102). By Theorem 1 of Section 7.1.7, $F(x)$ is invariant under r one-parameter groups of transformations T_{a^α}. Since, by Theorem 7.4.2, any transformation of G_r can be obtained as a composition of these one-parameter groups, it follows that $F(T_a(x)) = f(x)$ for all $T_a \in G_r$.

Theorem 2. The group G_r with infinitesimal generators (7.100) has $n - r_*$ functionally independent invariants $\psi_1(x), \ldots, \psi_{n-r_*}(x)$ (a *basis of invariants*), where r_* is the rank of the matrix of coefficients of (7.100) at generic points x:

$$r_* = \text{rank} \left\| \xi_\alpha^i(x) \right\|. \tag{7.103}$$

An arbitrary invariant $F(x)$ of G_r has the form $F = \Phi(\psi_1(x), \ldots, \psi_{n-r_*}(x))$.

Proof. The proof of the theorem and methods for calculating invariants are provided by Theorem 1 and Sections 4.5.2 and 4.5.3. Indeed, since the operators (7.100) span a Lie algebra, the system (7.102) is complete (cf. (4.96) and (7.76)).

Example. Consider the group G_3 of rotations in $\mathbb{R}^3$ with generators

$$X_1 = x^2 \frac{\partial}{\partial x^1} - x^1 \frac{\partial}{\partial x^2}, \quad X_2 = x^3 \frac{\partial}{\partial x^2} - x^2 \frac{\partial}{\partial x^3}, \quad X_3 = x^1 \frac{\partial}{\partial x^3} - x^3 \frac{\partial}{\partial x^1}.$$

Here $n = 3$ and $r_* = 2$. Consequently, a basis of invariants of the group G_3 is given by one invariant, e.g. by $|x|^2 = (x^1)^2 + (x^2)^2 + (x^3)^2$ (see Problem 7.24).

7.4.4 Regular and singular invariant manifolds

Definition 7.2.1 of invariant equations (7.47) and invariant manifolds $M \subset \mathbb{R}^n$ remains the same in the case of multi-parameter groups.

Theorem 1. The system of equations (7.47), $F_\sigma(x) = 0$ ($\sigma = 1, \ldots, s$), is invariant under the group G_r with generators X_α (7.100) if and only if

$$X_\alpha F_\sigma \big|_M = 0, \quad \sigma = 1, \ldots, s; \; \alpha = 1, \ldots, r, \tag{7.104}$$

170   7. INFINITESIMAL TRANSFORMATIONS AND LOCAL GROUPS

where $|_M$ means evaluated on the manifold M defined by equations (7.47).

Proof. Use Theorem 7.2.1 and proceed as in the proof of Theorem 1 of the preceding section.

Definition. An invariant manifold $M \subset \mathbb{R}^n$ for G_r is said to be *regular* (with respect to the group G_r) if

$$r_*|_M = r_* \qquad (7.105)$$

and *singular* otherwise, i.e. if

$$r_*|_M < r_*. \qquad (7.106)$$

Here r_* is the generic rank (7.103), and $r_*|_M$ is the rank evaluated on M.

Carrying over the proof of Theorem 7.2.2 to multi-parameter groups, we arrive at one of the key results of applied group analysis.

Theorem 2. Any regular invariant manifold M for G_r admits an invariant representation, i.e. can be given by equations of the form

$$\Phi_\sigma(\psi_1(x), \ldots, \psi_{n-r_*}(x)) = 0, \quad \sigma = 1, \ldots, s, \qquad (7.107)$$

where $\psi_1, \ldots, \psi_{n-r_*}$ is a basis of invariants of the group G_r. Equations (7.107) with arbitrary functions Φ_σ of $n - r_*$ variables furnish the general form of regular invariant manifolds for G_r.

Remark. Singular invariant manifolds are defined by (7.106) and (7.104):

$$\operatorname{rank} \left\| \xi_\alpha^i(x) \right\|_M < r_* \quad \text{and} \quad X_\alpha F_\sigma \big|_M = 0. \qquad (7.108)$$

They will be employed, e.g. in Chapter 10.

7.5 Approximate transformation groups

An *approximate transformation group*, or briefly an *approximate group* is a representation of a local group by what is called *approximate transformations*.

7.5.1 Motivation

The initiation and subsequent development of the theory of approximate transformation groups were inspired by the following two chief circumstances.

A variety of differential equations, recognized as mathematical models in engineering and physical sciences, involve empirical parameters or constitutive laws. Therefore the coefficients of model equations are defined approximately with an inevitable error. Consequently, differential equations depending on a small parameter occur frequently in applications. Unfortunately, Lie symmetries are unstable with respect to perturbations of coefficients of differential equations. This instability, reducing the practical value of group theoretic methods, was the first circumstance that led us [7.13](i) to the concept of approximate groups.

7.6. APPROXIMATE TRANSFORMATION GROUPS

The second factor is that, in practical applications, Lie group analysis may come across unjustified complexities. Consider, as an illustration, a de Sitter universe (cf. Example 3 in Section 5.5.4) with the metric

$$ds^2 = -\left(1+\varepsilon\sigma^2\right)^{-2}\sum_{\mu=1}^{4}(dx^\mu)^2, \quad \sigma^2 = \sum_{\mu=1}^{4}(x^\mu)^2.$$

Here $\varepsilon = K/4$ with K denoting the curvature of the de Sitter space-time. According to cosmological data, the curvature K is a small constant, and hence ε can be treated as a small parameter. The *de Sitter group* (i.e., the group of isometric motions in the de Sitter space-time) differs from the *Poincaré group* (the group of isometries in the Minkowski space-time, see Example 9 in Section 6.3.3) in that the usual translations of space-time coordinates x^μ are replaced by more complicated transformations, the so-called "generalized translations" in the de Sitter space-time. The generalized translation, e.g. along the x^1-axis, has the infinitesimal generator

$$X = \left(1 + \varepsilon[(x^1)^2 - (x^2)^2 - (x^3)^2 - (x^4)^2]\right)\frac{\partial}{\partial x^1}$$

$$+ 2\varepsilon x^1\left(x^2\frac{\partial}{\partial x^2} + x^3\frac{\partial}{\partial x^3} + x^4\frac{\partial}{\partial x^4}\right). \tag{7.109}$$

The corresponding group transformations have the form

$$\bar{x}^1 = 2\frac{x^1\cos(2a\sqrt{\varepsilon}) + (1-\varepsilon\sigma^2)/(2\sqrt{\varepsilon})\sin(2a\sqrt{\varepsilon})}{1+\varepsilon\sigma^2 + (1-\varepsilon\sigma^2)\cos(2a\sqrt{\varepsilon}) - 2x^1\sqrt{\varepsilon}\sin(2a\sqrt{\varepsilon})},$$

$$\bar{x}^j = 2\frac{x^j}{1+\varepsilon\sigma^2 + (1-\varepsilon\sigma^2)\cos(2a\sqrt{\varepsilon}) - 2x^1\sqrt{\varepsilon}\sin(2a\sqrt{\varepsilon})}, \tag{7.110}$$

where $j = 2, 3, 4$. It is assumed here that $\varepsilon \geq 0$.

Since ε is small, one can use an approximate expression of the transformations (7.110) obtained by expanding in powers of ε and considering only the leading terms of the first order. In other words, one can consider the de Sitter group as a perturbation of the Poincaré group by the curvature K. Then the result is rather simple:

$$\bar{x}^1 \approx x^1 + a + \varepsilon\Big([(x^1)^2 - (x^2)^2 - (x^3)^2 - (x^4)^2]a + x^1a^2 + \frac{1}{3}a^3\Big),$$

$$\bar{x}^j \approx x^j + \varepsilon(2ax^1 + a^2)x^j, \quad j = 2, 3, 4. \tag{7.111}$$

The theory of approximate groups provides a regular method of calculation, e.g. the perturbation (7.111) directly, without using the complicated group transformations (7.110). The calculation is based on *approximate Lie equations*.

The present chapter focuses on one-parameter approximate transformation groups treated in the first order of precision. For the general theory, see [7.13](ii).

7.5.2 Notation

In what follows, the functions $f(x,\varepsilon)$ of n variables $x = (x^1, \ldots, x^n)$ and a parameter ε are considered locally in a neighborhood of $\varepsilon = 0$. In the subsequent discussion, all functions are supposed to be analytic in the variables x and ε near $\varepsilon = 0$. Recall the definition and notation of infinitely small functions.

Definition 1. A function $f(x,\varepsilon)$ is said to be of order less than ε^p (with an integer $p \geq 1$) and written $f(x,\varepsilon) = o(\varepsilon^p)$ if the following holds:

$$\lim_{\varepsilon \to 0} \frac{f(x,\varepsilon)}{\varepsilon^p} = 0. \tag{7.112}$$

Equation (7.112) is equivalent to any one of the following conditions:

$$f(x,\varepsilon) = \varepsilon^{p+1} \varphi(x,\varepsilon), \quad \varphi(x,\varepsilon) \text{ is analytic near } \varepsilon = 0; \tag{7.113}$$

$$|f(x,\varepsilon)| \leq C |\varepsilon|^{p+1}, \quad C \text{ is a positive constant.} \tag{7.114}$$

Remark 1. For vector-functions $f(x,\varepsilon) = (f^1(x,\varepsilon), \ldots, f^m(x,\varepsilon))$, the relation $f(x,\varepsilon) = o(\varepsilon^p)$ is defined by imposing the condition (7.112) or (7.113) on all $f^i(x,\varepsilon)$, $i = 1, \ldots, m$, whereas the condition (7.114) can be conveniently written, by using the norm of vectors defined in Section 3.1.3, in the following form:

$$\|f(x,\varepsilon)\| \leq C |\varepsilon|^{p+1}. \tag{7.115}$$

Definition 2. Two functions, $f(x,\varepsilon)$ and $g(x,\varepsilon)$, are said to be *approximately equal* (with an error $o(\varepsilon^p)$) if

$$g(x,\varepsilon) - f(x,\varepsilon) = o(\varepsilon^p).$$

To designate the approximate equality, we use either the notation

$$g(x,\varepsilon) = f(x,\varepsilon) + o(\varepsilon^p),$$

or $g \approx f$ when there is no ambiguity.

The approximate equality defines an equivalence relation, and we join functions into equivalence classes by letting the functions $f(x,\varepsilon)$ and $g(x,\varepsilon)$ be members of the same class if $f \approx g$.

Given a function $f(x,\varepsilon)$, let $f_0(x) + \varepsilon f_1(x) + \cdots + \varepsilon^p f_p(x)$ be the approximating polynomial of degree p in ε obtained via the Taylor expansion of $f(x,\varepsilon)$ in powers of ε about $\varepsilon = 0$. Then any function $g \approx f$ (in particular, f itself) has the form

$$g(x,\varepsilon) = f_0(x) + \varepsilon f_1(x) + \cdots + \varepsilon^p f_p(x) + o(\varepsilon^p).$$

Hence, the following definition.

Definition 3. The equivalence class of functions containing f is the set of all functions $g(x,\varepsilon)$ such that

$$g(x,\varepsilon) \approx f_0(x) + \varepsilon f_1(x) + \cdots + \varepsilon^p f_p(x),$$

7.6. APPROXIMATE TRANSFORMATION GROUPS

the function $f_0(x) + \varepsilon f_1(x) + \cdots + \varepsilon^p f_p(x)$ being its *canonical representative*.

Remark 2. One can identify the equivalence class of functions $g(x,\varepsilon) \approx f(x,\varepsilon)$ as the ordered set of $p+1$ functions, $f_0(x), f_1(x), \ldots, f_p(x)$.

In the theory of approximate transformation groups, we consider ordered sets

$$f_0(x,a), \quad f_1(x,a), \quad \ldots, \quad f_p(x,a)$$

of vector-functions depending on x and a parameter a, with coordinates

$$f_0^i(x,a), \; f_1^i(x,a), \ldots f_p^i(x,a), \quad i=1,\ldots,n.$$

Definition 4. *An approximate transformation* $\bar{x} \approx f(x,a)$ *in* $\mathbb{R}^n$, *or*

$$\bar{x}^i \approx f_0^i(x,a) + \varepsilon f_1^i(x,a) + \cdots + \varepsilon^p f_p^i(x,a), \quad i=1,\ldots,n, \tag{7.116}$$

obeying the initial conditions

$$\bar{x}^i\big|_{a=0} \approx x^i, \quad i=1,\ldots,n, \tag{7.117}$$

is the set of all invertible transformations

$$\bar{x}^i = g^i(x,a,\varepsilon) \tag{7.118}$$

such that $g^i(x,a,\varepsilon) \approx f_0^i(x,a) + \varepsilon f_1^i(x,a) + \cdots + \varepsilon^p f_p^i(x,a)$. Furthermore, it is assumed that the representations (7.118) of (7.116) are defined in a neighborhood of $a=0$ and that, in this neighborhood, $g^i(x,a,\varepsilon) \approx x^i$ if and only if $a=0$.

7.5.3 One-parameter approximate transformation groups

Definition. We say that (7.116)–(7.117) define a one-parameter *approximate transformation group* if any representation (7.118) of (7.116) satisfies the group property (with an error $o(\varepsilon^p)$):

$$g(g(x,a,\varepsilon),b,\varepsilon) \approx g(x,a+b,\varepsilon). \tag{7.119}$$

Remark. In (7.119), unlike the usual group property (7.11), g does not necessarily denote the same function at each occurrence. It can be replaced by any function $f \approx g$. For example, let us take $n=1, p=1$ and consider two functions:

$$f(x,a,\varepsilon) = x + a(1+\varepsilon x + \varepsilon a/2) \quad \text{and} \quad g(x,a,\varepsilon) = x + a(1+\varepsilon x)(1+\varepsilon a/2).$$

They are equal in the first order of precision,

$$g(x,a,\varepsilon) = f(x,a,\varepsilon) + \varepsilon^2 \varphi(x,a), \quad \varphi(x,a) = a^2 x/2,$$

and satisfy the approximate group property (7.119), e. g.,

$$f(g(x,a,\varepsilon),b,\varepsilon) = f(x, a+b, \varepsilon) + \varepsilon^2 \psi(x,a,b,\varepsilon),$$

where $\psi(x,a,b,\varepsilon) = a(ax+ab+2bx+\varepsilon abx)/2$.

7.5.4 Approximate group generator

Definition. The generator of an approximate transformation group (7.116) is the set of all first-order linear differential operators

$$X = \xi^i(x,\varepsilon)\frac{\partial}{\partial x^i}$$

such that $\xi^i(x,\varepsilon) \approx \xi_0^i(x) + \varepsilon\xi_1^i(x) + \cdots + \varepsilon^p\xi_p^i(x)$, where

$$\xi_\nu^i(x) = \frac{\partial}{\partial a}\left[f_\nu^i(x,a)\right]_{a=0}, \quad i = 1,\ldots,n;\ \nu = 0,\ldots,p. \quad (7.120)$$

Thus, using the notation of Section 7.5.2, an approximate group generator is written

$$X \approx \left(\xi_0^i(x) + \varepsilon\xi_1^i(x) + \cdots + \varepsilon^p\xi_p^i(x)\right)\frac{\partial}{\partial x^i}.$$

In calculation, it is convenient to identify X with its *canonical representative*:

$$X = \left(\xi_0^i(x) + \varepsilon\xi_1^i(x) + \cdots + \varepsilon^p\xi_p^i(x)\right)\frac{\partial}{\partial x^i}. \quad (7.121)$$

7.5.5 A preparatory lemma

To carry over Lie's theorem (Theorem 7.1.5) to approximate transformation groups, we will need the following variant of the theorem on continuous dependence of a solution of Cauchy's problem on parameters.

Lemma. Let $f(x,\varepsilon)$ and $g(x,\varepsilon)$ be approximately equal vector-functions:

$$g(x,\varepsilon) = f(x,\varepsilon) + o(\varepsilon^p). \quad (7.122)$$

Let $x = x(t,\varepsilon)$ and $y = y(t,\varepsilon)$ be the solutions to the Cauchy problems (3.12):

$$\frac{dx}{dt} = f(x,\varepsilon), \quad x|_{t=0} = x_0(\varepsilon)$$

and

$$\frac{dy}{dt} = g(y,\varepsilon), \quad y|_{t=0} = y_0(\varepsilon),$$

where $y_0(\varepsilon) = x_0(\varepsilon) + o(\varepsilon^p)$. Then the solutions are approximately equal:

$$y(t,\varepsilon) = x(t,\varepsilon) + o(\varepsilon^p). \quad (7.123)$$

Proof. It is readily seen that $u(t,\varepsilon) = y(t,\varepsilon) - x(t,\varepsilon)$ satisfies the conditions:

$$\|u(0,\varepsilon)\| = \|y_0(\varepsilon) - x_0(\varepsilon)\| \leq M|\varepsilon|^{p+1}, \quad (7.124)$$

$$\left|\frac{du^i}{dt}\right| \leq |f^i(x,\varepsilon) - g^i(y,\varepsilon)| \leq \|f(x,\varepsilon) - g(x,\varepsilon)\| + \|g(x,\varepsilon) - g(y,\varepsilon)\|. \quad (7.125)$$

7.6. APPROXIMATE TRANSFORMATION GROUPS

By virtue of (7.122) rewritten in the form (7.115), $\|g(x,\varepsilon) - f(x,\varepsilon)\| \leq C|\varepsilon|^{p+1}$, and the Lipschitz condition (3.13), $\|g(x,\varepsilon) - g(y,\varepsilon)\| \leq K\|x - y\|$, (7.125) yields:

$$\left|\frac{du^i}{dt}\right| \leq K\|u\| + C|\varepsilon|^{p+1}, \quad i = 1, \ldots, n. \tag{7.126}$$

It follows from the continuity of $u(t,\varepsilon)$ that, given ε in a vicinity of $\varepsilon = 0$, there exists t_ε such that every function $u^i(t,\varepsilon)$ has a constant sign when t is within the interval from 0 to t_ε. Let us assume that $t_\varepsilon > 0$ and consider the inequality (7.126) in the interval $0 \leq t \leq t_\varepsilon$. Then it can be written:

$$\frac{d}{dt}|u^i| \leq K\|u\| + C|\varepsilon|^{p+1}, \quad i = 1, \ldots, n.$$

Hence, by definition of the norm (see Section 3.1.3),

$$\frac{d}{dt}\|u\| \leq n\left(K\|u\| + C|\varepsilon|^{p+1}\right) \leq B\left(\|u\| + |\varepsilon|^{p+1}\right),$$

where $B = \max\{nK, nC\}$ is a positive constant. Now one can divide the above inequality by $|u| + |\varepsilon|^{p+1}$ and integrate from 0 to t within the preceding interval, to get $\ln(\|u(t,\varepsilon)\| + |\varepsilon|^{p+1}) - \ln(\|u(0,\varepsilon)\| + |\varepsilon|^{p+1}) \leq Bt$, or

$$\|u(t,\varepsilon)\| \leq \left(e^{Bt} - 1\right)|\varepsilon|^{p+1} + u(0,\varepsilon)e^{Bt}. \tag{7.127}$$

Replacing in (7.127) $\exp(Bt)$ by $\exp(Bt_\varepsilon) \geq \exp(Bt)$ and invoking (7.124), one arrives at the inequality $\|u(t,\varepsilon)\| \leq N|\varepsilon|^{p+1}$ (with $N = (M+1)\exp(Bt_\varepsilon) - 1$) equivalent to the approximate equation (7.123). The reader can carry out similar calculations in the case $t_\varepsilon < 0$ (Problem 7.26*), thus completing the proof.

7.5.6 Lie equations in the first order of precision

In the general theory, approximate groups are considered with an accuracy of an arbitrary order $p \geq 1$. However, in practical applications we use a simplified approach by restricting the theory to the first order of precision, i.e. by assuming $p = 1$. This assumption is adopted in what follows.

Let

$$X = X_0 + \varepsilon X_1 \tag{7.128}$$

be a given approximate operator, where

$$X_0 = \xi_0^i(x)\frac{\partial}{\partial x^i}, \quad X_1 = \xi_1^i(x)\frac{\partial}{\partial x^i}.$$

The corresponding approximate group of transformations of points x into points $\bar{x} = \bar{x}_0 + \varepsilon\bar{x}_1$ with coordinates

$$\bar{x}^i = \bar{x}_0^i + \varepsilon\bar{x}_1^i, \quad i = 1, \ldots, n, \tag{7.129}$$

176   7. INFINITESIMAL TRANSFORMATIONS AND LOCAL GROUPS

is determined by the following equations:

$$\frac{d\bar{x}_0^i}{da} = \xi_0^i(\bar{x}_0), \quad \bar{x}_0^i\big|_{a=0} = x^i, \tag{7.130}$$

$$\frac{d\bar{x}_1^i}{da} = \sum_{k=1}^{n} \bar{x}_1^k \cdot \left[\frac{\partial \xi_0^i(x)}{\partial x^k}\right]_{x=\bar{x}_0} + \xi_1^i(\bar{x}_0), \quad \bar{x}_1^i\big|_{a=0} = 0. \tag{7.131}$$

Equations (7.130)–(7.131) are called the *approximate Lie equations*.

An approach to the solution of approximate Lie equations is illustrated by the following simple examples.

Example 1. Let $n = 1$ and let

$$X = (1 + \varepsilon x)\frac{d}{dx}.$$

Here $\xi_0(x) = 1$ and $\xi_1(x) = x$, and the approximate Lie equations (7.130)–(7.131) are written:

$$\frac{d\bar{x}_0}{da} = 1, \quad \bar{x}_0\big|_{a=0} = x,$$

$$\frac{d\bar{x}_1}{da} = \bar{x}_0, \quad \bar{x}_1\big|_{a=0} = 0.$$

Integration yields:

$$\bar{x}_0 = x + a, \quad \bar{x}_1 = ax + \frac{a^2}{2}.$$

Hence, the approximate group is given by

$$\bar{x} \approx x + a + \varepsilon\left(ax + \frac{a^2}{2}\right).$$

Example 2. Let $n = 2$ and let

$$X = (1 + \varepsilon x^2)\frac{\partial}{\partial x} + \varepsilon xy\frac{\partial}{\partial y}.$$

Here $\xi_0(x, y) = (1, 0)$ and $\xi_1(x, y) = (x^2, xy)$, and the approximate Lie equations (7.130)–(7.131) are written:

$$\frac{d\bar{x}_0}{da} = 1, \quad \frac{d\bar{y}_0}{da} = 0, \quad \bar{x}_0\big|_{a=0} = x, \quad \bar{y}_0\big|_{a=0} = y,$$

$$\frac{d\bar{x}_1}{da} = (\bar{x}_0)^2, \quad \frac{d\bar{y}_1}{da} = \bar{x}_0\bar{y}_0, \quad \bar{x}_1\big|_{a=0} = 0, \quad \bar{y}_1\big|_{a=0} = 0.$$

Integration yields:

$$\bar{x} \approx x + a + \varepsilon\left(ax^2 + a^2x + \frac{a^3}{3}\right), \quad \bar{y} \approx y + \varepsilon\left(axy + \frac{a^2}{2}y\right).$$

7.5.7 The approximate exponential map

Let X and Y be linear differential operators of the first order. Consider the exponential map

$$e^X = \sum_{k=0}^{\infty} \frac{1}{k!} X^k \equiv 1 + X + \frac{1}{2!}X^2 + \frac{1}{3!}X^3 + \cdots.$$

Recall that the *differential of the exponential map* is a linear mapping given by the following infinite sum (see, e.g. [7.14], Chap. 3, § 4.3):

$$\sum_{k=0}^{\infty} \frac{1}{(k+1)!} (\text{ad } X)^k,$$

where ad X is the *inner derivation* by X (known also as the *map adjoint to X*) and it is defined by the linear mapping

$$\text{ad } X(Y) = [X, Y]$$

with the usual commutator

$$[X, Y] = XY - YX.$$

Let us denote by $\langle\!\langle X, Y \rangle\!\rangle$ the differential operator of the infinite order given by the formal infinite sum:

$$\langle\!\langle X, Y \rangle\!\rangle = \sum_{k=0}^{\infty} \frac{1}{(k+1)!} (\text{ad } X)^k Y.$$

By substituting

$$\text{ad } X(Y) = [X, Y], \quad (\text{ad } X)^2(Y) = [X, [X, Y]], \ldots$$

one obtains:

$$\langle\!\langle X, Y \rangle\!\rangle = Y + \frac{1}{2!}[X, Y] + \frac{1}{3!}[X, [X, Y]] + \frac{1}{4!}[X, [X, [X, Y]]] + \cdots. \quad (7.132)$$

Theorem [7.15]. Given an approximate generator (7.128),

$$X = X_0 + \varepsilon X_1,$$

the approximate transformation group (7.129),

$$\overline{x}^i = \overline{x}_0^i + \varepsilon \overline{x}_1^i, \quad i = 1, \ldots, n,$$

is determined by the following:

$$\overline{x}_0^i = e^{aX_0}(x^i), \quad \overline{x}_1^i = \langle\!\langle aX_0, aX_1 \rangle\!\rangle(\overline{x}_0^i), \quad i = 1, \ldots, n, \quad (7.133)$$

where e^{aX_0} is the usual exponential map (7.19) and

$$\langle\!\langle aX_0, aX_1\rangle\!\rangle = aX_1 + \frac{a^2}{2!}[X_0, X_1] + \frac{a^3}{3!}[X_0, [X_0, X_1]] + \cdots. \tag{7.134}$$

In other words, the approximate operator $X = X_0 + \varepsilon X_1$ generates the following one-parameter approximate group of transformations in $\mathbb{R}^n$:

$$\bar{x}^i = \Big(1 + \varepsilon\langle\!\langle aX_0, aX_1\rangle\!\rangle\Big)e^{aX_0}(x^i), \quad i = 1, \ldots, n. \tag{7.135}$$

Proof. Substitute $X = X_0 + \varepsilon X_1$ into the exponent (7.19),

$$e^{a(X_0 + \varepsilon X_1)} = 1 + a(X_0 + \varepsilon X_1) + \frac{a^2}{2!}(X_0 + \varepsilon X_1)^2 + \frac{a^3}{3!}(X_0 + \varepsilon X_1)^3 + \cdots,$$

and single out the sum of terms of the first degree in ε, to obtain:

$$e^{a(X_0+\varepsilon X_1)} \approx 1 + aX_0 + \frac{a^2}{2!}X_0^2 + \frac{a^3}{3!}X_0^3 + \cdots$$

$$+ \varepsilon\bigg\{aX_1 + \frac{a^2}{2!}(X_0X_1 + X_1X_0) + \frac{a^3}{3!}(X_0^2X_1 + X_0X_1X_0 + X_1X_0^2)$$

$$+ \frac{a^4}{4!}(X_0^3X_1 + X_0^2X_1X_0 + X_0X_1X_0^2 + X_1X_0^3) + \cdots\bigg\}. \tag{7.136}$$

By using the identities

$$X_0X_1 = X_1X_0 + [X_0, X_1],$$

$$X_0^2X_1 + X_0X_1X_0 = 2X_1X_0^2 + 3[X_0, X_1]X_0 + [X_0, [X_0, X_1]], \ldots$$

one can rewrite (7.136) in the form:

$$e^{a(X_0+\varepsilon X_1)} \approx 1 + aX_0 + \frac{a^2}{2!}X_0^2 + \frac{a^3}{3!}X_0^3 + \cdots$$

$$+ \varepsilon\bigg\{aX_1\Big(1 + aX_0 + \frac{a^2}{2!}X_0^2 + \frac{a^3}{3!}X_0^3 + \cdots\Big)$$

$$+ \frac{a^2}{2!}[X_0, X_1]\Big(1 + aX_0 + \frac{a^2}{2!}X_0^2 + \frac{a^3}{3!}X_0^3 + \cdots\Big)$$

$$+ \frac{a^3}{3!}[X_0, [X_0, X_1]]\Big(1 + aX_0 + \frac{a^2}{2!}X_0^2 + \frac{a^3}{3!}X_0^3 + \cdots\Big) + \cdots\bigg\}.$$

Thus, invoking (7.132):

$$e^{a(X_0+\varepsilon X_1)} \approx \Big(1 + \varepsilon\langle\!\langle aX_0, aX_1\rangle\!\rangle\Big)e^{aX_0}. \tag{7.137}$$

Hence, the exponential map (7.19) written for the operator (7.128) in the first order of precision with respect to ε has the form (7.135). Taking into account (7.129), one obtains the formulae (7.133), thus proving the theorem.

7.6. APPROXIMATE TRANSFORMATION GROUPS

7.5.8 Examples on the exponential map

Example 1. Let us use Theorem 7.5.7 in Example 1 of Section 7.5.6. Here,

$$X_0 = \frac{d}{dx}, \quad X_1 = x\frac{d}{dx}.$$

Therefore,

$$X_0(x) = 1, \quad X_0^2(x) = X_0^3(x) = \cdots = 0,$$

and

$$[X_0, X_1] = \frac{d}{dx} = X_0, \quad [X_0, [X_0, X_1]] = [X_0, X_0] = 0, \ldots.$$

Consequently,

$$\bar{x}_0 = e^{aX_0}(x) = x + a,$$

and

$$\langle\!\langle aX_0, aX_1 \rangle\!\rangle = \left(ax + \frac{a^2}{2!}\right)\frac{d}{dx},$$

whence

$$\bar{x}_1 = \langle\!\langle aX_0, aX_1 \rangle\!\rangle(\bar{x}_0) = \left(ax + \frac{a^2}{2!}\right)\frac{d}{dx}(x + a) = ax + \frac{a^2}{2!}.$$

Hence,

$$\bar{x} \approx x + a + \varepsilon\left(ax + \frac{a^2}{2}\right).$$

Example 2. Let us use Theorem 7.5.7 in Example 2 of Section 7.5.6. Here,

$$X_0 = \frac{\partial}{\partial x}, \quad X_1 = x^2\frac{\partial}{\partial x} + xy\frac{\partial}{\partial y}.$$

Therefore,

$$\bar{x}_0 = e^{aX_0}(x) = x + a,$$

and

$$[X_0, X_1] = 2x\frac{\partial}{\partial x} + y\frac{\partial}{\partial y}, \quad [X_0, [X_0, X_1]] = 2\frac{\partial}{\partial x}, \quad [X_0, [X_0, [X_0, X_1]]] = 0, \ldots$$

Consequently,

$$\langle\!\langle aX_0, aX_1 \rangle\!\rangle = aX_1 + \frac{a^2}{2}\left(2x\frac{\partial}{\partial x} + y\frac{\partial}{\partial y}\right) + \frac{a^3}{3}\frac{\partial}{\partial x}$$

$$= \left(ax^2 + a^2 x + \frac{a^3}{3}\right)\frac{\partial}{\partial x} + \left(axy + \frac{a^2}{2}y\right)\frac{\partial}{\partial y}.$$

It follows that

$$\bar{x}_1 = \langle\!\langle aX_0, aX_1 \rangle\!\rangle(\bar{x}_0) = \left(ax^2 + a^2 x + \frac{a^3}{3}\right)\frac{\partial}{\partial x}(x + a),$$

180   7. INFINITESIMAL TRANSFORMATIONS AND LOCAL GROUPS

$$\bar{y}_1 = \langle\!\langle aX_0, aX_1\rangle\!\rangle(\bar{y}_0) = \left(axy + \frac{a^2}{2}y\right)\frac{\partial}{\partial y}(y).$$

Thus,

$$\bar{x}_1 = ax^2 + a^2x + \frac{a^3}{3}, \quad \bar{y}_1 = axy + \frac{a^2}{2}y,$$

hence the result given in Section 7.5.6:

$$\bar{x} \approx x + a + \varepsilon\left(ax^2 + a^2x + \frac{a^3}{3}\right), \quad \bar{y} \approx y + \varepsilon\left(axy + \frac{a^2}{2}y\right).$$

Exercise. Find the approximate transformation group generated by operator (7.109), $X = X_0 + \varepsilon X_1$, where

$$X_0 = \frac{\partial}{\partial x^1}, \quad X_1 = \left((x^1)^2 - (x^2)^2 - (x^3)^2 - (x^4)^2\right)\frac{\partial}{\partial x^1} + 2x^1\left(x^2\frac{\partial}{\partial x^2} + x^3\frac{\partial}{\partial x^3} + x^4\frac{\partial}{\partial x^4}\right).$$

Solution. Let us use the exponential map. The operator X_0 generates the translation group along the x^1-axis:

$$\bar{x}_0^1 = x^1 + a, \quad \bar{x}_0^j = x^j, \quad j = 2, 3, 4.$$

Now we calculate the differential of the exponential map by formula (7.132) applied to the operators X_0 and X_1. We have:

$$[X_0, X_1] = 2x^1\left(x^1\frac{\partial}{\partial x^1} + x^2\frac{\partial}{\partial x^2} + x^3\frac{\partial}{\partial x^3} + x^4\frac{\partial}{\partial x^4}\right),$$

$$[X_0, [X_0, X_1]] = 2\frac{\partial}{\partial x^1}, \quad [X_0, [X_0, [X_0, X_1]]] = 0, \ldots.$$

Consequently, formula (7.134) takes the form:

$$\langle\!\langle aX_0, aX_1\rangle\!\rangle = \left([(x^1)^2 - (x^2)^2 - (x^3)^2 - (x^4)^2]a + x^1a^2 + \frac{1}{3}a^3\right)\frac{\partial}{\partial x^1}$$

$$+(2ax^1 + a^2)\left(x^2\frac{\partial}{\partial x^2} + x^3\frac{\partial}{\partial x^3} + x^4\frac{\partial}{\partial x^4}\right).$$

Therefore (7.133) yields

$$\bar{x}_1^1 = \langle\!\langle aX_0, aX_1\rangle\!\rangle(\bar{x}_0^1) = [(x^1)^2 - (x^2)^2 - (x^3)^2 - (x^4)^2]a + x^1a^2 + \frac{1}{3}a^3,$$

$$\bar{x}_1^j = (2ax^1 + a^2)x^j, \quad j = 2, 3, 4.$$

We thus arrive at the approximate transformation given in Section 7.5.1:

$$\bar{x}^1 \approx \bar{x}_0^1 + \varepsilon\bar{x}_1^1 = x^1 + a + \varepsilon\left([(x^1)^2 - (x^2)^2 - (x^3)^2 - (x^4)^2]a + x^1a^2 + \frac{1}{3}a^3\right),$$

$$\bar{x}^j \approx \bar{x}_0^j + \varepsilon\bar{x}_1^j = x^j + \varepsilon(2ax^1 + a^2)x^j, \quad j = 2, 3, 4.$$

Problems

7.1. In the dilation group $\bar{x} = ax$, the identical transformation corresponds to $a_0 = 1$. Introduce a new parameter such that $a_0 = 0$ as mentioned in Section 7.1.1.

7.2. Deduce the composition law (7.7) and the parameter a^{-1} of the inverse transformation (7.6) for each of following three different representations of the dilation group (cf. Problem 7.1): (i) $\bar{x} = ax$, (ii) $\bar{x} = x + ax$, (iii) $\bar{x} = e^a x$.

7.3. Prove the properties (7.8) of the composition law.

7.4. Find the generator (7.15):
   (i) for the group of conformal transformation (6.36) on the plane;
   (ii) for the group of conformal transformation (6.46) in $\mathbb{R}^3$.

7.5. Show that $\bar{x} = x\sqrt{1-a^2} + ya$, $\bar{y} = y\sqrt{1-a^2} - xa$ ($|a| < 1$) is a one-parameter group. Find the composition law (7.7) and the canonical parameter (7.21).

7.6*. Deduce the formula (7.25), $F(\bar{x}) = e^{aX} F(x)$. Note that it can be written, invoking (7.19), in the form $F(e^{aX}(x)) = e^{aX} F(x)$ or, since x is arbitrary, $Fe^{aX} = e^{aX} F$.

7.7. Check that the change of variables $u = yx^{-2}$, $v = zx^2$, $w = \ln|x|$ reduces the group G of dilations $\bar{x} = xe^a$, $\bar{y} = ye^{2a}$, $\bar{z} = ze^{-2a}$ to the group of translations of w, and hence, u, v, w are canonical variables for the group G (cf. Example 1 in Section 7.1.8).

7.8. Find one-parameter groups of transformations in the (x, y) plane, by using all three methods of Section 7.1.9, for the following generators (k and l are any constants):

(i) $X = l\dfrac{\partial}{\partial x} - k\dfrac{\partial}{\partial y}$, (ii) $X = x\dfrac{\partial}{\partial x} + ky\dfrac{\partial}{\partial y}$, (iii) $X = y\dfrac{\partial}{\partial x}$, (iv) $X = y\dfrac{\partial}{\partial x} - x\dfrac{\partial}{\partial y}$,

(v) $X = y\dfrac{\partial}{\partial x} + x\dfrac{\partial}{\partial y}$, (vi) $X = x^2\dfrac{\partial}{\partial x} + (3xy + x^2)\dfrac{\partial}{\partial y}$, (vii) $X = x\ln x\dfrac{\partial}{\partial x} - 2(1 + \ln x)\dfrac{\partial}{\partial y}$.

7.9. In the (t, x, u)-space, find the one-parameter group with the generator (use all three methods of Section 7.1.9) $X = 2t\,\partial/\partial x - xu\,\partial/\partial u$. Is this group local or global?

7.10*. Prove the statement of Remark 7.2.1. In other words, derive equations (7.56), $XF_\sigma(x) = \lambda_\sigma^\nu(x)F_\nu(x)$, from (7.50).

7.11. Check the infinitesimal test (7.50) for the invariance of the spiral (7.65) under the operator (7.66). Give an invariant representation of the spiral.

7.12. Apply the exponential map and the method of canonical variables to the operator (7.66) to calculate the group transformations (7.69).

7.13. Deduce the operator (7.67) from (7.66).

7.14*. Check that Newton's equations (1.34), $w^i + \alpha x^i/r^3 = 0$ ($i = 1, 2, 3$), where $w^i = d^2 x^i/dt^2$, define a regular invariant manifold for the group G_3 with generators

$$X_1 = x^2 \frac{\partial}{\partial x^1} - x^1 \frac{\partial}{\partial x^2} + w^2 \frac{\partial}{\partial w^1} - w^1 \frac{\partial}{\partial w^2},$$

$$X_2 = x^3 \frac{\partial}{\partial x^2} - x^2 \frac{\partial}{\partial x^3} + w^3 \frac{\partial}{\partial w^2} - w^2 \frac{\partial}{\partial w^3},$$

$$X_3 = x^1 \frac{\partial}{\partial x^3} - x^3 \frac{\partial}{\partial x^1} + w^1 \frac{\partial}{\partial w^3} - w^3 \frac{\partial}{\partial w^1}.$$

Find an invariant representation of Newton's equations.

7.15. Prove that the derived algebra $L^{(1)}$ is a Lie algebra and that it is an ideal of L.

7.16. Prove that any two-dimensional Lie algebra is solvable.

7.17. Show that the following operators in $\mathbb{R}^1, \mathbb{R}^2$ and $\mathbb{R}^3$ span non-solvable algebras:

(i) $\quad X_1 = \dfrac{\partial}{\partial x}, \quad X_2 = x\dfrac{\partial}{\partial x}, \quad X_3 = x^2\dfrac{\partial}{\partial x};$

(ii) $\quad X_1 = (1+x^2)\dfrac{\partial}{\partial x} + xy\dfrac{\partial}{\partial y}, \quad X_2 = xy\dfrac{\partial}{\partial x} + (1+y^2)\dfrac{\partial}{\partial y}, \quad X_3 = y\dfrac{\partial}{\partial x} - x\dfrac{\partial}{\partial y};$

(iii) $\quad X_1 = y\dfrac{\partial}{\partial x} - x\dfrac{\partial}{\partial y}, \quad X_2 = z\dfrac{\partial}{\partial y} - y\dfrac{\partial}{\partial z}, \quad X_3 = x\dfrac{\partial}{\partial z} - z\dfrac{\partial}{\partial x}.$

7.18. Let X_1 and X_2 span a non-Abelian algebra L_2, i.e. $[X_1, X_2] = \alpha_1 X_1 + \alpha_2 X_2 \neq 0$. Show that there exists a basis X_1', X_2' of L_2 such that $[X_1', X_2'] = X_1'$.

7.19*. Read Lie's proof of Theorem 1 of Section 7.3.8 given in [5.14], Chap. 21, §§2, 3.

7.20*. Show that the Lie algebras $L_2, \overline{L}_3$, and $\widetilde{L}_3$ spanned, respectively, by the operators:

(i) $\quad X_1 = \dfrac{\partial}{\partial x} + \dfrac{\partial}{\partial y}, \quad X_2 = x\dfrac{\partial}{\partial x} + y\dfrac{\partial}{\partial y}, \quad X_3 = x^2\dfrac{\partial}{\partial x} + y^2\dfrac{\partial}{\partial y},$

(ii) $\quad \overline{X}_1 = \dfrac{\partial}{\partial \overline{x}}, \quad \overline{X}_2 = \overline{x}\dfrac{\partial}{\partial \overline{x}} + \overline{y}\dfrac{\partial}{\partial \overline{y}}, \quad \overline{X}_3 = \overline{x}^2\dfrac{\partial}{\partial \overline{x}} + (2\overline{x}\,\overline{y} + \overline{y}^2)\dfrac{\partial}{\partial \overline{y}},$

(iii) $\quad \widetilde{X}_1 = \dfrac{\partial}{\partial \widetilde{x}} + \widetilde{x}\dfrac{\partial}{\partial \widetilde{y}}, \quad \widetilde{X}_2 = \widetilde{x}\dfrac{\partial}{\partial \widetilde{x}} + 2\widetilde{y}\dfrac{\partial}{\partial \widetilde{y}}, \quad \widetilde{X}_3 = (\widetilde{x}^2 - \widetilde{y})\dfrac{\partial}{\partial \widetilde{x}} + \widetilde{x}\,\widetilde{y}\dfrac{\partial}{\partial \widetilde{y}},$

are similar. Find the changes of variables transforming (ii) and (iii) to (i), i.e. to the canonical form 1) of Theorem 2 of Section 7.3.8.

7.21*. Show that the Lie algebras $L_3, \overline{L}_3$, and $\widetilde{L}_3$ spanned, respectively, by the operators:

(i) $\quad X_1 = \dfrac{\partial}{\partial x}, \quad X_2 = 2x\dfrac{\partial}{\partial x} + y\dfrac{\partial}{\partial y}, \quad X_3 = x^2\dfrac{\partial}{\partial x} + xy\dfrac{\partial}{\partial y},$

(ii) $\quad \overline{X}_1 = \dfrac{\partial}{\partial \overline{x}}, \quad \overline{X}_2 = \overline{x}\dfrac{\partial}{\partial \overline{x}} + \overline{y}\dfrac{\partial}{\partial \overline{y}}, \quad \overline{X}_3 = \overline{x}^2\dfrac{\partial}{\partial \overline{x}} + 2\overline{x}\,\overline{y}\dfrac{\partial}{\partial \overline{y}},$

(iii) $\quad \widetilde{X}_1 = \widetilde{x}\dfrac{\partial}{\partial \widetilde{y}}, \quad \widetilde{X}_2 = \widetilde{x}\dfrac{\partial}{\partial \widetilde{x}} - \widetilde{y}\dfrac{\partial}{\partial \widetilde{y}}, \quad \widetilde{X}_3 = \widetilde{y}\dfrac{\partial}{\partial \widetilde{x}},$

are similar. Find the changes of variables transforming (ii) and (iii) to (i), i.e. to the canonical form 2) of Theorem 2 of Section 7.3.8.

7.22. Check that the vector space spanned by the operators $X_2 = \partial/\partial y$, $X_3 = y\partial/\partial x$ (see Example 7.4.2) is not a Lie algebra, and show that the composition of two one-parameter groups generated by these operators is not a two-parameter group.

7.23. Find by Theorem 7.4.2 the six-parameter group generated by the operators (7.77).

7.24. Find the invariants of rotations in $\mathbb{R}^3$ (Example 7.4.3) by solving equations (7.102).

7.25. Construct by Theorem 7.4.2 the groups G_3 for the algebras given in Problem 7.17.

7.26*. Complete the proof of Lemma 7.5.5 by considering the case $t_\varepsilon < 0$.

7.27. Find one-parameter approximate transformation groups (in the first order of precision) for the following operators (see [5.39](i)):

$$X_1 = \dfrac{2\varepsilon}{3}x^3\dfrac{\partial}{\partial x} + \left[1 + \varepsilon\left(yx^2 + \dfrac{11}{60}x^5\right)\right]\dfrac{\partial}{\partial y}, \quad X_2 = \dfrac{\varepsilon}{6}x^4\dfrac{\partial}{\partial x} + \left[x + \varepsilon\left(\dfrac{yx^3}{3} + \dfrac{7x^6}{180}\right)\right]\dfrac{\partial}{\partial y}.$$

Chapter 8
Calculus of differential algebra

Differential algebra [8.1] furnishes us with a *convenient language, basic devices* and the *universal space* of modern groups analysis.

In mathematical analysis, it is customary to deal with *functions* $u^\alpha(x)$. Their derivatives $u_i^\alpha(x) \equiv \partial u^\alpha(x)/\partial x^i$, $u_{ij}^\alpha(x) \equiv \partial^2 u^\alpha(x)/\partial x^i \partial x^j, \ldots$ are also regarded as functions of independent variables $x = (x^1, \ldots, x^n)$. In differential algebra, it is expedient to treat $u^\alpha, u_i^\alpha, u_{ij}^\alpha, \ldots$ as variables. Furthermore, composite functions $f(x, u(x), \partial u(x)/\partial x, \ldots)$ of x are treated in differential algebra as *differential functions* $f(x, u, u_{(1)}, \ldots)$ of the variables $x, u, u_{(1)}, \ldots$.

8.1 Main variables and total derivatives

In differential algebra we deal with an infinite number of variables:

$$x = \{x^i\}, \; u = \{u^\alpha\}, \; u_{(1)} = \{u_i^\alpha\}, \; u_{(2)} = \{u_{i_1 i_2}^\alpha\}, \; u_{(3)} = \{u_{i_1 i_2 i_3}^\alpha\}, \ldots, \quad (8.1)$$

where $\alpha = 1, \ldots, m$; $i, i_1, \ldots = 1, \ldots, n$. The variables $u_{i_1 i_2}^\alpha$, $u_{i_1 i_2 i_3}^\alpha, \ldots$ are symmetric in the subscripts $i_1 i_2, \; i_1 i_2 i_3, \ldots$.

The main operation in the calculus of differential algebra is the *total differentiation* given by the following formal infinite sums (cf. (4.39)):

$$D_i = \frac{\partial}{\partial x^i} + u_i^\alpha \frac{\partial}{\partial u^\alpha} + u_{ii_1}^\alpha \frac{\partial}{\partial u_{i_1}^\alpha} + u_{ii_1 i_2}^\alpha \frac{\partial}{\partial u_{i_1 i_2}^\alpha} + \cdots, \quad i = 1, \ldots, n. \quad (8.2)$$

The total derivatives D_i act on functions involving any finite number of variables (8.1), e.g.

$$D_i\left(f(x, u, u_{(1)})\right) = \frac{\partial f}{\partial x^i} + u_i^\alpha \frac{\partial f}{\partial u^\alpha} + u_{ii_1}^\alpha \frac{\partial f}{\partial u_{i_1}^\alpha}.$$

In particular, letting $f = u^\alpha$, $f = u_j^\alpha, \ldots$, one obtains:

$$u_i^\alpha = D_i(u^\alpha), \; u_{ij}^\alpha = D_i(u_j^\alpha) = D_i D_j(u^\alpha), \ldots. \quad (8.3)$$

Consequently, x^i are called *independent variables* and u^α *differential variables* with the successive *derivatives* $u_{(1)}, u_{(2)}, \ldots$.

8.2 The universal space $\mathcal{A}$ of modern group analysis

The prolongation theory of Lie point and Lie contact transformation groups requires the introduction of functions depending not only on the independent variables x and differential variables u (often called dependent variables), but also on derivatives of finite orders.

This prolongation is sufficient in the context of classical Lie theory. However, it is insufficient for the natural generalization of the classical theory given by Lie–Bäcklund transformation groups. In this generalization one deals with transformations acting on intrinsically infinite-dimensional spaces. This new approach mandates the space $\mathcal{A}$ of differential functions as the universal space of modern group analysis.

8.2.1 Differential functions and the space $\mathcal{A}$

Let us denote by z the sequence

$$z = (x,\, u,\, u_{(1)},\, u_{(2)}, \ldots) \tag{8.4}$$

with elements z^ν, $\nu \geq 1$, where, e.g.,

$$z^i = x^i\ (1 \leq i \leq n),\quad z^{n+\alpha} = u^\alpha\ (1 \leq \alpha \leq m),$$

with the remaining elements representing the derivatives of u. However, in applications one invariably utilizes only finite subsequences of z which are denoted by $[z]$. In the case of one independent variable x and one differential variable y, we will use the common notation and write $z = (x, y, y', y'', \ldots, y^{(s)}, \ldots)$.

Definition. A *differential function* $f([z])$ is a locally analytic function (i.e. locally expandable in a Taylor series with respect to all arguments) of a finite number of variables (8.4). The highest order of derivatives appearing in f is called the order of the differential function and is denoted by $\mathrm{ord}(f)$, e.g. if $f([z]) = f(x, u, u_{(1)}, \ldots, u_{(s)})$ then $\mathrm{ord}(f) = s$. The set of all differential functions of finite order is denoted by $\mathcal{A}$ [8.2].

Theorem. The set $\mathcal{A}$ is a vector space endowed with the usual multiplication of functions. In other words, if $f([z]) \in \mathcal{A}$ and $g([z]) \in \mathcal{A}$ are differential functions and a and b any constants, then

$$a f + b g \in \mathcal{A},\quad \mathrm{ord}(a f + b g) \leq \max\{\mathrm{ord}(f), \mathrm{ord}(g)\},$$

$$f g \in \mathcal{A},\quad \mathrm{ord}(f g) = \max\{\mathrm{ord}(f), \mathrm{ord}(g)\}.$$

Furthermore, the space $\mathcal{A}$ is closed under the total derivation: if $f \in \mathcal{A}$, then

$$D_i(f) \in \mathcal{A},\quad \mathrm{ord}\left(D_i(f)\right) = \mathrm{ord}(f) + 1,$$

Proof. The first part of the theorem is evident since a linear combination and the product of analytic functions are analytic. To prove the second part, we note

8.2. THE UNIVERSAL SPACE OF MODERN GROUP ANALYSIS

that the total differential operators defined as formal sums (8.2) truncate when they act on differential functions. Specifically, if $\mathrm{ord}(f) = s$, then $D_i(f)$ is a finite sum, the terminal term of which contains derivatives of u of order $s+1$. Since a partial derivative of any analytic function is analytic, each term in the finite sum representing the total derivative $D_i(f)$, and hence the total derivative itself, is analytic. Thus, total differentiation converts any differential function of order s into a differential function of order $s+1$.

8.2.2 Successive derivatives of differential functions

If $f \in \mathcal{A}$, then $D_i(f)$ are themselves differential functions and can, in turn, be differentiated. This can be repeated to obtain derivatives of higher order. Thus, given a differential function $f([z])$ of order s, its successive derivatives

$$D_i(f), \quad D_j D_i(f) = D_j(D_i(f)), \ldots$$

are differential functions of respective orders $s+1, s+2, \ldots$. Furthermore, one can prove using the supposition $u_{ij}^\alpha = u_{ji}^\alpha, \ldots$, that the order of differentiating successively with respect to D_i and D_j is immaterial, i.e.

$$D_j D_i(f) = D_i D_j(f). \tag{8.5}$$

8.2.3 One independent and one differential variable. Faà de Bruno's formula

For one independent variable x and one differential variable y with the successive derivatives $y', y'', \ldots, y^{(s)}, \ldots$, the total differentiation (8.2) has the form (1.14):

$$D_x = \frac{\partial}{\partial x} + y'\frac{\partial}{\partial y} + y''\frac{\partial}{\partial y'} + \cdots + y^{(s+1)}\frac{\partial}{\partial y^{(s)}} + \cdots. \tag{8.6}$$

Example 1. This example illustrates the difference between the total and partial derivations. Let us take $f = x$, $f = y$, and $f = xy'$. Then (8.6) yields $D_x(x) = 1$, $D_x(y) = y'$, $D_x(xy') = y' + xy''$, whereas the partial derivatives are:

$$\frac{\partial x}{\partial x} = 1, \quad \frac{\partial y}{\partial x} = 0, \quad \frac{\partial(xy')}{\partial x} = y'.$$

In analysis one often deals with a function $u = f(y)$ of a function $y = \varphi(x)$. Then u is said to be a *composite function* of x through y and written $u = f(\varphi(x))$. Its first derivative is given by the commonly known *chain rule*:

$$\frac{du}{dx} = \frac{df}{dy}\frac{dy}{dx}.$$

However, in the theory of differential equations, one also encounters the problem of calculation of higher derivatives of composite functions.

8. CALCULUS OF DIFFERENTIAL ALGEBRA

Let us use the language of differential algebra and discuss the calculation of total derivatives of higher order, $D_x^k(f)$, of differential functions of the form $f(y)$.

Example 2. Let $k = 3$. We denote $f' = df/dy, \ldots$ and calculate successively to obtain:

$$D_x(f) = f'y', \quad D_x^2(f) = f''y'^2 + f'y'', \quad D_x^3(f) = f'''y'^3 + 3f''y'y'' + f'y'''.$$

One might reasonably ask whether it is possible to find $D_x^k(f)$ directly, without calculating all preceding derivatives, $D_x(f), \ldots, D_x^{(k-1)}(f)$. The following result, due to Faà de Bruno [8.3], gives affirmative answer to this question.

Theorem. The kth-order derivative of a function $f(y)$ is given, in terms of $y', \ldots, y^{(k)}$ and $f' = df/dy, \ldots, f^{(k)} = d^k f/dy^k$, by the following formula:

$$D_x^k(f) = \sum \frac{k!}{l_1! l_2! \cdots l_k!} f^{(p)} \left(\frac{y'}{1!}\right)^{l_1} \left(\frac{y''}{2!}\right)^{l_2} \cdots \left(\frac{y^{(s)}}{s!}\right)^{l_s} \cdots \left(\frac{y^{(k)}}{k!}\right)^{l_k}, \quad (8.7)$$

where the sum runs through all non-negative integers $l_1, \ldots, l_k$ such that

$$l_1 + 2l_2 + \cdots + kl_k = k, \quad (8.8)$$

and p is the positive integer defined, for every solution set $l_1, \ldots, l_k$ of (8.7), by

$$p = l_1 + l_2 + \cdots + l_k. \quad (8.9)$$

Example 3. If $k = 1$, then (8.8) reduces to $l_1 = 1$, and (8.9) yields $p = 1$. Hence, formula (8.7) reduces to the chain rule, $D_x(f) = f'y'$.

Exercise 1. Find $D_x^5(f)$ by using Faà de Bruno's formula (8.7).

Solution. In this case, equations (8.8) and (8.9) are written:

$$l_1 + 2l_2 + 3l_3 + 4l_4 + 5l_5 = 5, \quad p = l_1 + l_2 + l_3 + l_4 + l_5.$$

It is easy to see from the first equation that all possible candidates for its solutions are provided by the following sets of whole numbers: $l_1 = \{5, 3, 2, 1, 0\}$, $l_2 = \{2, 1, 0\}$, $l_3 = \{1, 0\}$, $l_4 = \{1, 0\}$, $l_5 = \{1, 0\}$. By testing all the possibilities, one obtains the following values (different from 0) of l and p:

1) $l_1 = 5$, $p = 5$; 2) $l_1 = 3$, $l_2 = 1$, $p = 4$; 3) $l_1 = 2$, $l_3 = 1$, $p = 3$;

4) $l_1 = 1$, $l_2 = 2$, $p = 3$; 5) $l_1 = 1$, $l_4 = 1$, $p = 2$;

6) $l_2 = 1$, $l_3 = 1$, $p = 2$; 7) $l_5 = 1$, $p = 1$.

Hence, Faà de Bruno's formula for $D_x^5(f)$ contains seven terms and has the form:

$$D_x^5(f) = f^{(5)} y'^5 + 10 f^{(4)} y'^3 y'' + 10 f''' y'^2 y''' + 15 f''' y' y''^2 + 5 f'' y' y^{(4)} + 10 f'' y'' y''' + f' y^{(5)}.$$

Note that the Leibnitz formula for higher derivatives of the product of functions applies to arbitrary differential functions as well. Namely, if $f, g \in \mathcal{A}$ then

$$D_x^k(fg) = D_x^k(f)g + \sum_{s=1}^{k-1} \frac{k!}{(k-s)!s!} D_x^{k-s}(f) D_x^s(g) + f D_x^k(g). \qquad (8.10)$$

Exercise 2. Let t and z be new independent and differential variables defined by

$$t = \phi(x), \quad y = \sigma(x) z. \qquad (8.11)$$

Let $z' = D_t(z), z'' = D_t^2(z), \ldots$ denote the successive derivatives of z with respect to t. Single out in $y^{(k)} = D_x^k[\sigma(x)z]$ the terms containing $z^{(k)}$, $z^{(k-1)}$ and $z^{(k-2)}$.

Solution. One obtains by (8.10) the leading terms (containing $z^{(k)}$, $z^{(k-1)}$ $z^{(k-2)}$) :

$$y^{(k)} = D_x^k[\sigma(x)z] \approx D_x^k(z)\sigma + k D_x^{k-1}(z)\sigma' + \frac{k(k-1)}{2} D_x^{k-2}(z)\sigma''. \qquad (8.12)$$

Let us calculate $D_x^k(z)$ by Faà de Bruno's formula (8.7), where f and y are replaced by z and t. The terms with $z^{(k)}$ are found by letting $p = k$ in (8.9) and solving the equations $l_1 + 2l_2 + \cdots + k l_k = k$, $l_1 + l_2 + \cdots + l_k = k$. Elimination of l_1 from these equations yields $l_2 + 2l_3 + \cdots + (k-1)l_k = 0$, whence $l_2 = l_3 = \cdots = l_k = 0$, and $l_1 = k$.

Likewise, one obtains the terms with $z^{(k-1)}$ by letting $p = k-1$ and solving the equations $l_1 + 2l_2 + \cdots + k l_k = k$, $l_1 + l_2 + \cdots + l_k = k - 1$. Elimination of l_1 yields $l_2 + 2l_3 + \cdots + (k-1)l_k = 1$, whence $l_2 = 1$, $l_3 = \cdots = l_k = 0$, and hence $l_1 = k-2$.

For the terms with $z^{(k-2)}$, one has to solve the equations $l_1 + 2l_2 + \cdots + k l_k = k$, $l_1 + l_2 + \cdots + l_k = k - 2$. The equation $l_2 + 2l_3 + \cdots + (k-1)l_k = 2$ then follows. It is satisfied by $l_2 = 2$ and then $l_1 = k - 4$, or by $l_3 = 1$ and then $l_1 = k - 3$.

Summing up, one obtains: 1) $l_1 = k$, $p = k$, 2) $l_1 = k-2$, $l_2 = 1$, $p = k-1$, 3) $l_1 = k-4$, $l_2 = 2$, $p = k-2$, 4) $l_1 = k-3$, $l_3 = 1$, $p = k-2$. Hence, (8.7) yields:

$$D_x^k(z) \approx \phi'^k z^{(k)} + \frac{k! \phi'^{k-2} \phi''}{(k-2)! 2!} z^{(k-1)} + \left(\frac{k! \phi'^{k-4} \phi''^2}{(k-4)!(2!)^3} + \frac{k! \phi'^{k-3} \phi'''}{(k-3)! 3!} \right) z^{(k-2)}. \qquad (8.13)$$

Replacing here k by $k-1$ and $k-2$, one also obtains:

$$D_x^{k-1}(z) \approx \phi'^{k-1} z^{(k-1)} + \frac{(k-1)! \phi'^{k-3} \phi''}{(k-3)! 2!} z^{(k-2)}, \quad D_x^{k-2}(z) \approx \phi'^{k-2} z^{(k-2)}. \qquad (8.14)$$

Substituting the expressions (8.13)–(8.14) into (8.12), one ultimately obtains the terms of $y^{(k)}$ containing $z^{(k)}$, $z^{(k-1)}$, $z^{(k-2)}$.

8.3 Extended point transformation groups. Contact transformations

8.3.1 Transformations of the plane

Point transformations in the (x, y) plane are given by

$$\bar{x} = f(x, y, a) \approx x + a\xi(x, y), \quad \bar{y} = \varphi(x, y, a) \approx y + a\eta(x, y). \qquad (8.15)$$

If one treats x as an independent variable and y as a differential one, then a transformation (8.15) carries with it a transformation of derivatives $y', y'', \ldots$. Namely, invoking the equation $\mathrm{d}y = y'\mathrm{d}x$, one obtains the transformation of y':

$$\bar{y}' \equiv \frac{\mathrm{d}\bar{y}}{\mathrm{d}\bar{x}} = \frac{\varphi_x \mathrm{d}x + \varphi_y \mathrm{d}y}{f_x \mathrm{d}x + f_y \mathrm{d}y} = \frac{\varphi_x + y'\varphi_y}{f_x + y'f_y} = \frac{D_x(\varphi)}{D_x(f)}. \tag{8.16}$$

Thus, starting from a point transformation (8.15) and then adding the formula (8.16), one obtains the transformation in the space of three independent variables, x, y, y'. This is what is known as an *extended point transformation;* specifically, the extension of a point transformation (8.15) to the first derivative, sometimes called the *first prolongation* [8.4].

Likewise, the action of a transformation (8.15) can be enlarged upon higher derivatives via equations $\bar{y}'' = \mathrm{d}\bar{y}'/\mathrm{d}\bar{x}, \ldots$ to obtain an s-times extended point transformation affecting the variables $x, y, y', y'', \ldots y^{(s)}$.

Theorem. If transformations (8.15) form a one-parameter group, then their extension to derivatives $y', y'', \ldots y^{(s)}$ of any order is again a one-parameter group and is called an *extended point transformation group*.

Proof. It suffices to prove the theorem for the first extension. Let a be a canonical parameter. Then the group property (7.11) for transformations (8.15) is written $\bar{\bar{x}} \equiv f(\bar{x}, \bar{y}, b) = f(x, y, a+b)$, $\bar{\bar{y}} \equiv \varphi(\bar{x}, \bar{y}, b) = \varphi(x, y, a+b)$. Let us rewrite (8.16) in the form

$$\bar{y}' = \psi(x, y, y', a) \equiv \frac{D_x(\varphi(x, y, a))}{D_x(f(x, y, a))}.$$

To prove the theorem, we have to show that

$$\bar{\bar{y}}' \equiv \psi(\bar{x}, \bar{y}, \bar{y}', b) = \psi(x, y, y', a+b).$$

The latter can be obtained by using the chain rule[1] $\overline{D}_x = D_x(f(x, y, a))\overline{D}_x$ and the following simple calculations. We have

$$\bar{\bar{y}}' = \psi(\bar{x}, \bar{y}, \bar{y}', b) = \frac{\overline{D}_x(\varphi(\bar{x}, \bar{y}, b))}{\overline{D}_x(f(\bar{x}, \bar{y}, b))}.$$

Now we rewrite the last expression by multiplying its numerator and denominator by $D_x(f(x, y, a))$ and invoking the group property of f and φ, as follows:

$$\frac{D_x(f(x, y, a))\overline{D}_x(\varphi(\bar{x}, \bar{y}, b))}{D_x(f(x, y, a))\overline{D}_x f(\bar{x}, \bar{y}, b))} = \frac{D_x(\varphi(x, y, a+b))}{D_x(f(x, y, a+b))} = \psi(x, y, y', a+b),$$

thus completing the proof.

---

[1] $\overline{D}_x$ denotes the total derivation with respect to $\bar{x}$, see also formula (8.28).

8.3. EXTENDED POINT AND CONTACT TRANSFORMATIONS

The group analysis employs extensions of infinitesimal transformation (8.15), $\bar{x} \approx x + a\xi(x,y)$, $\bar{y} \approx y + a\eta(x,y)$. Substituting this into (8.16), one obtains the infinitesimal transformation $\bar{y}' \approx y' + a\zeta_1(x,y,y')$, viz.

$$\bar{y}' \approx \frac{D_x(y + a\eta)}{D_x(x + a\xi)} = \frac{y' + aD_x(\eta)}{1 + aD_x(\xi)} \approx [y' + aD_x(\eta)][1 - aD_x(\xi)],$$

or $\bar{y}' \approx y' + a[D_x(\eta) - y'D_x(\xi)]$. Hence,

$$\zeta_1 = D_x(\eta) - y'D_x(\xi). \tag{8.17}$$

Thus, given a one-parameter group with the generator (7.88),

$$X = \xi(x,y)\frac{\partial}{\partial x} + \eta(x,y)\frac{\partial}{\partial y},$$

the *extended generator* $X_{(1)}$ affecting the variables x, y, and y' has the form [8.5]:

$$X_{(1)} = \xi\frac{\partial}{\partial x} + \eta\frac{\partial}{\partial y} + \zeta_1\frac{\partial}{\partial y'}, \tag{8.18}$$

where ζ_1 is determined by (8.17) often called the *first prolongation formula*.

The $(s+1)$-times extended infinitesimal transformations, and hence their symbols $X_{(s+1)}$, are obtained recursively:

$$\bar{y}^{(s+1)} \approx [y^{(s+1)} + aD_x(\zeta_s)][1 - aD_x(\xi)] \approx y^{(s+1)} + a[D_x(\zeta_s) - y^{(s+1)}D_x(\xi)],$$

whence, by setting $\bar{y}^{(s+1)} \approx y^{(s+1)} + a\zeta_{s+1}$,

$$\zeta_{s+1} = D_x(\zeta_s) - y^{(s+1)}D_x(\xi), \quad s = 1, 2, \ldots. \tag{8.19}$$

In practical applications, one often needs the expanded forms of the coordinates ζ_1, ζ_2 and ζ_3 of the second and third extensions, $X_{(2)}$ and $X_{(3)}$. Namely,

$$\begin{aligned}
\zeta_1 &= D(\eta) - y'D(\xi) = \eta_x + (\eta_y - \xi_x)y' - \xi_y y'^2, \\
\zeta_2 &= D(\zeta_1) - y''D(\xi) = \eta_{xx} + (2\eta_{xy} - \xi_{xx})y' + (\eta_{yy} - 2\xi_{xy})y'^2 \\
&\quad -\xi_{yy}y'^3 + (\eta_y - 2\xi_x - 3\xi_y y')y'', \\
\zeta_3 &= D(\zeta_2) - y'''D(\xi) = \eta_{xxx} + (3\eta_{xxy} - \xi_{xxx})y' + 3(\eta_{xyy} - \xi_{xxy})y'^2 \\
&\quad +(\eta_{yyy} - 3\xi_{xyy})y'^3 - \xi_{yyy}y'^4 + 3[\eta_{xy} - \xi_{xx} + (\eta_{yy} - 3\xi_{xy})y' \\
&\quad -2\xi_{yy}y'^2]y'' - 3\xi_y y''^2 + (\eta_y - 3\xi_x - 4\xi_y y')y'''.
\end{aligned} \tag{8.20}$$

Thus, the first, second, and third extensions of the generator X (7.88) are given, respectively, by (8.18) and by

$$X_{(2)} = \xi\frac{\partial}{\partial x} + \eta\frac{\partial}{\partial y} + \zeta_1\frac{\partial}{\partial y'} + \zeta_2\frac{\partial}{\partial y''},$$

$$X_{(3)} = \xi\frac{\partial}{\partial x} + \eta\frac{\partial}{\partial y} + \zeta_1\frac{\partial}{\partial y'} + \zeta_2\frac{\partial}{\partial y''} + \zeta_3\frac{\partial}{\partial y'''}. \quad (8.21)$$

Example. The first extension of the infinitesimal rotation has the form

$$X_{(1)} = y\frac{\partial}{\partial x} - x\frac{\partial}{\partial y} - (1 + y'^2)\frac{\partial}{\partial y'}.$$

8.3.2 One independent and several differential variables

Consider point transformations

$$\overline{x} = f(x, u, a) \approx x + a\xi(x, u), \quad \overline{u}^\alpha = \varphi^\alpha(x, u, a) \approx u^\alpha + a\eta^\alpha(x, u).$$

with one independent variable x and m differential variables $u = (u^1, \ldots, u^m)$. Then (8.16) is replaced by

$$\overline{u}_1^\alpha \equiv \frac{d\,\overline{u}^\alpha}{d\,\overline{x}} = \frac{D_x(\varphi^\alpha)}{D_x(f)}.$$

In precisely the same way as in the previous section, we are led to the following *prolongation formulas* (cf. (8.17) and (8.19)):

$$\zeta_1^\alpha = D_x(\eta^\alpha) - u_1^\alpha D_x(\xi), \quad \zeta_{s+1}^\alpha = D_x(\zeta_s^\alpha) - u_{s+1}^\alpha D_x(\xi), \quad s = 1, 2, \ldots, \quad (8.22)$$

where $u_1^\alpha, u_2^\alpha, \ldots$ denote the first, second, etc. derivatives. Hence, given a one-parameter group with the generator

$$X = \xi(x, u)\frac{\partial}{\partial x} + \eta^\alpha(x, u)\frac{\partial}{\partial u^\alpha},$$

we obtain, e.g., the twice-extended generator by the formula (summation in α):

$$X_{(2)} = \xi(x, u)\frac{\partial}{\partial x} + \eta^\alpha(x, u)\frac{\partial}{\partial u^\alpha} + \zeta_1^\alpha\frac{\partial}{\partial u_1^\alpha} + \zeta_2^\alpha\frac{\partial}{\partial u_2^\alpha}. \quad (8.23)$$

8.3.3 Point transformations involving many variables

Let G be a one-parameter group of transformations of independent variables $x = (x^1, \ldots, x^n)$ and differential variables $u = (u^1, \ldots, u^m)$:

$$\overline{x}^i = f^i(x, u, a), \quad f^i|_{a=0} = x^i, \quad (8.24)$$

$$\overline{u}^\alpha = \varphi^\alpha(x, u, a), \quad \varphi^\alpha|_{a=0} = u^\alpha. \quad (8.25)$$

The generator of the group G is written in the form

$$X = \xi^i(x, u)\frac{\partial}{\partial x^i} + \eta^\alpha(x, u)\frac{\partial}{\partial u^\alpha}, \quad (8.26)$$

8.3. EXTENDED POINT AND CONTACT TRANSFORMATIONS

where

$$\xi^i(x,u) = \left.\frac{\partial f^i(x,u,a)}{\partial a}\right|_{a=0}, \quad \eta^\alpha(x,u) = \left.\frac{\partial \varphi^\alpha(x,u,a)}{\partial a}\right|_{a=0}. \tag{8.27}$$

Let $\overline{D}_i$ denote total derivations in new variables $\overline{x}^i$. The chain rule yields:

$$D_i = \left(\frac{\partial f^j}{\partial x^i} + u_i^\beta \frac{\partial f^j}{\partial u^\beta}\right)\overline{D}_j \equiv D_i(f^j)\overline{D}_j. \tag{8.28}$$

Equations (8.3) are written in the new variables:

$$\overline{u}_i^\alpha = \overline{D}_i(\overline{u}^\alpha), \quad \overline{u}_{ij}^\alpha = \overline{D}_j(\overline{u}_i^\alpha), \ldots$$

and therefore equations (8.25), (8.28) yield the *rule for the change of derivatives*:

$$\overline{u}_j^\alpha D_i(f^j) = D_i(\varphi^\alpha). \tag{8.29}$$

Upon solving (8.29) with respect to $\overline{u}_i^\alpha$, one obtains the transformation of the first derivatives, $\overline{u}_i^\alpha = \psi_i^\alpha(x, u, u_{(1)}, a)$. Extensions to the second- and higher-order derivatives are obtained by further differentiating equation (8.29) by means of (8.28). An extension of a group of transformations (8.24)–(8.25) to derivatives $u_{(1)}, \ldots, u_{(s)}$ of any order is again a one-parameter group and is called an *extended point transformation group* (see Problem 8.8*).

By letting $f^i = x^i + a\xi^i$, $\varphi^\alpha = u^\alpha + a\eta^\alpha$ and setting $\overline{u}_i^\alpha = u_i^\alpha + a\zeta_i^\alpha$, one readily obtains from equation (8.29) the *first prolongation formula*:

$$\zeta_i^\alpha = D_i(\eta^\alpha) - u_j^\alpha D_i(\xi^j). \tag{8.30}$$

Hence, the extended generator (8.26):

$$X_{(1)} = \xi^i \frac{\partial}{\partial x^i} + \eta^\alpha \frac{\partial}{\partial u^\alpha} + \zeta_i^\alpha \frac{\partial}{\partial u_i^\alpha}. \tag{8.31}$$

Similarly, the twice-extended generator is written:

$$X_{(2)} = \xi^i \frac{\partial}{\partial x^i} + \eta^\alpha \frac{\partial}{\partial u^\alpha} + \zeta_i^\alpha \frac{\partial}{\partial u_i^\alpha} + \zeta_{i_1 i_2}^\alpha \frac{\partial}{\partial u_{i_1 i_2}^\alpha}, \tag{8.32}$$

where $\zeta_{i_1 i_2}^\alpha$ are defined by the *second prolongation formula*:

$$\begin{aligned}\zeta_{i_1 i_2}^\alpha &= D_{i_2}(\zeta_{i_1}^\alpha) - u_{ji_1}^\alpha D_{i_2}(\xi^j) \\ &\equiv D_{i_2} D_{i_1}(\eta^\alpha) - u_j^\alpha D_{i_2} D_{i_1}(\xi^j) - u_{ji_1}^\alpha D_{i_2}(\xi^j).\end{aligned} \tag{8.33}$$

The higher-order prolongations are defined recursively:

$$\zeta_{i_1\ldots i_s}^\alpha = D_{i_s}(\zeta_{i_1\ldots i_{s-1}}^\alpha) - u_{ji_1\ldots i_{s-1}}^\alpha D_{i_s}(\xi^j). \tag{8.34}$$

Introducing the functions W^α (a generalization to many variables of Lie's characteristic function W, see Section 5.3.2):

$$W^\alpha = \eta^\alpha - \xi^j u_j^\alpha, \tag{8.35}$$

we unify the above prolongation formulas as follows [8.6]:

$$\zeta_{i_1\ldots i_s}^\alpha = D_{i_1}\cdots D_{i_s}(W^\alpha) + \xi^j u_{ji_1\ldots i_s}^\alpha, \quad s = 1, 2, \ldots. \tag{8.36}$$

8.3.4 Properties of extended generators

The following properties manifest that the extension (8.30)–(8.31) of group generators complies with all operations in Lie algebras.

Lemma. The operation of extension is invariant under any change of variables x^i, u^α $(i = 1, \ldots, n; \ \alpha = 1, \ldots, m)$:
$$x'^i = x'^i(x, u), \quad u'^\alpha = u'^\alpha(x, u). \tag{8.37}$$

Namely, $\overline{X}_{(1)} = \overline{X_{(1)}}$, where $\overline{X}_{(1)}$ is obtained by first rewriting the operator X (8.26) in the new coordinates x'^i AND u'^α (cf. (7.30)):
$$\overline{X} = X(x'^i)\frac{\partial}{\partial x'^i} + X(u'^\alpha)\frac{\partial}{\partial u'^\alpha}, \tag{8.38}$$

and then extending $\overline{X}$ to the derivatives u'^α_i, whereas $\overline{X_{(1)}}$ is the operator obtained by first extending X by formula (8.30) and then rewriting the extended operator $X_{(1)}$ (8.31) in the new coordinates x'^i, u'^α, and u'^α_i.

Theorem. Let $X_1, \ldots, X_r$ be r linearly independent operators of the form (8.26). If they span an r-dimensional Lie algebra L_r, then the s-times extended operators span a Lie algebra with the same structure constants as the algebra L_r. This statement results from the following equations:
$$(kX + lY)_{(1)} = kX_{(1)} + lY_{(1)}, \tag{8.39}$$
$$[X, Y]_{(1)} = [X_{(1)}, Y_{(1)}], \tag{8.40}$$

where $k, l = \text{const.}$, and X and Y denote any two operators of the form (8.26).

It follows that if the point transformations (8.24)–(8.25), where the parameter a is replaced by a vector-parameter $a = (a^1, \ldots, a^r)$, form an r-parameter group, then their extension to derivatives of any order is also an r-parameter group with the same group composition law.

8.3.5 Differential invariants. Invariant differentiation

Definition 1. Let G be a point transformation group (8.24)–(8.25). An invariant $F(x, u, u_{(1)}, \ldots, u_{(s)})$ of the s-times extended transformation group is called a *differential invariant of order s* of the group G, provided that $\text{ord}(F) = s \geq 1$ (if $s = 0$, F is the usual invariant). Invoking the invariant test (7.24) and denoting the infinite-order extension of the generator (8.26) again by X (see [8.5]):
$$X = \xi^i\frac{\partial}{\partial x^i} + \eta^\alpha\frac{\partial}{\partial u^\alpha} + \zeta^\alpha_{i_1}\frac{\partial}{\partial u^\alpha_{i_1}} + \cdots + \zeta^\alpha_{i_1 \ldots i_s}\frac{\partial}{\partial u^\alpha_{i_1 \ldots i_s}} + \cdots, \tag{8.41}$$

we can define differential invariants as differential functions $F \in \mathcal{A}$, $\text{ord}(F) = s$, satisfying the partial differential equation $X(F) = 0$,
$$X(F) \equiv \xi^i\frac{\partial F}{\partial x^i} + \eta^\alpha\frac{\partial F}{\partial u^\alpha} + \zeta^\alpha_{i_1}\frac{\partial F}{\partial u^\alpha_{i_1}} + \cdots + \zeta^\alpha_{i_1 \ldots i_s}\frac{\partial F}{\partial u^\alpha_{i_1 \ldots i_s}} = 0.$$

8.3. EXTENDED POINT AND CONTACT TRANSFORMATIONS

Consider the particular case when G is a group (8.15) in the (x,y) plane. By Theorem 2 of Section 7.1.7, G has precisely one independent invariant. Let us denote it by $u = u(x,y)$. The extension of G to the first derivative y' adds another invariant, $v = v(x,y,y')$, which necessarily depends upon y' and is therefore a differential invariant of the first order. Likewise, the second extension adds one more invariant which should contain the second derivative y'', and hence, it is a second-order differential invariant of the group G, etc.

Example. Let G be the group of Galilean transformations $\bar{x} = x + ay$, $\bar{y} = y$ (see Table 7.1.10). Its generator $X = y\partial/\partial x$, after extending to all derivatives by (8.41), yields the equation:

$$X(F) \equiv y\frac{\partial F}{\partial x} - y'^2\frac{\partial F}{\partial y'} - 3y'y''\frac{\partial F}{\partial y''} - \left(3y''^2 + 4y'y'''\right)\frac{\partial F}{\partial y'''} - \cdots = 0. \quad (8.42)$$

Letting $\mathrm{ord}(F) = 0$, i.e. $F = F(x,y)$, we reduce equation (8.42) to $yF_x = 0$, whence $F = F(y)$. Therefore, one can take, as an independent invariant, $u = y$. To obtain differential invariants, e.g. of the first and second orders, we truncate the operator X in (8.42) by letting $\mathrm{ord}(F) = 2$, i.e. $F = F(x,y,y',y'')$. Then X acts as the twice-extended operator $X_{(2)}$ [8.5], and equation (8.42) is written

$$X_{(2)}(F) \equiv y\frac{\partial F}{\partial x} - y'^2\frac{\partial F}{\partial y'} - 3y'y''\frac{\partial F}{\partial y''} = 0.$$

Solving the characteristic system,

$$\frac{dx}{y} = -\frac{dy'}{y'^2} = -\frac{dy''}{3y'y''},$$

we obtain, along with the invariant $u = y$, two independent differential invariants (of the first and second orders): $v = (y/y') - x$ and $w = y''/y'^3$.

Now one can find, by Theorem 7.2.2, equations invariant under the group of extended Galilean transformations. The invariant equations, e.g. for the first and second extensions, are given by $v = \Phi(u)$ and $w = \Phi(u,v)$. Hence, the following differential equations admit the group G in the sense of Section 5.2.2:

$$y' = \frac{y}{x + \Phi(y)}, \quad y'' = y'^3 \Phi\left(y, \frac{y}{y'} - x\right).$$

One can continue the process and obtain higher-order differential invariants by substituting in equation (8.42) differential functions F of orders 3, 4, etc. However, Lie [8.7] showed that all differential invariants of orders 2, 3, ... can be found merely by differentiation, provided that u and v are known.

Theorem 1. Let G be a one-parameter point transformation group in the plane with the generator X. Let $u(x,y)$ and $v(x,y,y')$ be an invariant and a first-order differential invariant of G. Then

$$w = \frac{dv}{du} = \frac{v_x + y'v_y + y''v_{y'}}{u_x + y'u_y} = \frac{D_x(v)}{D_x(u)} \quad (8.43)$$

is a second-order differential invariant of the group G.

Proof. The differential equation $v(x, y, y') - ku(x, y) - l = 0$ with arbitrary constants k and l is invariant under G since its left-hand term is a differential invariant. Hence, the totality of the integral curves of this equation is invariant under G. This is true also when we hold k fixed and let l alone vary. Then the totality of the integral curves is the set of solutions of the second-order equation obtained by eliminating l through differentiation, $dv - kdu = 0$, or $w - k = 0$ in notation (8.43). It follows from the invariance of the set of solutions of the latter equation that it admits the twice-extended generator, i.e. $X_{(2)}(w - k) \equiv X_{(2)}(w) = 0$, whenever $w - k = 0$. Since k is arbitrary and $X_{(2)}(w)$ does not involve k, it follows that $X_{(2)}(w) = 0$ identically. Hence, $w = dv/du$ is a differential invariant of the second order.

Remark. Since $X_{(2)}F(x, y, y', y'') = 0$ has precisely three independent solutions, all differential invariants F, $\text{ord}(F) \leq 2$, are given by $F = \Phi(u, v, w)$.

Let us rewrite (8.43) in the form $w = \mathcal{D}(v)$ with the operator $\mathcal{D}$ defined by

$$\mathcal{D} = \lambda D_x, \quad \text{where} \quad \lambda = \frac{1}{D_x(u)}. \tag{8.44}$$

Definition 2. The operator $\mathcal{D}$ (8.44), where $u(x, y)$ is any invariant of G, is called an *invariant differentiation* for the group G.

The significance of this definition is disclosed by following statement. It can be proved by repeating the proof of Theorem 1.

Theorem 2. The invariant differentiation $\mathcal{D}$ converts any differential invariant $F \in \mathcal{A}$ of the group G into a differential invariant $\mathcal{D}(F) \in \mathcal{A}$ of G. Furthermore, any differential invariant F, $\text{ord}(F) = s + 1$, can be expressed as a function of $u(x, y)$, $v(x, y, y')$ and successive invariant derivatives of the first-order differential invariant v:

$$F = \Phi\left(u, v, \mathcal{D}(v), \mathcal{D}^2(v), \ldots, \mathcal{D}^s(v)\right).$$

Similar results hold also in the case of many variables and multi-parameter groups [8.8]. To formulate this generalization, let us first note that the coefficient $\lambda = 1/D_x(u)$ of the invariant differentiation (8.44) satisfies the equation

$$X_{(1)}(\lambda) = \lambda D_x(\xi). \tag{8.45}$$

Indeed, one can verify by straightforward computation that the following operator identity holds for the infinite-times extended generator X (8.41):

$$XD_i - D_iX = -D_i(\xi^j)D_j. \tag{8.46}$$

Let us consider the one-dimensional case and apply (8.46), written in the form $XD_x = D_xX - D_x(\xi)D_x$, to $\lambda(x, y, y') = 1/D_x(u)$. Invoking that $u(x, y)$ is an invariant, i.e. $X(u) = 0$, we arrive at (8.45) as follows:

$$X_{(1)}(\lambda) = -\frac{X(D_x(u))}{(D_x(u))^2} = -\frac{D_xX(u) - D_x(\xi)D_x(u)}{(D_x(u))^2} = \frac{D_x(\xi)}{D_x(u)} = \lambda D_x(\xi).$$

8.3. EXTENDED POINT AND CONTACT TRANSFORMATIONS

The generalization of Definition 2 and Theorem 2 is given by the following invariant derivations. Let G_r be any r-parameter group of point transformations of the form (8.24)–(8.25) with infinitesimal generators

$$X_\nu = \xi^i_\nu(x,u)\frac{\partial}{\partial x^i} + \eta^\alpha_\nu(x,u)\frac{\partial}{\partial u^\alpha}, \quad \nu = 1,\ldots,r.$$

Then there exist n independent invariant derivations [8.9]

$$\mathcal{D} = \lambda^i D_i, \quad \lambda^i \in \mathcal{A}, \tag{8.47}$$

where $\lambda^i(x, u, u_{(1)}, u_{(2)}, \ldots)$ are differential functions determined from the equations similar to (8.45):

$$X_\nu(\lambda^i) = \lambda^j D_j(\xi^i_\nu), \quad i = 1,\ldots,n;\ \nu = 1,\ldots,r.$$

8.3.6 Change of derivatives under differential substitutions

The rule for changes of derivatives (8.29) holds also in a more general setting, when point transformations (8.24) are replaced by *differential substitutions* [8.10]:

$$\bar{x}^i = f^i(x, u, u_{(1)}, \ldots, u_{(p)}), \quad \bar{u}^\alpha = \varphi^\alpha(x, u, u_{(1)}, \ldots, u_{(q)}), \tag{8.48}$$

where $f^i, \varphi^\alpha \in \mathcal{A}$ are any differential functions, $\mathrm{ord}(f^i) = p$, $\mathrm{ord}(\varphi^\alpha) = q$. If $p = q = 0$, then (8.48) reduces to the usual change of variables x, u.

The differential substitution (8.48) is accompanied by the conversions

$$D_i \mapsto \overline{D}_i, \quad u^\alpha_i \mapsto \bar{u}^\alpha_i = \overline{D}_i(\bar{u}^\alpha), \quad u^\alpha_{ij} \mapsto \bar{u}^\alpha_{ij} = \overline{D}_j(\bar{u}^\alpha_i), \ldots$$

obtained as follows. The first equation (8.48) provides the relations (cf. (8.28))

$$D_i = D_i(f^j)\overline{D}_j, \quad i = 1, \ldots, n, \tag{8.49}$$

(summation in j) for total derivatives in the variables x and $\bar{x}$. Then its application to the second equation (8.48) yields[2] the transformation law (8.29) for the first derivatives. One can again differentiate the latter equation by means of (8.49) to obtain $\bar{u}^\alpha_{ij}$, etc. Summing up, we have the following general rule.

The differential substitution (8.48) carries with it the change of derivatives $u_{(1)}, u_{(2)}, \ldots$ determined by the equations

$$\bar{u}^\alpha_j D_i(f^j) = D_i(\varphi^\alpha), \quad \bar{u}^\alpha_{jl} D_k(f^l) D_i(f^j) + \bar{u}^\alpha_j D_k D_i(f^j) = D_k D_i(\varphi^\alpha), \ldots \tag{8.50}$$

provided that equations (8.50) can be solved for $\bar{u}^\alpha_i$, $\bar{u}^\alpha_{ij}, \ldots$.

Example 1. The simplest example is provided by the rule of differentiation of *inverse functions* used in differential calculus. Recall that inverse functions, $g(x)$

---

[2] Specifically, the right-hand term of (8.49) acts on the left-hand side of the second equation (8.48), and vice versa.

and $h(y)$, are obtained by solving an equation $y = g(x)$ with respect to x to obtain $x = h(y)$. Then one can prove that the derivative of the inverse function is equal to the reciprocal of the derivative of the direct function, $h'(y) = 1/g'(x)$, where x is replaced by $x = h(y)$. This rule is also written $x' = 1/y'$.

In differential algebra, we use the change of variables $\bar{x} = y$, $\bar{y} = x$, i.e. (8.48) with $f = y$, $\varphi = x$. Here, (8.49) is written $D_{\bar{x}} = y'D_y$. Furthermore,

$$\bar{y}' = \frac{d\bar{y}}{d\bar{x}} = \frac{dx}{dy} = x'.$$

The first equation (8.50), $\bar{y}'D_{\bar{x}}(f) = D_x(\varphi)$, yields $x'D_x(y) = D_x(x)$, or $x'y' = 1$. Hence, the first derivative $x' = 1/y'$. Differentiating the latter by means of $y'D_y = D_x$ yields $y'D_y(x') = D_x(1/y')$, or $y'x'' = -y''/y'^2$. Hence, the second derivative $x'' = -y''/y'^3$. It can be obtained also from the second equation (8.50).

Example 2. Consider, in the case of n independent variables x^i and one differential variable u, the following example of (8.48): $\bar{x}^i = u_i$, $\bar{u} = -u + x^j u_j$. Formula (8.49) is written $D_i = u_{ij}\bar{D}_j$, and the first equation (8.50) yields

$$\sum_{j=1}^{n} u_{ij}(\bar{u}_j - x^j) = 0, \quad i = 1, \ldots, n.$$

Since u_{ij} are arbitrary variables, $\det\|u_{ij}\| \neq 0$. Hence, $\bar{u}_j - x^j = 0$, $j = 1, \ldots, n$, and we have the Legendre transformation (5.12),

$$\bar{x}^i = u_i, \quad \bar{u} = -u + x^j u_j, \quad \bar{u}_i = x^i, \quad i = 1, \ldots, n.$$

8.3.7 Infinitesimal contact transformations

A group of *contact transformations* can be defined by differential substitutions (8.48), involving n independent variables $x = (x^1, \ldots, x^n)$, one differential variable u, and its first derivatives $u_{(1)} = (u_1, \ldots, u_n)$:

$$\bar{x}^i = f^i(x, u, u_{(1)}, a) \approx x^i + a\xi^i(x, u, u_{(1)}),$$

$$\bar{u} = \varphi(x, u, u_{(1)}, a) \approx u + a\eta(x, u, u_{(1)}), \qquad (8.51)$$

provided that the extension of (8.51) to the first derivatives has the form

$$\bar{u}_i = \psi_i(x, u, u_{(1)}, a) \approx u_i + a\zeta_i(x, u, u_{(1)}), \qquad (8.52)$$

i.e., it does not involve higher derivatives (cf. Section 5.3.1).

Theorem. Given an operator

$$X = \xi^i(x, u, u_{(1)})\frac{\partial}{\partial x^i} + \eta(x, u, u_{(1)})\frac{\partial}{\partial u} + \zeta_i(x, u, u_{(1)})\frac{\partial}{\partial u_i}, \qquad (8.53)$$

8.3. EXTENDED POINT AND CONTACT TRANSFORMATIONS

it is a symbol of a group of contact transformations (8.51)–(8.52) if and only if its coordinates can be represented in the form (5.15):

$$\xi^i = -\frac{\partial W}{\partial u_i}, \quad \eta = W - u_i \frac{\partial W}{\partial u_i}, \quad \zeta_i = \frac{\partial W}{\partial x^i} + u_i \frac{\partial W}{\partial u} \tag{8.54}$$

with a differential function $W \in \mathcal{A}$ of the first order, i.e. $W = W(x, u, u_{(1)})$.

Proof. The extended transformation (8.52) is obtained from the first equation (8.50), $\overline{u}_i D_i(f^j) = D_i(\varphi)$. Hence, one can apply the calculations of Section 8.3.3, e.g. represent ζ_i in (8.52) in the form (8.36), $\zeta_i = D_i(W) + \xi^j u_{ij}$, or

$$\zeta_i = \frac{\partial W}{\partial x^i} + u_i \frac{\partial W}{\partial u} + u_{ij}\left(\xi^j + \frac{\partial W}{\partial u_i}\right), \tag{8.55}$$

where W is given by (8.35), $W = \eta - \xi^j u_j$. It follows that ζ_i do not contain the second derivatives u_{ij} if and only if the last term in (8.55) vanishes, i.e.

$$\xi^j + \frac{\partial W}{\partial u_i} = 0, \quad j = 1, \ldots, n.$$

Hence, the first equation (8.54). Furthermore, (8.55) provides the third equation (8.54). The expression for η given in (8.54) is obtained from $W = \eta - \xi^j u_j$ after substituting $\xi^j = -\partial W/\partial u_i$.

8.3.8 Irreducible contact transformation groups in the plane

One can obtain contact transformation groups, e.g. by subjecting an arbitrary point transformation group to any contact transformation (5.10). However, such contact transformation groups are regarded as trivial ones since they are reducible to point transformation groups. In the case of contact transformations in the plane, all irreducible groups were enumerated by Lie [8.11].

Definition. An r-parameter group of contact transformations (8.51)–(8.52) is said to be *reducible* if it can be reduced, by a contact transformation

$$x'^i = x'^i(x, u, u_{(1)}), \quad u' = u'(x, u, u_{(1)}), \quad u'_i = u'_i(x, u, u_{(1)}) \quad i = 1, \ldots, n,$$

to a group of extended point transformations of x'^i, u', and *irreducible* otherwise.

Theorem. Any finite continuous [8.12] irreducible group of contact transformations in the (x, y) plane is a 6- or 7- or 10-parameter group. Its infinitesimal generators can be mapped, by an appropriate contact transformation, to one of the following forms (in accordance with the dimension of the group):

$$X_1 = \frac{\partial}{\partial x}, \quad X_2 = \frac{\partial}{\partial y}, \quad X_3 = x\frac{\partial}{\partial y} + \frac{\partial}{\partial y'}, \quad X_4 = \frac{1}{2}x^2\frac{\partial}{\partial y} + x\frac{\partial}{\partial y'},$$

$$X_5 = x\frac{\partial}{\partial x} - y'\frac{\partial}{\partial y'}, \quad X_6 = y'\frac{\partial}{\partial x} + \frac{1}{2}y'^2\frac{\partial}{\partial y}; \tag{8.56}$$

or
$$X_1 = \frac{\partial}{\partial x},\ X_2 = \frac{\partial}{\partial y},\ X_3 = x\frac{\partial}{\partial y} + \frac{\partial}{\partial y'},\ X_4 = \frac{1}{2}x^2\frac{\partial}{\partial y} + x\frac{\partial}{\partial y'},$$
$$X_5 = x\frac{\partial}{\partial x} - y'\frac{\partial}{\partial y'},\ X_6 = y'\frac{\partial}{\partial x} + \frac{1}{2}y'^2\frac{\partial}{\partial y},\ X_7 = x\frac{\partial}{\partial x} + 2y\frac{\partial}{\partial y} + y'\frac{\partial}{\partial y'};\quad (8.57)$$

or
$$X_1 = \frac{\partial}{\partial x},\ X_2 = \frac{\partial}{\partial y},\ X_3 = x\frac{\partial}{\partial y} + \frac{\partial}{\partial y'},\ X_4 = \frac{1}{2}x^2\frac{\partial}{\partial y} + x\frac{\partial}{\partial y'},$$
$$X_5 = x\frac{\partial}{\partial x} - y'\frac{\partial}{\partial y'},\ X_6 = y'\frac{\partial}{\partial x} + \frac{1}{2}y'^2\frac{\partial}{\partial y},\ X_7 = x\frac{\partial}{\partial x} + 2y\frac{\partial}{\partial y} + y'\frac{\partial}{\partial y'},$$
$$X_8 = (y - xy')\frac{\partial}{\partial x} - \frac{1}{2}xy'^2\frac{\partial}{\partial y} - \frac{1}{2}y'^2\frac{\partial}{\partial y'},\ X_9 = \frac{1}{2}x^2\frac{\partial}{\partial x} + xy\frac{\partial}{\partial y} + y\frac{\partial}{\partial y'},$$
$$X_{10} = \left(xy - \frac{1}{2}x^2y'\right)\frac{\partial}{\partial x} + \left(y^2 - \frac{1}{4}x^2y'^2\right)\frac{\partial}{\partial y} + \left(yy' - \frac{1}{2}xy'^2\right)\frac{\partial}{\partial y'}. \quad (8.58)$$

Thus, the maximal finite irreducible group is similar to the 10-parameter group generated by (8.58). Two other irreducible groups are similar to its 6- and 7-parameter subgroups generated by (8.56) and (8.57), respectively.

8.4 Operators and identities in $\mathcal{A}$

The operators presented in this section play a central role in the calculus of variations and in the study of symmetries and conservation laws of differential equations. These operators are correctly defined in the space $\mathcal{A}$, namely the formal sums representing them truncate while acting on differential functions.

The *fundamental identity*, connecting three main operators, furnishes the basis of a differential algebraic approach to conservation theorems.

8.4.1 The Euler-Lagrange operator. Test for a total derivative

For the motivation of the following formal definition, see Section 9.7.1 (cf. also equation (5.26)) [8.13].

Definition. The *Euler-Lagrange operator* in $\mathcal{A}$ is defined by the formal sum
$$\frac{\delta}{\delta u^\alpha} = \frac{\partial}{\partial u^\alpha} + \sum_{s=1}^{\infty}(-1)^s D_{i_1}\cdots D_{i_s}\frac{\partial}{\partial u^\alpha_{i_1\cdots i_s}},\quad \alpha = 1,\ldots,m, \quad (8.59)$$

where, for every s, the summation is presupposed over the repeated indices $i_1\ldots i_s$ running from 1 to n (cf. (8.65)).

Lemma 1. The total derivative D_i commutes with the partial differentiation with respect to u^α:
$$\frac{\partial}{\partial u^\alpha}D_i = D_i\frac{\partial}{\partial u^\alpha}.$$

8.4. OPERATORS AND IDENTITIES IN $\mathcal{A}$

In the theory of ordinary differential equations, one deals with the one-dimensional version (one independent variable x) of the Euler-Lagrange operator:

$$\frac{\delta}{\delta u^\alpha} = \frac{\partial}{\partial u^\alpha} - D_x \frac{\partial}{\partial u_x^\alpha} + D_x^2 \frac{\partial}{\partial u_{xx}^\alpha} - D_x^3 \frac{\partial}{\partial u_{xxx}^\alpha} + \cdots. \qquad (8.60)$$

In the (x, y) plane, where y is a differential variable, (8.60) is written:

$$\frac{\delta}{\delta y} = \sum_{s=0}^{\infty} (-1)^s D_x^s \frac{\partial}{\partial y^{(s)}} = \frac{\partial}{\partial y} - D_x \frac{\partial}{\partial y'} + D_x^2 \frac{\partial}{\partial y''} - D_x^3 \frac{\partial}{\partial y'''} + \cdots. \qquad (8.61)$$

Lemma 2. The following operator identity holds for every α:

$$\frac{\delta}{\delta u^\alpha} D_x = 0. \qquad (8.62)$$

Proof. Straightforward calculations and Lemma 1 yield:

$$\frac{\delta}{\delta u^\alpha} D_x = \left(\frac{\partial}{\partial u^\alpha} - D_x \frac{\partial}{\partial u_x^\alpha} + D_x^2 \frac{\partial}{\partial u_{xx}^\alpha} - \cdots \right) \left(\frac{\partial}{\partial x} + u_x^\beta \frac{\partial}{\partial u^\beta} + u_{xx}^\beta \frac{\partial}{\partial u_x^\beta} + \cdots \right)$$

$$= \frac{\partial}{\partial u^\alpha} D_x - D_x \frac{\partial}{\partial u^\alpha} - D_x^2 \frac{\partial}{\partial u_x^\alpha} + D_x^2 \frac{\partial}{\partial u_x^\alpha} + D_x^3 \frac{\partial}{\partial u_{xx}^\alpha} - D_x^3 \frac{\partial}{\partial u_{xx}^\alpha} - \cdots = 0.$$

It is useful to supply the integration method of Section 2.24 with the following *test for total derivatives*.

Theorem. A necessary and sufficient condition that a differential function $f(x, u, \ldots, u_{(s)}) \in \mathcal{A}$ be a total derivative,

$$f = D_x(g), \quad g(x, u, \ldots, u_{(s-1)}) \in \mathcal{A}, \qquad (8.63)$$

is that the following equations hold identically in $x, u, u_{(1)}, \ldots$:

$$\frac{\delta f}{\delta u^\alpha} = 0, \quad \alpha = 1, \ldots, m. \qquad (8.64)$$

Proof. The necessity, i.e. that (8.63) implies (8.64), is evident from the identity (8.62). For the proof of sufficiency, see, e.g., [8.14]. See also the next exercise and solve Problem 8.15.

Exercise 1. Verify the above theorem for $m = 2$ and $s = 1$. In other words, show that $f(x, u, v, u', v') = D_x(g(x, u, v))$ if and only if

$$\frac{\delta f}{\delta u} \equiv f_u - D_x(f_{u'}) = 0, \quad \frac{\delta f}{\delta v} \equiv f_v - D_x(f_{v'}) = 0. \qquad (8.65)$$

Solution. If $f(x, u, v, u', v') = D_x(g(x, u, v)) \equiv g_x + u'g_u + v'g_v$, then equations (8.65) follow from Lemma 1:

$$\frac{\delta f}{\delta u} = \frac{\partial}{\partial u} D_x(g) - D_x(g_u) = \left(\frac{\partial}{\partial u} D_x - D_x \frac{\partial}{\partial u} \right)(g) = 0,$$

$$\frac{\delta f}{\delta v} = \frac{\partial}{\partial v} D_x(g) - D_x(g_v) = \left(\frac{\partial}{\partial v} D_x - D_x \frac{\partial}{\partial v}\right)(g) = 0.$$

Conversely, let $f(x, u, v, u', v')$ satisfy equations (8.65):

$$f_u - f_{xu'} - u' f_{uu'} - v' f_{vu'} - u'' f_{u'u'} - v'' f_{u'v'} = 0,$$

$$f_v - f_{xv'} - u' f_{uv'} - v' f_{vv'} - u'' f_{u'v'} - v'' f_{v'v'} = 0.$$

Since f does not involve u'' and v'', it follows that $f_{u'u'} = f_{u'v'} = f_{v'v'} = 0$. Hence, $f = a(x, u, v)u' + b(x, u, v)v' + c(x, u, v)$, and the above equations take the form:

$$a_v = b_u, \quad a_x = c_u, \quad b_x = c_v.$$

These equations furnish the integrability conditions for the system (cf. Section 2.1.4)

$$\frac{\partial g}{\partial u} = a(x, u, v), \quad \frac{\partial g}{\partial v} = b(x, u, v), \quad \frac{\partial g}{\partial x} = c(x, u, v).$$

Letting $g(x, u, v)$ be its solution, we get $f = g_u u' + g_v v' + g_x \equiv D_x(g)$.

Exercise 2. Let $f(x, y, y', \ldots, y^{(s)}) \in \mathcal{A}$. Prove that if $D_x(f) = 0$ identically in all variables $x, y, y', \ldots, y^{(s)}$, and $y^{(s+1)}$, then $f = C = \mathrm{const}$.

Solution. We have: $D_x(f) = f_x + y' f_y + y'' f_{y'} + \cdots + y^{(s)} f_{y^{(s-1)}} + y^{(s+1)} f_{y^{(s)}} = 0$. Since f does not involve $y^{(s+1)}$, the last term shows that $f_{y^{(s)}} = 0$. Likewise, the term next to the last shows that $f_{y^{(s-1)}} = 0$. Proceeding in this manner, one ultimately reaches the conditions $f_{y^{(s)}} = 0, \ldots, f_{y'} = 0, f_y = 0, f_x = 0$, meaning that $f = \mathrm{const}$.

In the case of several variables, the test for total derivatives is replaced by conditions for differential functions to be the divergence. For the sake of brevity, consider differential functions of the first order with two independent variables, x and y, and one differential variable u.

Exercise 3. Prove that a differential function $f = f(x, y, u, u_x, u_y)$ is the divergence if and only if its variational derivative vanishes [8.15]:

$$f = \mathrm{div}\, \boldsymbol{H} \quad \text{if and only if} \quad \frac{\delta f}{\delta u} = 0, \tag{8.66}$$

where $\boldsymbol{H} = (g(x, y, u), h(x, y, u))$, $\mathrm{div}\, \boldsymbol{H} = D_x(g) + D_y(h)$, and $\delta/\delta u$ is the Euler-Lagrange operator (8.60) with independent variables x, y and a differential variable u:

$$\frac{\delta}{\delta u} \equiv \frac{\partial}{\partial u} - D_x \frac{\partial}{\partial u_x} - D_y \frac{\partial}{\partial u_y} + D_x^2 \frac{\partial}{\partial u_{xx}} D_y D_x \frac{\partial}{\partial u_{xy}} + D_y^2 \frac{\partial}{\partial u_{yy}} - \cdots. \tag{8.67}$$

Solution. Let $f(x, u, v, u', v')$ be the divergence,

$$f = D_x(g) + D_y(h) \equiv g_x + u_x g_u + h_y + u_y h_u. \tag{8.68}$$

Then, invoking Lemma 1, one obtains:

$$\frac{\delta f}{\delta u} = \frac{\partial}{\partial u} D_x(g) - D_x(g_u) + \frac{\partial}{\partial u} D_y(h) - D_y(h_u)$$

$$= \left(\frac{\partial}{\partial u} D_x - D_x \frac{\partial}{\partial u}\right)(g) + \left(\frac{\partial}{\partial u} D_y - D_y \frac{\partial}{\partial u}\right)(h) = 0.$$

8.4. OPERATORS AND IDENTITIES IN $\mathcal{A}$

Conversely, let $\delta f/\delta u = f_u - D_x(f_{u_x}) - D_y(f_{u_y}) = 0$:

$$f_u - f_{xu_x} - u_x f_{uu_x} - u_{xx} f_{u_x u_x} - u_{xy} f_{u_x u_y} - f_{yu_y} - u_y f_{uu_y} - u_{xy} f_{u_x u_y} - u_{yy} f_{u_y u_y} = 0.$$

Since f does not contain the variables u_{xx}, u_{xy}, and u_{yy}, it follows from this equation that $f_{u_x u_x} = f_{u_x u_y} = f_{u_y u_y} = 0$. Hence,

$$f = a(x, y, u)u_x + b(x, y, u)u_y + c(x, y, u). \qquad (8.69)$$

The equation $\delta f/\delta u = 0$ gives one condition for the coefficients of (8.69):

$$c_u = a_x + b_y, \quad \text{or} \quad c = \int (a_x + b_y) du + p(x, y). \qquad (8.70)$$

Comparing (8.69) with (8.68) and taking into account the condition (8.70), we conclude that the function (8.69) takes the divergence form (8.68) with $g_u = a$ and $h_u = b$, i.e.

$$g = \int a(x, y, u) du + k(x, y), \quad h = \int b(x, y, u) du + l(x, y), \qquad (8.71)$$

where $k(x, y)$ and $l(x, y)$ should comply with the condition (8.70), i.e. $k_x + l_y = p$. For example, one can take $l = 0$ and $k(x, y) = \int p(x, y) dx$.

Example. The function $f = 2(u/x)u_x + 2(u/y)u_y + 2xy - u^2(x^{-2} + y^{-2})$ has the form (8.69) with $a = 2(u/x)$, $b = 2(u/y)$, $c = -u^2(x^{-2} + y^{-2}) + 2xy$. The condition (8.70) is satisfied. Taking the functions (8.71) with $l = 0$ and $k = \int 2xy dx = x^2 y$, one represents f as the divergence:

$$2u\left(\frac{u_x}{x} + \frac{u_y}{y}\right) - u^2\left(\frac{1}{x^2} + \frac{1}{y^2}\right) + 2xy = D_x\left(\frac{u^2}{x} + x^2 y\right) + D_y\left(\frac{u^2}{y}\right).$$

8.4.2 Lie-Bäcklund operators

Lie-Bäcklund operators provide a significant generalization of Lie point and contact symmetries [8.16], and have application, e.g. in accounting for the hidden symmetry associated with Kepler's first law (Section 5.5.2).

Definition 1. A *Lie-Bäcklund operator* is defined by the formal sum

$$X = \xi^i \frac{\partial}{\partial x^i} + \eta^\alpha \frac{\partial}{\partial u^\alpha} + \zeta_i^\alpha \frac{\partial}{\partial u_i^\alpha} + \zeta_{i_1 i_2}^\alpha \frac{\partial}{\partial u_{i_1 i_2}^\alpha} + \cdots, \qquad (8.72)$$

where $\xi^i([z]), \eta^\alpha([z]) \in \mathcal{A}$ are any differential functions (for the notation, see Section 8.2.1), and the other coefficients are determined by the prolongation formula (8.36) with $W^\alpha = \eta^\alpha - \xi^j u_j^\alpha \in \mathcal{A}$:

$$\zeta_i^\alpha = D_i(W^\alpha) + \xi^j u_{ij}^\alpha, \quad \zeta_{i_1 i_2}^\alpha = D_{i_1} D_{i_2}(W^\alpha) + \xi^j u_{j i_1 i_2}^\alpha, \ldots.$$

The operator (8.72) is in fact the infinite-order extension of

$$X = \xi^i \frac{\partial}{\partial x^i} + \eta^\alpha \frac{\partial}{\partial u^\alpha}, \quad \xi^i, \eta^\alpha \in \mathcal{A}. \qquad (8.73)$$

The abbreviated operator (8.73) is also referred to as a Lie–Bäcklund operator and is denoted by the same symbol X, provided that its extended action given by (8.72) is implied [8.5]. An alternative form of the operator (8.72) is

$$X = \xi^i D_i + W^\alpha \frac{\partial}{\partial u^\alpha} + D_i(W^\alpha)\frac{\partial}{\partial u_i^\alpha} + D_{i_1}D_{i_2}(W^\alpha)\frac{\partial}{\partial u_{i_1 i_2}^\alpha} + \cdots. \quad (8.74)$$

Let

$$X_\nu = \xi_\nu^i \frac{\partial}{\partial x^i} + \eta_\nu^\alpha \frac{\partial}{\partial u^\alpha}, \quad \xi_\nu^i, \eta_\nu^\alpha \in \mathcal{A}, \quad \nu = 1, 2,$$

be two Lie–Bäcklund operators (8.73). Their commutator (written in the form (8.73)) is defined by the usual formula (7.74):

$$[X_1, X_2] = X_1 X_2 - X_2 X_1 = (X_1(\xi_2^i) - X_2(\xi_1^i))\frac{\partial}{\partial x^i} + (X_1(\eta_2^\alpha) - X_2(\eta_1^\alpha))\frac{\partial}{\partial u^\alpha}.$$

Consequently, the set of all Lie–Bäcklund operators is an infinite-dimensional Lie algebra. It will be denoted by $L_\mathcal{B}$. Let us outline the basic properties of the algebra $L_\mathcal{B}$ used in modern group analysis.

Theorem. Any operator of the form $X_* = \xi^i D_i$ is a Lie–Bäcklund operator for arbitrary differential functions $\xi^i = \xi^i([z])$, $i = 1, \ldots, n$, i.e.

$$X_* = \xi^i D_i \in L_\mathcal{B} \quad \text{for any} \quad \xi^i \in \mathcal{A}. \quad (8.75)$$

Furthermore, the set L_* of all operators X_* (8.75) is an ideal of $L_\mathcal{B}$, i.e.

$$[X, X_*] \in L_* \quad \text{for any} \quad X \in L_\mathcal{B}. \quad (8.76)$$

Proof. For the operator $X_* = \xi^i D_i$, we have $\eta^\alpha = \xi^i u_i^\alpha$ and therefore $W^\alpha = 0$. Thus, X_* is a particular case of the Lie–Bäcklund operator (8.74). Hence, (8.75). Property (8.76) follows from the identity (cf. (8.46))

$$[X, X_*] = \Big(X(\xi_*^i) - X_*(\xi^i)\Big) D_i.$$

In accordance with property (8.76), two operators $X_1, X_2 \in L_\mathcal{B}$ are said to be *equivalent* (cf. Section 7.3.4) if $X_1 - X_2 \in L_*$. The equivalence relation is written $X_1 \sim X_2$. In particular, every $X \in L_\mathcal{B}$ is equivalent to an operator (8.73) with $\xi^i = 0, i = 1, \ldots, n$. Namely, according to (8.74),

$$X \sim Y = X - \xi^i D_i = W^\alpha \frac{\partial}{\partial u^\alpha}, \quad W^\alpha = (\eta^\alpha - \xi^i u_i^\alpha). \quad (8.77)$$

Definition 2. Lie–Bäcklund operators (8.73) of the form

$$X = \eta^\alpha \frac{\partial}{\partial u^\alpha}, \quad \eta^\alpha \in \mathcal{A},$$

are called *canonical operators*. Thus, any $X \in L_\mathcal{B}$ is equivalent to a canonical Lie–Bäcklund operator Y (8.77) known as a *canonical representation* of X [8.17].

Example. Consider the case of one independent variable x and one differential variable u. Let X be the generator of the translation group, $X = \partial/\partial x$. Here $W = -u_x$, and (8.77) gives the canonical representation $Y = u_x \partial/\partial u$ of X.

8.4. OPERATORS AND IDENTITIES IN $\mathcal{A}$

8.4.3 Operators N^i associated with Lie-Bäcklund operators

Definition. Given a Lie-Bäcklund operator X (8.74), let us define n operators N^i $(i = 1, \ldots, n)$ by the formal sums [5.31]:

$$\mathsf{N}^i = \xi^i + W^\alpha \frac{\delta}{\delta u_i^\alpha} + \sum_{s=1}^{\infty} D_{i_1} \cdots D_{i_s}(W^\alpha) \frac{\delta}{\delta u_{ii_1 \cdots i_s}^\alpha}, \qquad (8.78)$$

where the Euler-Lagrange operators with respect to derivatives of u^α are obtained from (8.59) by replacing u^α by the corresponding derivatives, e.g.

$$\frac{\delta}{\delta u_i^\alpha} = \frac{\partial}{\partial u_i^\alpha} + \sum_{s=1}^{\infty}(-1)^s D_{j_1} \cdots D_{j_s} \frac{\partial}{\partial u_{ij_1 \cdots j_s}^\alpha}. \qquad (8.79)$$

Exercise. For one independent variable x, operators (8.74) and (8.79) are written:

$$X = \xi D_x + W^\alpha \frac{\partial}{\partial u^\alpha} + D_x(W^\alpha) \frac{\partial}{\partial u_1^\alpha} + D_x^2(W^\alpha) \frac{\partial}{\partial u_2^\alpha} + \cdots,$$

$$\mathsf{N} = \xi + W^\alpha \frac{\delta}{\delta u_1^\alpha} + D_x(W^\alpha) \frac{\delta}{\delta u_2^\alpha} + D_x^2(W^\alpha) \frac{\delta}{\delta u_3^\alpha} + \cdots,$$

where u_s^α denotes the sth derivative of u^α, and $u_0^\alpha = u^\alpha$. Prove that $\mathsf{N} = X/D_x$, i.e.

$$X = \mathsf{N} D_x. \qquad (8.80)$$

In other words, prove that any Lie-Bäcklund operator with one independent variable is divisible by the total derivation D_x.

Solution. Equation (8.80) follows from the identities

$$\frac{\delta}{\delta u_{s+1}^\alpha} D_x = \frac{\partial}{\partial u_s^\alpha}, \quad s = 0, 1, 2, \ldots. \qquad (8.81)$$

Indeed:

$$\mathsf{N} D_x = \xi D_x + W^\alpha \frac{\delta}{\delta u_1^\alpha} D_x + D_x(W^\alpha) \frac{\delta}{\delta u_2^\alpha} D_x + D_x^2(W^\alpha) \frac{\delta}{\delta u_3^\alpha} D_x + \cdots$$

$$= \xi D_x + W^\alpha \frac{\partial}{\partial u^\alpha} + D_x(W^\alpha) \frac{\partial}{\partial u_1^\alpha} + D_x^2(W^\alpha) \frac{\partial}{\partial u_2^\alpha} D_x + \cdots = X.$$

8.4.4 The fundamental identity

Theorem. The Euler-Lagrange (8.59), Lie-Bäcklund (8.72) and the associated operators (8.78) are connected by the identity [5.31]

$$X + D_i(\xi^i) = W^\alpha \frac{\delta}{\delta u^\alpha} + D_i \mathsf{N}^i. \qquad (8.82)$$

Remark. Here, $D_i(\xi^i)$ is a differential function since it is a finite sum of differential functions obtained by total derivations of the differential functions $\xi^i([z])$. It is the divergence, $D_i(\xi^i) = \mathrm{div}\,\boldsymbol{\xi}$, of the vector $\boldsymbol{\xi} = (\xi^1, \ldots, \xi^n)$. The expression $D_i \mathsf{N}^i$ is an operator obtained as the sum of the products of operators D_i and N^i.

8.5 The frame of differential equations

A differential equation is made up of two distinct ingredients. The first component of a differential equation is its *frame*, and the second one is a *class of solutions*. The frame is an object of study for differential algebra, whereas a class of solutions is provided by functional analysis [8.18].

Consider, e.g. ordinary differential equations of the first order (2.1),

$$y' = f(x, y), \quad \text{where} \quad y' = dy/dx. \tag{8.83}$$

The frame of equation (8.83) is the surface in the space of three independent variables, x, y, and p, given by the equation

$$p = f(x, y). \tag{8.84}$$

It is obtained from the differential equation (8.83) merely by setting $p = y'$. In other words, the expression $F(x, y, y') = y' - f(x, y)$ is treated as a differential function of the first order, $\text{ord}(F) = 1$. For example, the frame of a Riccati equation, considered in Example 2 of Section 5.2.2, is a hyperbolic paraboloid $p + y^2 - 2/x^2 = 0$ (see Fig. 8.1). *Isoclines* to equation (8.83) (that is, the curves $f(x, y) = c$ with different values of a constant c) are projections of cuts $p = c$ of the frame (8.84) to the (x, y) plane (see Fig. 8.2).

A *class of solutions* is defined in accordance with certain "natural" mathematical assumptions or from the physical significance of a differential equation. For example, classical solutions (Definition 3.1.1) are continuously differentiable functions $y = \phi(x)$ such that $\phi'(x) = f(x, \phi(x))$ identically in x. In terms of the frame, it means that the curve $y = \phi(x)$, $p = \phi'(x)$ in the (x, y, p) space belongs to the frame (8.84). Going over to discontinuous or generalized solutions (keeping the same frame) drastically changes the situation [3.6].

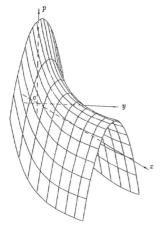

Figure 8.1 The frame of the Riccati equation $y' + y^2 - 2/x^2 = 0$.

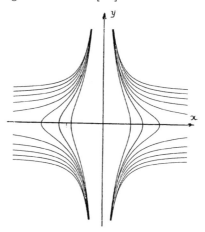

Figure 8.2 Isoclines to $y' + y^2 - 2/x^2 = 0$.

8.5. THE FRAME OF DIFFERENTIAL EQUATIONS

In integrating ordinary differential equations, a decisive step is that of simplifying the frame by a change of variables. The Lie group analysis furnishes a method for determining a suitable change of variables. Provided that an infinitesimal symmetry is known, we merely introduce canonical variables. This simplifies the equation by converting its frame into a cylinder (see Fig. 8.3), i.e. the explicit dependence of one of the variables x or y has been eliminated.

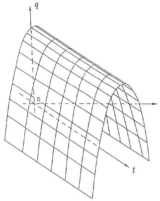

Figure 8.3 Equation $y' + y^2 - 2/x^2 = 0$ admits $X = x\partial/\partial x - y\partial/\partial y$. Canonical variables for X are $t = \ln x$, $u = xy$. In these variables the above Riccati equation is written $u' + u^2 - u - 2 = 0$. Its frame, obtained by setting $u' = q$, is given here. It is a parabolic cylinder, $q + u^2 - u - 2 = 0$, protracted along t. Hence, the hyperbolic paraboloid of Fig. 8.1 is straightened out by passing to canonical variables.

A similar approach is fruitful for all (ordinary and partial) differential equations with known symmetries. For this purpose, we use the following definition.

Definition. Given any differential function $F \in \mathcal{A}$, $\text{ord}(F) = s$, the equation

$$F(x, u, u_{(1)}, \ldots, u_{(s)}) = 0 \tag{8.85}$$

defines a manifold in the space of variables $x, u, u_{(1)}, \ldots, u_{(s)}$. This manifold is called the frame of the sth-order partial differential equation

$$F\left(x, u, \frac{\partial u}{\partial x}, \ldots, \frac{\partial^s u}{\partial x^s}\right) = 0. \tag{8.86}$$

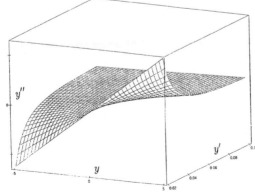

Figure 8.4 The frame of the second-order equation

$$y'' = \frac{y'}{y^2} - \frac{1}{xy} \tag{8.87}$$

is a three-dimensional manifold in the four-dimensional space of variables x, y, y', y''. The figure shows a two-dimensional cut of the frame obtained by setting $x = 1$ in (8.87).

Problems

8.1*. Prove (8.5), i.e. $D_j D_i(f) = D_i D_j(f)$ for any $f \in \mathcal{A}$.

8.2. Calculate the partial and total derivatives, $\partial f/\partial x$ and $D_x(f)$, where D_x is given by (8.6), of the following differential functions:

(i) $f = \ln|xy|$, (ii) $f = xyy'''$, (iii) $f = \sin(x^2 + y')$, (iv) $f = x^2 e^{y''}$, (v) $f = x^2 y^3 + \sin y'$.

8.3. Continue Example 2 of Section 8.2.3 and find the derivatives $D_x^4(f)$ and $D_x^5(f)$.

8.4. Find $D_x^5(f)$ by using Faà de Bruno's formula (8.7).

8.5. Find the extension to the first derivative y' of the rotation group in the (x, y) plane by two methods: first, use formula (8.16); second, solve the Lie equations (7.17) for the extended generator $X_{(1)}$ of the rotation group given in Example 8.3.1.

8.6. Find the extension (8.18) of the following operators in the (x, y) plane:

$$X_1 = \frac{\partial}{\partial x}, \quad X_2 = \frac{\partial}{\partial y}, \quad X_3 = \frac{\partial}{\partial x} + \frac{\partial}{\partial y}, \quad X_4 = x\frac{\partial}{\partial x} + y\frac{\partial}{\partial y},$$

$$X_5 = 2x\frac{\partial}{\partial x} + y\frac{\partial}{\partial y}, \quad X_6 = (x-y)\frac{\partial}{\partial x} + (x+y)\frac{\partial}{\partial y}, \quad X_7 = x^2\frac{\partial}{\partial x} + y^2\frac{\partial}{\partial y},$$

$$X_8 = (1+x^2)\frac{\partial}{\partial x} + xy\frac{\partial}{\partial y}, \quad X_9 = y\frac{\partial}{\partial x} + x\frac{\partial}{\partial y}, \quad X_{10} = (x^2 - y^2)\frac{\partial}{\partial x} + xy\frac{\partial}{\partial y}.$$

8.7. Find the extension (8.18) of the approximate generators X_1, X_2 from Problem 7.27.

8.8*. Prove that if (8.24)-(8.25) is a one-parameter group, then the extended transformations, given by (8.24)-(8.25) and (8.29), also form a one-parameter group.

8.9. Find the second extension $X_{(2)}$ of the infinitesimal rotation, $X = y\partial/\partial x - x\partial/\partial y$.

8.10. Deduce the prolongation formula (8.30) from equation (8.29).

8.11*. Prove the properties (8.39) and (8.40) of extended generators.

8.12. Find a third-order differential invariant of the Galilean transformation by solving equation (8.42) for $F = F(x, y, y', y'', y''')$.

8.13. Deduce differential invariants of orders 2 and 3 for the group of Galilean transformations, considered in Example 8.3.5, by means of the invariant differentiation (8.44).

8.14. Prove Lemma 1 of Section 8.4.1.

8.15. Verify Theorem 8.4.1 for $m = 1$ and for differential functions of the first order, i.e. show that $f(x, y, y') = D_x(g(x, y))$ if and only if

$$\frac{\delta f}{\delta y} \equiv \frac{\partial f}{\partial y} - D_x\left(\frac{\partial f}{\partial y'}\right) = 0.$$

8.16. Find the canonical representation Y (8.77) of the infinitesimal generators of: (i) dilation and (ii) Galilean transformation (u is a single differential variable),

(i) $X = x\frac{\partial}{\partial x} + ky\frac{\partial}{\partial y} + lu\frac{\partial}{\partial u}$, $(k, l = \text{const.})$; (ii) $X = t\frac{\partial}{\partial x} + \frac{\partial}{\partial u}$.

8.17*. Prove the identities (8.81).

8.18. Draw the frame of the first-order differential equation $xy' = 0$ given in [3.6].

Chapter 9

Symmetry of differential equations

The purpose of this chapter is to guide the reader through the totality of infinitesimal methods, needed for determining symmetry groups, with a minimum number of theoretical constructions, and to assist the reader in developing skills in applying these methods [9.1]. The reader whose intention is to find and use symmetry groups in his own particular field of interest, may find that the simplest way is to work through the illustrative examples.

9.1 Notation and assumptions

Throughout the book we consider only classical solutions of differential equations (Definition 3.1.1). The class of solutions being fixed, a system of kth-order differential equations can be identified with its frame:

$$F_\sigma(x, u, u_{(1)}, \ldots, u_{(k)}) = 0, \quad \sigma = 1, \ldots, s, \tag{9.1}$$

where $F_\sigma \in \mathcal{A}$, and the order k refers to the highest derivative appearing in (9.1).

Definition. A system of differential equations (9.1) is said to be *locally solvable* if, for any generic point $z^0 = (x^0, u^0, u^0_{(1)}, \ldots, u^0_{(k)})$ lying on its frame,

$$F_\sigma(x^0, u^0, u^0_{(1)}, \ldots, u^0_{(k)}) = 0, \quad \sigma = 1, \ldots, s,$$

there exists a solution $u = \phi(x)$ of the system passing through z^0, i.e. such that $\phi(x)$ and its derivatives at x^0 assume the respective values $u^0, u^0_{(1)}, \ldots, u^0_{(k)}$.

The differential equations commonly used as mathematical models are locally solvable. In what follows, we consider locally solvable equations with n independent variables $x = (x^1, \ldots, x^n)$ and m differential (dependent) variables $u = (u^1, \ldots, u^m)$. Note that in most applications the number of equations (9.1) coincides with the number of differential variables, $s = m$. If $s = m = 1$, we have a single equation, also termed a *scalar equation*.

9.2 Determination of infinitesimal symmetries

The concept of a symmetry group of a differential equation can be specified either in terms of its solutions or its frame. Both definitions are useful and are given below.

The first definition, dealing with the totality of solutions of a given differential equation, does not employ the theory of extended transformations. This natural approach depends, however, upon knowledge of the general solution of the equation in question. Since solutions are not known in advance, the first definition does not provide a practical method for finding symmetries.

The second definition treats a symmetry group of a differential equation as a group of transformations whose extension to derivatives leaves invariant the frame of the differential equation in question. This differential algebraic definition does not assume knowledge of solutions. It is central to the infinitesimal method used to test for symmetry via *determining equations*.

9.2.1 Two definitions of a symmetry group

By way of introduction, consider point transformations (8.15) of the plane,

$$\bar{x} = f(x, y, a), \qquad \bar{y} = \varphi(x, y, a). \tag{9.2}$$

Invariant manifolds (Section 7.2.1) of a one-parameter group G of transformations (9.2) may have the dimension one (curves) or zero (points). Here, we drop the latter case and consider curves in the (x, y) plane. It is clear from Definition 7.2.1, that curves invariant under G coincide with path curves of the group G.

Now we take a further step and introduce the notion of *invariant families of curves*. Consider a family of curves (1.11) depending on n arbitrary parameters:

$$\Phi(x, y, C_1, \ldots, C_n) = 0. \tag{9.3}$$

The family of curves (9.2) is said to be invariant under the group G if the curves of the family are permuted among themselves by every transformation (9.2) of G. Now recall that a family of curves (9.3) is described by a differential equation (1.13) and provides its general solution. Hence, we have the natural definition.

Definition 1. A system of differential equations (9.1) is said to be *invariant* under a group G of point transformations (8.24)-(8.25) if the solutions of the system are merely permuted among themselves (or are individually unaltered) by every transformation of the group G. The group G is also termed a *symmetry group* for a system (9.1), or a group *admitted* by the system. We shall speak of the generator X of a symmetry group G as an *infinitesimal symmetry*, or an *admitted operator* for a system of differential equations (9.1).

Example 1. The family of straight lines $y = C_1 x + C_2$ is invariant under an arbitrary rotation about the origin, $\bar{x} = x\cos a + y\sin a$, $\bar{y} = y\cos a - x\sin a$.

9.2. DETERMINATION OF INFINITESIMAL SYMMETRIES

Indeed, substituting $x = \bar{x}\cos a - \bar{y}\sin a$, $y = \bar{y}\cos a + \bar{x}\sin a$ into the equation of straight lines, one obtains $\bar{y}\cos a + \bar{x}\sin a = C_1(\bar{x}\cos a - \bar{y}\sin a) + C_2$. Hence, any straight line $y = C_1 x + C_2$ is mapped into a straight line $\bar{y} = \bar{C}_1 \bar{x} + \bar{C}_2$ with

$$\bar{C}_1 = \frac{C_1 \cos a - \sin a}{C_1 \sin a + \cos a}, \quad \bar{C}_2 = \frac{C_2}{C_1 \sin a + \cos a}.$$

Thus, according to Definition 1, the differential equation of straight lines (1.16), $y'' = 0$, admits the rotation group. The latter is a subgroup of a more general symmetry group given in the next example.

Example 2. It was shown in Exercise 6.3.2 that the family of straight lines is invariant under any projective transformation (1.32). Hence, the second-order differential equation $y'' = 0$ admits the eight-parameter group of projective transformations (1.32) of the plane.

Example 3. Neither the rotation $\bar{x} = x\cos a + y\sin a$, $\bar{y} = y\cos a - x\sin a$ nor the projective transformation (6.33),

$$\bar{x} = \frac{x}{1 - ax}, \quad \bar{y} = \frac{y}{1 - ax},$$

leaves invariant the family of parabolas $y = C_1 x^2 + C_2 x + C_3$. But the latter is invariant under the group (cf. Problem 9.3)

$$\bar{x} = \frac{x}{1 - ax}, \quad \bar{y} = \frac{y}{(1 - ax)^2}. \tag{9.4}$$

Indeed, the rotation maps a parabola into an algebraic curve

$$\bar{y}\cos a + \bar{x}\sin a = C_1(\bar{x}\cos a - \bar{y}\sin a)^2 + C_2(\bar{x}\cos a - \bar{y}\sin a) + C_3,$$

which obviously is not a parabola. Likewise, one can easily check that a curve, obtained from a parabola by the projective transformation, is not a parabola. Let us turn now to transformation (9.4). It is convenient to use the inverse transformation, $x = \bar{x}/(1 + a\bar{x})$, $y = \bar{y}/(1 + a\bar{x})^2$. Then it is easily seen that a parabola $y = C_1 x^2 + C_2 x + C_3$ is mapped into a parabola:

$$\frac{\bar{y}}{(1 + a\bar{x})^2} = C_1 \frac{\bar{x}^2}{(1 + a\bar{x})^2} + C_2 \frac{\bar{x}}{1 + a\bar{x}} + C_3,$$

or

$$\bar{y} = \left(C_1 + aC_2 + a^2 C_3\right)\bar{x}^2 + (C_2 + 2aC_3)\bar{x} + C_3 \equiv +\bar{C}_1 \bar{x}^2 + \bar{C}_2 \bar{x} + \bar{C}_3.$$

Hence, the differential equation (1.17) $y''' = 0$ admits the group (9.4).

Definition 2. A system of kth-order differential equations is said to be invariant under a group G if the frame of the system is an invariant manifold for the extension of the group G to the kth-order derivatives.

Theorem [9.2]. For locally solvable systems, Definitions 1 and 2 are equivalent. In other words, both definitions provide the same symmetry group for a given system of differential equations.

9.2.2 Determining equations

Definition 2 of the previous section and Theorem 7.2.1 provide the following infinitesimal criterion for symmetry groups of differential equations.

Theorem 1. The system of differential equations (9.1) is invariant under the group with an infinitesimal generator X if and only if

$$XF_\sigma\Big|_{(9.1)} = 0, \quad \sigma = 1, \ldots, s, \qquad (9.5)$$

where X is extended to all derivatives involved in $F_\sigma(x, u, u_{(1)}, \ldots, u_{(k)})$ (cf. [8.5]), and the symbol $|_{(9.1)}$ means evaluated on the frame (9.1).

Definition. Equations (9.5) determine all infinitesimal symmetries of a system (9.1) and therefore they are known as *determining equations*.

The determining equations can also be written in the form (7.56),

$$XF_\sigma = \phi_\sigma^\nu F_\nu, \quad \sigma = 1, \ldots, s, \qquad (9.6)$$

where the indeterminate coefficients $\phi_k^l(x, u, u_{(1)}, \ldots) \in \mathcal{A}, \updownarrow, \| = \infty, \ldots, \int$, are assumed to be differential functions that are bounded on the frame (9.1).

Determining equations (9.5) seem, at first glance, to be more complicated than the differential equations (9.1) in question. However, this is an apparent complexity. Indeed, the left-hand sides of the determining equations involve the derivatives $u_{(1)}, u_{(2)}, \ldots$, along with the variables x, u and the functions ξ^i and η^α of x, u. Since equations (9.5) should be satisfied identically with respect to all the variables involved, the determining equations split into a system of several equations. As a rule, this is an overdetermined system. Therefore, in many practical applications, the determining equations can be solved analytically. The solution of the determining equation can be carried out either "by hand" or, in simple cases, by using modern symbolic manipulation programs [9.3].

As mentioned in the preamble to Part II, the set of all transformations leaving a given object unaltered is a group. It follows that, upon solving the determining equations, one obtains a Lie algebra. This property of the determining equations is central to the whole subject. Let us prove it.

Theorem 2. The solutions of any determining equations form a Lie algebra.

Proof. Let us consider the generators (8.26) of point transformation groups and apply Theorem 8.3.4 to the extended generators.

The determining equations (9.5) are linear homogeneous partial differential equations for unknown functions $\xi^i(x, u)$ and $\eta^i(x, u)$. Hence, the set of all solutions to (9.5) is a vector space. It remains to verify that this vector space has the property of closure with respect to the commutator, i.e. that if X_1 and X_2 satisfy equations (9.5) then the same is true for the commutator $X = [X_1, X_2]$. This property can be easily deduced from (9.6). Indeed, let

$$X_1 F_\sigma = \phi_\sigma^\nu F_\nu, \quad X_2 F_\sigma = \varphi_\sigma^\nu F_\nu, \quad \text{where} \quad \phi_\sigma^\nu, \varphi_\sigma^\nu \in \mathcal{A}.$$

9.2. DETERMINATION OF INFINITESIMAL SYMMETRIES

Then, invoking the definition of the commutator, $X = [X_1, X_2] \equiv X_1 X_2 - X_2 X_1$, and using equation (8.40), one obtains

$$XF_\sigma = X_1\left(\varphi^\nu_\sigma F_\nu\right) - X_2\left(\phi^\nu_\sigma F_\nu\right),$$

and eventually arrives at equations (9.5) for the commutator:

$$XF_\sigma = \psi^\nu_\sigma F_\nu$$

with coefficients $\psi^\nu_\sigma \in \mathcal{A}$ defined by $\quad \psi^\nu_\sigma = X_1(\varphi^\nu_\sigma) - X_2(\phi^\nu_\sigma) + \varphi^\mu_\sigma \phi^\nu_\mu - \phi^\mu_\sigma \varphi^\nu_\mu.$

9.2.3 Intrinsic symmetries of mathematical models

Mathematical models usually inherit fundamental symmetries which lie at the core of natural phenomena. These intrinsic symmetry groups comprise *translations, rotations, scaling transformations, Galilean and Lorentz boosts,* and *conformal transformations* (see Section 6.3). Consequently, the main concern of the group analysis, e.g. in problems of mathematical physics, is not so much with calculating groups admitted by differential equations. Rather a physicist is interested in possible ways of using "natural symmetry groups" of the physical laws for investigating mathematical models.

Therefore, it is reasonable to begin the group analysis of a mathematical model with testing the differential equations in question for simple symmetries. The simplest symmetry is provided by translations. Indeed, translational invariance of a differential equation, e.g. under translations of an independent variable x, means that the equation does not involve x (see Example 1 in Section 5.2.2 and Problem 9.1). Rotational symmetry is usually evident from geometry (see, e.g. Example 4 in Section 7.2.3).

The scaling invariance of a differential equation may be obtained by simple algebraic calculations in cases when the frame of the equation is given by a rational fraction. The approach is illustrated by the following examples.

Example 1. Consider again the Riccati equation (see Section 8.5, Fig. 8.1)

$$y' + y^2 - \frac{2}{x^2} = 0. \tag{9.7}$$

Its left-hand side is a rational fraction in the variables $x, y, y' = dy/dx$. Setting

$$\bar{x} = kx, \quad \bar{y} = ly, \tag{9.8}$$

we transform the left-hand side of equation (9.7)

$$\bar{y}' + \bar{y}^2 - \frac{2}{\bar{x}^2} = \frac{l}{k}y' + l^2 y^2 - \frac{1}{k^2}\frac{2}{x^2},$$

and obtain from the invariance condition, $\bar{y}' + \bar{y}^2 - 2/\bar{x}^2 = \lambda(y' + y^2 - 2/x^2)$, that $l/k = l^2 = 1/k^2$. The latter equations yield $l = 1/k$, where k is an arbitrary positive parameter. Setting $k = e^a$, we get the group $\bar{x} = xe^a$, $\bar{y} = ye^{-a}$. Hence, the Riccati equation (9.7) admits the operator (cf. Example 2 in Section 5.2.2)

$$X = x\frac{\partial}{\partial x} - y\frac{\partial}{\partial y}. \tag{9.9}$$

Example 2. Let us apply a similar approach to the equation

$$y'' = \frac{y'}{y^2} - \frac{1}{xy}. \tag{9.10}$$

After dilation (9.8), we have:

$$\bar{y}'' - \frac{\bar{y}'}{\bar{y}^2} + \frac{1}{\bar{x}\bar{y}} = \frac{l}{k^2}y'' - \frac{1}{kl}\frac{y'}{y^2} + \frac{1}{kl}\frac{1}{xy}.$$

The invariance condition yields $l/k^2 = 1/kl$, whence $k = l^2$. Setting $l = e^a$, we have $\bar{x} = xe^{2a}$, $\bar{y} = ye^a$. Hence, the second-order equation (9.10) admits

$$X = 2x\frac{\partial}{\partial x} + y\frac{\partial}{\partial y}. \tag{9.11}$$

Example 3. Consider a partial differential equation from gas dynamics [7.6]:

$$u_x u_{xx} + u_{yy} = 0. \tag{9.12}$$

In new variables $\bar{x} = kx$, $\bar{y} = ly$, $\bar{u} = mu$, it is written

$$\bar{u}_{\bar{x}}\bar{u}_{\bar{x}\bar{x}} + \bar{u}_{\bar{y}\bar{y}} = \frac{m^2}{k^3}u_x u_{xx} + \frac{m}{l^2}u_{yy}.$$

The invariance condition, $\bar{u}_{\bar{x}}\bar{u}_{\bar{x}\bar{x}} + \bar{u}_{\bar{y}\bar{y}} = \lambda(u_x u_{xx} + u_{yy})$, yields $m^2/k^3 = m/l^2$. Hence $m = k^3/l^2$, where k and l are two arbitrary positive parameters. Letting $k = e^a, l = 0$, and then $k = 0, l = e^b$, we obtain two independent scaling symmetries. Consequently, we have two admitted operators (cf. (7.77)):

$$X_1 = x\frac{\partial}{\partial x} + 3u\frac{\partial}{\partial u}, \quad X_2 = y\frac{\partial}{\partial x} - 2u\frac{\partial}{\partial u}.$$

Thus, one can find simple symmetries, such as translations, rotations and dilations, without using general determining equations. However, not all symmetries of the equations of mathematical physics are transparent, e.g. the projective symmetry of the monatomic gas and shallow water (Section 5.4.4), the hidden symmetry of Newton's gravitation law (Section 5.5.2), etc. Moreover, many various differential equations appearing as approximations to universal natural laws (e.g. in fluid mechanics), in constructing invariant solutions or integrating ordinary differential equations, may not inherit the natural symmetries of the basic model, or possess additional groups. Consequently, one often encounters equations admitting sophisticated transformations that do not have a simple geometrical significance and therefore cannot be easily seen (see, e.g. Problem 9.5). In these cases, one should solve the determining equations (9.5).

9.3 Samples for solution of determining equations

This section provides a concise practical guide for the reader interested in using the determining equations in his own problems. We begin with typical situations encountered in determining point symmetries for a single ordinary differential equation. Then we consider point symmetries for a single partial differential equation and a system of ordinary differential equations. Calculations for systems of partial differential equations are similar and can be carried out by the reader while solving problems to this chapter. Finally, a simple example on contact symmetries provides an alternative approach to the problem via Lie's characteristic function W.

9.3.1 Ordinary differential equations: Theorems on the maximum number of symmetries and examples

A general statement about point symmetries of ordinary differential equations is the following result due to Lie [9.4].

Theorem 1. Consider the ordinary differential equations

$$y^{(n)} = f(x, y, y', \ldots, y^{(n-1)}).$$

If $n = 1$, the symmetry group contains an arbitrary function, i.e. first-order equations admit an infinite continuous group. If the order n of the equation is equal to or greater than two, then the continuous symmetry group contains at most arbitrary constants, i.e. it is a finite continuous group (see Definition 1 in Section 6.3.1). Furthermore, an ordinary differential equation of order $n \geq 3$ admits at most $(n+4)$-dimensional Lie algebra, this maximum being reached for $y^{(n)} = 0$.

Let us restrict consideration to second-order equations written in the form

$$y'' = f(x, y, y'). \tag{9.13}$$

We look for an admissible infinitesimal generator

$$X = \xi(x, y) \frac{\partial}{\partial x} + \eta(x, y) \frac{\partial}{\partial y} \tag{9.14}$$

with coefficients ξ and η to be found from the determining equation (9.5):

$$X\left(y'' - f(x, y, y')\right)\Big|_{(9.13)} \equiv \left(\zeta_2 - \zeta_1 f_{y'} - \xi f_x - \eta f_y\right)\Big|_{y''=f} = 0. \tag{9.15}$$

After substituting ζ_1 and ζ_2 from (8.20), equation (9.15) assumes the form

$$\eta_{xx} + (2\eta_{xy} - \xi_{xx})y' + (\eta_{yy} - 2\xi_{xy})y'^2 - y'^3 \xi_{yy} - \xi f_x - \eta f_y \\ + (\eta_y - 2\xi_x - 3y'\xi_y)f - [\eta_x + (\eta_y - \xi_x)y' - y'^2 \xi_y]f_{y'} = 0. \tag{9.16}$$

Here $f(x, y, y')$ is a known function when one deals with a given differential equation (9.13). Since equation (9.16) involves all three variables x, y and y', but y' does not occur in ξ and η, the determining equation (9.16) decomposes into several equations, thus becoming an overdetermined system of differential equations for ξ and η. After solving this system, one finds all generators of point transformations admitted by equation (9.13).

The following theoretical result due to Lie [9.5] supplements Theorem 1 and aids us considerably in the integration of second-order equations.

Theorem 2. A second-order ordinary differential equation (9.13) admits at most an eight-dimensional Lie algebra. This maximum is reached by $y'' = 0$.

Proof. Recall that, according to assumptions made in Section 9.1, $f \in \mathcal{A}$, i.e. it is a differential function and hence expandable in a Taylor series with respect to all its arguments. Therefore one can replace f and its partial derivatives $f_x, f_y, f_{y'}$, appearing in (9.16), by their expansion in a series with respect to y'. Then the left-hand side of (9.16) becomes a power series in y' with coefficients depending on x and y. Since y' is an independent variable, this series must vanish identically, i.e. the coefficient of each power of y' must be zero. Equating to zero the terms free of y' and the coefficients of y', y'^2, y'^3, we get

$$\eta_{xx} = h_1, \quad 2\eta_{xy} - \xi_{xx} = h_2, \quad \eta_{yy} - 2\xi_{xy} = h_3, \quad \xi_{yy} = h_4, \qquad (9.17)$$

where h_i are linear homogeneous functions of

$$\xi, \quad \eta, \quad \xi_x, \quad \xi_y, \quad \eta_x, \quad \eta_y, \qquad (9.18)$$

whose coefficients are known functions of x, y. Assigning the values of six functions (9.18) and ξ_{xx}, ξ_{xy} at any point (x_0, y_0), one can evaluate at this point all second and higher derivatives of ξ and η using the equations (9.17) together with their differential consequences. Proceeding in this manner, one ultimately constructs power series representations of functions $\xi(x, y)$ and $\eta(x, y)$ near (x_0, y_0). The functions ξ and η depend upon eight arbitrary constants, namely, arbitrary numerical values of (9.18) and ξ_{xx}, ξ_{xy} at (x_0, y_0). Thus far we have satisfied only equations (9.17) and ignored the further relations to be obtained by annulling the coefficients of y'^4, y'^5, etc. Consequently, the number of arbitrary parameters in the solution of the determining equation (9.16) may be fewer than eight. From the start, we knew that $y'' = 0$ admits the eight-parameter group of projective transformations (Example 2 of Section 9.2.1). This completes the proof.

Remark. Lie's group classification [5.17] shows that a Lie algebra of the maximal dimension eight is admitted if and only if (9.13) either is linear or can be linearized by a change of variables x, y, and hence, can be transformed to $y'' = 0$.

The following examples represent typical situations encountered in the search for infinitesimal symmetries (*symmetry algebras*) via the determining equations.

Example 1. Consider again the equation $y'' = 0$ and deduce its symmetry Lie algebra without appealing to geometry (cf. Example 2 in Section 9.2.1). Since here $f = 0$, the determining equation (9.16),

$$\eta_{xx} + (2\eta_{xy} - \xi_{xx})y' + (\eta_{yy} - 2\xi_{xy})y'^2 - y'^3 \xi_{yy} = 0,$$

9.3. SOLUTION OF DETERMINING EQUATIONS

is equivalent to the system (9.17) with $h_i = 0$:

$$\eta_{xx} = 0, \quad 2\eta_{xy} - \xi_{xx} = 0, \quad \eta_{yy} - 2\xi_{xy} = 0, \quad \xi_{yy} = 0. \tag{9.19}$$

The first and last equations yield $\xi = \phi_1(x)y + \phi_2(x)$ and $\eta = \psi_1(y)x + \psi_2(y)$. Differentiating the second equation (9.19) with respect to x and the third one with respect to y, we obtain $\xi_{xxx} = 0$, $\eta_{yyy} = 0$. It follows that $\phi_i(x)$ and $\psi_i(y)$ are quadratic functions of x and y, respectively. Hence,

$$\xi = C_1 + C_2 x + C_3 y + C_4 x^2 + C_5 xy + Ax^2 y, \quad \eta = C_6 + C_7 y + C_8 x + C_9 y^2 + C_{10} xy + Bx^2 y.$$

Substitution into the second and third equations (9.19) yields $Ay + C_4 = 2By + C_{10}$, $2Ax + C_5 = Bx + C_9$, whence $A = B = 0$, $C_{10} = C_4$, $C_9 = C_5$. As a result, we get the general solution with eight arbitrary constant coefficients C_i [9.6]:

$$\xi = C_1 + C_2 x + C_3 y + C_4 x^2 + C_5 xy, \quad \eta = C_6 + C_7 y + C_8 x + C_4 xy + C_5 y^2. \tag{9.20}$$

The operator obtained by substituting (9.20) into (9.14) is a linear combination of the following independent infinitesimal symmetries of the equation y'':

$$X_1 = \frac{\partial}{\partial x}, \quad X_2 = \frac{\partial}{\partial y}, \quad X_3 = x\frac{\partial}{\partial x}, \quad X_4 = y\frac{\partial}{\partial x}, \quad X_5 = x\frac{\partial}{\partial y},$$

$$X_6 = y\frac{\partial}{\partial y}, \quad X_7 = x^2\frac{\partial}{\partial x} + xy\frac{\partial}{\partial y}, \quad X_8 = xy\frac{\partial}{\partial x} + y^2\frac{\partial}{\partial y}. \tag{9.21}$$

Thus, $y'' = 0$ admits the eight-dimensional Lie algebra spanned by (9.21).

Example 2. Let us find all operators (9.14) admitted by (cf. Problem 9.3(xi))

$$y'' + \frac{y'}{x} - e^y = 0. \tag{9.22}$$

Here $f = e^y - y'/x$ and equation (9.16) becomes

$$\eta_{xx} + (2\eta_{xy} - \xi_{xx})y' + (\eta_{yy} - 2\xi_{xy})y'^2 - y'^3 \xi_{yy} - \xi\frac{y'}{x^2} - \eta e^y$$

$$+ (\eta_y - 2\xi_x - 3y'\xi_y)\left(e^y - \frac{y'}{x}\right) + \frac{1}{x}[\eta_x + (\eta_y - \xi_x)y' - y'^2 \xi_y] = 0.$$

The left-hand side of this equation is a polynomial of third degree in y'. Therefore the determining equation is split into the following four equations (cf. (9.17)):

$$\eta_{xx} + \frac{1}{x}\eta_x + (\eta_y - 2\xi_x - \eta)e^y = 0, \quad 2\eta_{xy} - \xi_{xx} + \left(\frac{\xi}{x}\right)_x - 3\xi_y e^y = 0, \tag{9.23}$$

$$\eta_{yy} - 2\xi_{xy} + \frac{2}{x}\xi_y = 0, \quad \xi_{yy} = 0. \tag{9.24}$$

It follows from equations (9.24) that

$$\xi = p(x)y + a(x), \quad \eta = \left(p'(x) - \frac{1}{x}p(x)\right)y^2 + q(x)y + b(x).$$

We substitute these expressions for ξ and η into equations (9.23). Since ξ and η are polynomials in y and equations (9.23) involve e^y, it follows from (9.23) that $\xi_y = 0$ and $\eta_y - 2\xi_x - \eta = 0$, whence $\xi = a(x)$, $\eta = -2a'(x)$. Now the second equation (9.23) becomes

$$\left(a' - \frac{a}{x}\right)' = 0$$

and yields $a = C_1 x \ln x + C_2 x$. Then the first equation (9.23) is valid identically. Thus, the general solution of the determining equations (9.23)-(9.24) is given by

$$\xi = C_1 x \ln x + C_2 x, \quad \eta = -2[C_1(1 + \ln x) + C_2]$$

with constant coefficients C_1 and C_2. By virtue of the linearity of the determining equations the general solution is represented as a linear combination of two independent solutions, $\xi_1 = x \ln x$, $\eta_1 = -2(1 + \ln x)$ and $\xi_2 = x$, $\eta_2 = -2$. Hence, equation (9.22) admits the two-dimensional Lie algebra spanned by

$$X_1 = x \ln x \frac{\partial}{\partial x} - 2(1 + \ln x)\frac{\partial}{\partial x}, \quad X_2 = x\frac{\partial}{\partial x} - 2\frac{\partial}{\partial y}. \quad (9.25)$$

Example 3. Consider the equation $y'' = e^{y'} + xy$. With $f = e^{y'} + xy$, the determining equation (9.16) involves terms free of y', the powers y', y'^2, y'^3, and the exponent $e^{y'}$. By setting the coefficient of the latter equal to zero, we get

$$\eta_y - 2\xi_x - 3y'\xi_y = \eta_x + (\eta_y - \xi_x)y' - y'^2\xi_y,$$

whence $\xi_y = 0$ and $\eta_y = \xi_x, \eta_x = -\xi_x$. Hence, $\xi = C_1 x + C_2$, $\eta = C_1(y-x) + C_3$. The remaining equation, $xy\xi_x + y\xi + x\eta = 0$, yields $C_i = 0$, or $\xi = \eta = 0$. Thus, $y'' = e^{y'} + xy$ is invariant under no infinitesimal point transformation [9.7].

9.3.2 Extended generators and calculation of symmetries of partial differential equations with two independent variables

For partial differential equations the construction of a symmetry group parallels the one-dimensional development of the previous section. To amplify this parallel, let us consider *second-order* equations with two independent variables.

Let x and y be independent variables, and u a differential variable. In the present case, the total derivatives (8.2) are written

$$D_x = \frac{\partial}{\partial x} + u_x \frac{\partial}{\partial u} + u_{xx} \frac{\partial}{\partial u_x} + u_{xy} \frac{\partial}{\partial u_y} + \cdots,$$

$$D_y = \frac{\partial}{\partial y} + u_y \frac{\partial}{\partial u} + u_{xy} \frac{\partial}{\partial u_x} + u_{yy} \frac{\partial}{\partial u_y} + \cdots.$$

9.3. SOLUTION OF DETERMINING EQUATIONS

We will use a generator of a point transformation group,

$$X = \xi^1(x,y,u)\frac{\partial}{\partial x} + \xi^2(x,y,u)\frac{\partial}{\partial y} + \eta(x,y,u)\frac{\partial}{\partial u}, \qquad (9.26)$$

in the following twice-extended form (see Section 8.3.3):

$$X = \xi^1\frac{\partial}{\partial x}+\xi^2\frac{\partial}{\partial y}+\eta\frac{\partial}{\partial u}+\zeta_1\frac{\partial}{\partial u_x}+\zeta_2\frac{\partial}{\partial u_y}+\zeta_{11}\frac{\partial}{\partial u_{xx}}+\zeta_{12}\frac{\partial}{\partial u_{xy}}+\zeta_{22}\frac{\partial}{\partial u_{yy}}. \quad (9.27)$$

The coefficients ζ_i and ζ_{ij} are given by (8.30):

$$\zeta_1 = D_x(\eta) - u_x D_x(\xi^1) - u_y D_x(\xi^2), \quad \zeta_2 = D_y(\eta) - u_x D_y(\xi^1) - u_y D_y(\xi^2), \quad (9.28)$$

and (8.33):

$$\begin{aligned}\zeta_{11} &= D_x(\zeta_1) - u_{xx} D_x(\xi^1) - u_{xy} D_x(\xi^2), \\ \zeta_{12} &= D_y(\zeta_1) - u_{xx} D_y(\xi^1) - u_{xy} D_y(\xi^2), \\ \zeta_{22} &= D_y(\zeta_2) - u_{xy} D_y(\xi^1) - u_{yy} D_y(\xi^2).\end{aligned} \qquad (9.29)$$

Invoking the definitions of D_x and D_y, one obtains:

$$\zeta_1 = \eta_x + u_x\eta_u - u_x\xi^1_x - (u_x)^2\xi^1_u - u_y\xi^2_x - u_x u_y\xi^2_u, \qquad (9.30)$$

$$\zeta_2 = \eta_y + u_y\eta_u - u_x\xi^1_y - u_x u_y\xi^1_u - u_y\xi^2_y - (u_y)^2\xi^2_u, \qquad (9.31)$$

$$\begin{aligned}\zeta_{11} &= \eta_{xx} + 2u_x\eta_{xu} + u_{xx}\eta_u + (u_x)^2\eta_{uu} - 2u_{xx}\xi^1_x - u_x\xi^1_{xx} \\ &\quad -2(u_x)^2\xi^1_{xu} - 3u_x u_{xx}\xi^1_u - (u_x)^3\xi^1_{uu} - 2u_{xy}\xi^2_x - u_y\xi^2_{xx} \\ &\quad -2u_x u_y\xi^2_{xu} - (u_y u_{xx} + 2u_x u_{xy})\xi^2_u - (u_x)^2 u_y\xi^2_{uu},\end{aligned} \qquad (9.32)$$

$$\begin{aligned}\zeta_{12} &= \eta_{xy} + u_y\eta_{xu} + u_x\eta_{yu} + u_{xy}\eta_u + u_x u_y\eta_{uu} - u_{xy}(\xi^1_x + \xi^2_y) \\ &\quad -u_x\xi^1_{xy} - u_{xx}\xi^1_y - u_x u_y(\xi^1_{xu} + \xi^2_{yu}) - (u_x)^2\xi^1_{yu} \\ &\quad -(2u_x u_{xy} + u_y u_{xx})\xi^1_u - (u_x)^2 u_y\xi^1_{uu} - u_y\xi^2_{xy} - u_{yy}\xi^2_x \\ &\quad -(u_y)^2\xi^2_{xu} - (2u_y u_{xy} + u_x u_{yy})\xi^2_u - u_x(u_y)^2\xi^2_{uu},\end{aligned} \qquad (9.33)$$

$$\begin{aligned}\zeta_{22} &= \eta_{yy} + 2u_y\eta_{yu} + u_{yy}\eta_u + (u_y)^2\eta_{uu} - 2u_{yy}\xi^2_y - u_y\xi^2_{yy} \\ &\quad -2(u_y)^2\xi^2_{yu} - 3u_y u_{yy}\xi^2_u - (u_y)^3\xi^2_{uu} - 2u_{xy}\xi^1_y - u_x\xi^1_{yy} \\ &\quad -2u_x u_y\xi^1_{yu} - (u_x u_{yy} + 2u_y u_{xy})\xi^1_u - u_x(u_y)^2\xi^1_{uu}.\end{aligned} \qquad (9.34)$$

Example 1. Consider again equation (9.12) from transonic gas dynamics:

$$u_x u_{xx} + u_{yy} = 0.$$

We seek a symmetry operator in the form (9.26) with unknown coefficients ξ^i and η to be found from the determining equation (9.5):

$$u_{xx}\zeta_1 + u_x\zeta_{11} + \zeta_{22} = 0, \qquad (9.35)$$

where we substitute $\zeta_1, \zeta_{11}, \zeta_{22}$ from (9.30), (9.32), (9.34) and set $u_{yy} = -u_x u_{xx}$. Then equation (9.35) contains the variables $x, y, u, u_x, u_y, u_{xx}, u_{xy}$, whereas ξ^1, ξ^2, η depend only upon x, y, u. Accordingly, we isolate the terms containing u_{xy}, u_{xx}, u_x, u_y, and those free of these variables, and set each term equal to zero. So, the terms containing u_{xy} lead to the equation

$$\xi_y^1 + u_x \xi_x^2 + u_y \xi_u^1 + (u_x)^2 \xi_u^2 = 0,$$

whence

$$\xi_y^1 = 0, \quad \xi_u^1 = 0, \quad \xi_x^2 = 0, \quad \xi_u^2 = 0. \tag{9.36}$$

The same argument applied to the terms containing u_{xx} yields

$$\eta_x = 0, \quad \eta_u - 3\xi_x^1 + 2\xi_y^2 = 0. \tag{9.37}$$

Then equation (9.35) reduces to

$$\eta_{yy} = 0, \quad 2\eta_{yu} - \xi_{yy}^2 = 0. \tag{9.38}$$

Thus, equation (9.35) is split into the overdetermined system of linear partial differential equations (9.36)–(9.38). The reckoning shows that the general solution of this system is given by

$$\xi^1 = C_1 x + C_2, \quad \xi^2 = C_3 y + C_4, \quad \eta = (3C_1 - 2C_3)u + C_5 y + C_6 \tag{9.39}$$

with six arbitrary constants C_i. Thus, we arrive at the six-dimensional Lie algebra L_6 spanned by the operators (7.77):

$$X_1 = \frac{\partial}{\partial x}, \quad X_2 = \frac{\partial}{\partial y}, \quad X_3 = \frac{\partial}{\partial u}, \quad X_4 = \frac{\partial}{\partial u},$$

$$X_5 = x\frac{\partial}{\partial x} + 3u\frac{\partial}{\partial u}, \quad X_6 = y\frac{\partial}{\partial y} - 2u\frac{\partial}{\partial u}.$$

Example 2. Consider the equation

$$u_{xx} + u_{yy} = e^u. \tag{9.40}$$

The determining equation has the form

$$\zeta_{11} + \zeta_{22} - \eta e^u = 0,$$

where we substitute ζ_{11}, ζ_{22} from (9.32), (9.34) and set $u_{yy} = e^u - u_{xx}$. The general solution of the determining equation is given by the functions $\xi^1(x,y)$ and $\xi^2(x,y)$ satisfying the Cauchy-Riemann system

$$\xi_x^1 - \xi_y^2 = 0, \quad \xi_y^1 + \xi_x^2 = 0, \tag{9.41}$$

and by $\eta = -2\xi_x^1$. Therefore, the nonlinear equation (9.40) admits an infinite continuous group (cf. Section 6.3.4) generated by

$$X = \xi^1(x,y)\frac{\partial}{\partial x} + \xi^2(x,y)\frac{\partial}{\partial y} - 2\xi_x^1(x,y)\frac{\partial}{\partial u}, \tag{9.42}$$

where $\xi^1(x,y)$ and $\xi^2(x,y)$ solve the Cauchy-Riemann system (9.41).

9.3.3 Solution of the determining equations for a system

All algorithms of the previous two sections can be naturally generalized to higher-order equations as well as to systems of ordinary and partial differential equations. Here, we illustrate this by considering a simple system of second-order ordinary differential equations. Accordingly, we adopt the notation of Section 8.3.2.

Example. Let us find the Lie algebra admitted by the following system:

$$\frac{d^2 u^\alpha}{dx^2} = 0, \quad \alpha = 1, 2, 3. \tag{9.43}$$

In analytic geometry of three dimensions, this system describes straight lines in a parametric representation, $u^\alpha = u^\alpha(x)$. In classical mechanics, (9.43) has the meaning of equations of the free motion of a particle, where u^α are coordinates of a position of a particle, and x is the time.

We seek a symmetry operator in the form (summation in α)

$$X = \xi(x, u^1, u^2, u^3) \frac{\partial}{\partial x} + \eta^\alpha(x, u^1, u^2, u^3) \frac{\partial}{\partial u^\alpha}.$$

Using the twice-extended operator (8.23), we obtain the determining equations:

$$\zeta_2^\alpha = 0, \quad \alpha = 1, 2, 3. \tag{9.44}$$

Bearing in mind that in (9.44) we set the second derivatives of u^α equal to zero, and invoking the prolongation formulas (8.22), we get:

$$\frac{\partial^2 \eta^\alpha}{\partial x^2} + 2u_1^\beta \frac{\partial^2 \eta^\alpha}{\partial x \partial u^\beta} + u_1^\beta u_1^\gamma \frac{\partial^2 \eta^\alpha}{\partial u^\beta \partial u^\gamma} - u_1^\alpha \frac{\partial^2 \xi}{\partial x^2} - 2 u_1^\alpha u_1^\beta \frac{\partial^2 \xi}{\partial x \partial u^\beta} - u_1^\alpha u_1^\beta u_1^\gamma \frac{\partial^2 \xi}{\partial u^\beta \partial u^\gamma} = 0,$$

where u_1^α denotes the first derivative (see Section 8.3.2). We again split this system by equating to zero the terms free of first derivatives u_1^α, then linear, quadratic and cubic in u_1^α. For example, the cubic terms give us $\partial^2 \xi / \partial u^\beta \partial u^\gamma = 0$, and the determining equations reduce to

$$\frac{\partial^2 \eta^\alpha}{\partial x^2} + 2u_1^\beta \frac{\partial^2 \eta^\alpha}{\partial x \partial u^\beta} + u_1^\beta u_1^\gamma \frac{\partial^2 \eta^\alpha}{\partial u^\beta \partial u^\gamma} - u_1^\alpha \frac{\partial^2 \xi}{\partial x^2} - 2 u_1^\alpha u_1^\beta \frac{\partial^2 \xi}{\partial x \partial u^\beta} = 0, \quad \alpha = 1, 2, 3.$$

After further splitting we obtain the following system:

$$\frac{\partial^2 \eta^\alpha}{\partial x^2} = 0, \quad \frac{\partial}{\partial x}\left(2\frac{\partial \eta^\alpha}{\partial u^\beta} - \delta_\beta^\alpha \frac{\partial \xi}{\partial x}\right) = 0, \quad \frac{\partial}{\partial u^\gamma}\left(\frac{\partial \eta^\alpha}{\partial u^\beta} - 2\delta_\beta^\alpha \frac{\partial \xi}{\partial x}\right) = 0, \tag{9.45}$$

where δ_β^α, known as the Kronecker deltas, are defined by

$$\delta_\beta^\alpha = 1 \text{ or } 0, \quad \text{as} \quad \alpha = \beta \text{ or } \alpha \neq \beta. \tag{9.46}$$

Solution of the system of equations (9.45) depends on 24 arbitrary constants. Consequently, equations (9.43) possess 24 independent infinitesimal symmetries. They form a Lie algebra L_{24} spanned by the following operators (cf. (9.21)):

$$X_0 = \frac{\partial}{\partial x}, \quad X_\alpha = \frac{\partial}{\partial u^\alpha}, \quad S = x\frac{\partial}{\partial x}, \quad P_\alpha = u^\alpha \frac{\partial}{\partial x}, \quad Q_\alpha = x\frac{\partial}{\partial u^\alpha},$$

$$Y_{\alpha\beta} = u^\alpha \frac{\partial}{\partial u^\beta}, \quad Z_0 = x^2 \frac{\partial}{\partial x} + xu^\beta \frac{\partial}{\partial u^\beta}, \quad Z_\alpha = u^\alpha x \frac{\partial}{\partial x} + u^\alpha u^\beta \frac{\partial}{\partial u^\beta}. \tag{9.47}$$

9.3.4 Determination of equations admitting a given group

The task of determining the differential equations admitting a given group (or its Lie algebra) is a simple matter. An approach based on differential invariants [9.8] and Theorem 7.2.2 was illustrated by Example 8.3.5. Consider now an alternative method based on the determining equation (9.5).

Let us fix our attention on second-order ordinary differential equations (9.13) solved for the second derivative, $y'' = f(x, y, y')$. The construction of all equations (9.13) admitting a given Lie algebra L requires the solution of the determining equation (9.16) with respect to the unknown function $f(x, y, y')$ for known coordinates ξ and η of basis operators of the algebra L.

Example. Consider the three-dimensional Lie algebra of type 1) in Lie's classification presented in Theorem 2 of Section 7.3.8, i.e. L_3 spanned by

$$X_1 = \frac{\partial}{\partial x} + \frac{\partial}{\partial y}, \quad X_2 = x\frac{\partial}{\partial x} + y\frac{\partial}{\partial y}, \quad X_3 = x^2\frac{\partial}{\partial x} + y^2\frac{\partial}{\partial x}.$$

For the operator X_1 we have $\xi = 1$, $\eta = 1$. On substituting these values of ξ and η, equation (9.16) assumes the form

$$\frac{\partial f}{\partial x} + \frac{\partial f}{\partial y} = 0, \quad \text{whence} \quad f = f(x - y, y').$$

Now we substitute into the determining equation (9.16) this expression for f and the coordinates $\xi = x$ and $\eta = y$ of X_2 to obtain $zf_z + f = 0$, where $z = x - y$. It follows that

$$f = \frac{g(y')}{x - y}.$$

Finally, we use the determining equation (9.16) with the coordinates $\xi = x^2$ and $\eta = y^2$ of X_3 to obtain the following differential equation:

$$2y'\frac{dg}{dy'} - 3g + 2(y'^2 - y') = 0.$$

Integration yields

$$g = -2(y' + Cy'^{3/2} + y'^2), \quad C = \text{const}.$$

Thus, the most general equation (9.13) admitting the algebra L_3 is

$$y'' + 2\frac{y' + Cy'^{3/2} + y'^2}{x - y} = 0.$$

Many other examples of ordinary differential equations of the first and second orders are collected in Tables 9.1 and 9.2. They are obtained by calculating the differential invariants of infinitesimal generators presented in the tables and using Theorem 7.2.2. Most of the generators are taken from Section 7.1.10.

9.3. SOLUTION OF DETERMINING EQUATIONS

Table 9.1 First-order equations with a known symmetry

No.	Equation	Symmetry
1	$y' = F(y)$	$X = \dfrac{\partial}{\partial x}$
	$y' = F(x)$	$X = \dfrac{\partial}{\partial y}$
	$y' = F(kx + ly)$	$X = l\dfrac{\partial}{\partial x} - k\dfrac{\partial}{\partial y}$
2	$y' = \dfrac{y + xF(\sqrt{x^2 + y^2})}{x - yF(\sqrt{x^2 + y^2})}$	$X = y\dfrac{\partial}{\partial x} - x\dfrac{\partial}{\partial y}$
3	$y' = F(y/x)$	$X = x\dfrac{\partial}{\partial x} + y\dfrac{\partial}{\partial y}$
4	$y' = x^{k-1}F(y/x^k)$	$X = x\dfrac{\partial}{\partial x} + ky\dfrac{\partial}{\partial y}$
5	$xy' = F(xe^{-y})$	$X = x\dfrac{\partial}{\partial x} + \dfrac{\partial}{\partial y}$
6	$y' = yF(ye^{-x})$	$X = \dfrac{\partial}{\partial x} + y\dfrac{\partial}{\partial y}$
7	$y' = \dfrac{y}{x} + xF(y/x)$	$X = \dfrac{\partial}{\partial x} + \dfrac{y}{x}\dfrac{\partial}{\partial y}$
8	$xy' = y + F(y/x)$	$X = x^2\dfrac{\partial}{\partial x} + xy\dfrac{\partial}{\partial y}$
9	$y' = \dfrac{y}{x + F(y/x)}$	$X = xy\dfrac{\partial}{\partial x} + y^2\dfrac{\partial}{\partial y}$
10	$y' = \dfrac{y}{x + F(y)}$	$X = y\dfrac{\partial}{\partial x}$
11	$xy' = y + F(x)$	$X = x\dfrac{\partial}{\partial y}$
12	$xy' = \dfrac{y}{\ln x + F(y)}$	$X = xy\dfrac{\partial}{\partial x}$
13	$xy' = y[\ln y + F(x)]$	$X = xy\dfrac{\partial}{\partial y}$
14	$y' = P(x)y + Q(x)$	$X = e^{\int P(x)dx}\dfrac{\partial}{\partial y}$
15	$y' = P(x)y + Q(x)y^n$	$X = y^n e^{(1-n)\int P(x)dx}\dfrac{\partial}{\partial y}$
16	$y' = P(x)y$	$X = y\dfrac{\partial}{\partial y}$

Table 9.2 Second-order equations with a known symmetry

No.	Equation	Symmetry
1	$y'' = F(y, y')$	$X = \dfrac{\partial}{\partial x}$
	$y'' = F(x, y')$	$X = \dfrac{\partial}{\partial y}$
	$y'' = F(kx + ly, y')$	$X = l\dfrac{\partial}{\partial x} - k\dfrac{\partial}{\partial y}$
2	$y'' = (1 + y'^2)^{3/2} F\left(\sqrt{x^2 + y^2},\ \dfrac{y - xy'}{x + yy'}\right)$	$X = y\dfrac{\partial}{\partial x} - x\dfrac{\partial}{\partial y}$
3	$y'' = F\left(y,\ \dfrac{y - xy'}{y'}\right)$	$X = y\dfrac{\partial}{\partial x}$
4	$y'' + \dfrac{q''(y)}{q(y)} xy'^3 = y'^3 F\left(y,\ \dfrac{1}{y'} - \dfrac{q'(y)}{q(y)}x\right)$	$X = q(y)\dfrac{\partial}{\partial x}$
5	$y'' = F(x,\ y - xy')$	$X = x\dfrac{\partial}{\partial y}$
6	$p(x)y'' - p''(x)y = F\Big(x,\ p(x)y' - p'(x)y\Big)$	$X = p(x)\dfrac{\partial}{\partial y}$
7	$p^2(x)y'' + p(x)p'(x)y' = F\Big(y,\ p(x)y'\Big)$	$X = p(x)\dfrac{\partial}{\partial x}$
8	$xy'' = F\left(\dfrac{y}{x},\ y'\right)$	$X = x\dfrac{\partial}{\partial x} + y\dfrac{\partial}{\partial y}$
9	$y'' = x^{k-2} F(x^{-k}y,\ x^{1-k}y')$	$X = x\dfrac{\partial}{\partial x} + ky\dfrac{\partial}{\partial y}$
10	$y'' = yF(ye^{-x},\ y'/y)$	$X = \dfrac{\partial}{\partial x} + y\dfrac{\partial}{\partial y}$
11	$y'' = yF(x,\ y'/y)$	$X = y\dfrac{\partial}{\partial x}$
12	$y'' = \dfrac{q'(y)}{q(y)}y'^2 + q(y)F\left(x,\ \dfrac{y'}{q(y)}\right)$	$X = q(y)\dfrac{\partial}{\partial y}$
13	$yy'' = y'^2 + y^2 F\left(x,\ \dfrac{xy'}{y} - \ln y\right)$	$X = xy\dfrac{\partial}{\partial y}$
14	$xy'' + y' = x^2 y'^3 F\left(y,\ \dfrac{y}{xy'} - \ln x\right)$	$X = xy\dfrac{\partial}{\partial x}$
15	$x^3 y'' = F\left(\dfrac{y}{x},\ y - xy'\right)$	$X = x^2\dfrac{\partial}{\partial x} + xy\dfrac{\partial}{\partial y}$
16	$x^3 y'' = y'^3 F\left(\dfrac{y}{x},\ \dfrac{y - xy'}{y'}\right)$	$X = xy\dfrac{\partial}{\partial x} + y^2\dfrac{\partial}{\partial y}$

9.3. SOLUTION OF DETERMINING EQUATIONS

9.3.5 Contact symmetries

Let us restrict the discussion of contact symmetries to ordinary differential equations. Then a contact transformation group acts, from the start, in the three-dimensional space of variables x, y, p, where $p = y'$. According to Theorem 8.3.7, its infinitesimal generator has the form

$$X = -W_p \frac{\partial}{\partial x} + (W - pW_p)\frac{\partial}{\partial y} + (W_x + pW_y)\frac{\partial}{\partial p}, \quad (9.48)$$

where $W = W(x, y, p)$ is Lie's characteristic function (see Section 5.3.2).

Thus, construction of infinitesimal contact symmetries reduces to the determination of the characteristic function $W(x, y, p)$ from a determining equation.

Example. Let us find the operators (9.48) admitted by

$$y''' = 0. \quad (9.49)$$

The extension of (9.48) to y'' and y''' by the usual prolongation procedure yields

$$X = -W_p \frac{\partial}{\partial x} + (W - pW_p)\frac{\partial}{\partial y} + (W_x + pW_y)\frac{\partial}{\partial p} + \zeta_2 \frac{\partial}{\partial y''} + \zeta_3 \frac{\partial}{\partial y'''}, \quad (9.50)$$

where (cf. (8.20))

$$\zeta_2 = D_x(W_x + pW_y) - y'' D_x(-W_p), \quad \zeta_3 = D_x(\zeta_2) - y''' D_x(-W_p).$$

The expanded forms of these expressions are:

$$\zeta_2 = W_{xx} + 2pW_{xy} + p^2 W_{yy} + y''[W_y + 2W_{xp} + 2pW_{yp}] + y''^2 W_{pp}, \quad (9.51)$$

$$\begin{aligned}\zeta_3 &= W_{xxx} + 3pW_{xxy} + 3p^2 W_{xyy} + p^3 W_{yyy} + 3y''[W_{xy} + W_{xxp} \\ &\quad + pW_{yy} + 2pW_{xyp} + p^2 W_{yyp}] + 3y''^2[W_{yp} + W_{xpp} + pW_{ypp}] \\ &\quad + y''^3 W_{ppp} + y'''[W_y + 3W_{xp} + 3pW_{yp} + 3y'' W_{pp}].\end{aligned} \quad (9.52)$$

The determining equation has the form $X(y''')|_{y'''=0} \equiv \zeta_3|_{y'''=0} = 0$ and, after substituting (9.52), splits into the following four equations:

$$\begin{aligned}(y'')^3 &: \quad W_{ppp} = 0, \\ (y'')^2 &: \quad W_{yp} + W_{xpp} + pW_{ypp} = 0, \\ (y'')^1 &: \quad W_{xy} + W_{xxp} + pW_{yy} + 2pW_{xyp} + p^2 W_{yyp} = 0, \\ (y'')^0 &: \quad W_{xxx} + 3pW_{xxy} + 3p^2 W_{xyy} + p^3 W_{yyy} = 0.\end{aligned}$$

The first two equations yield $W = f(x)p^2 + [g(x) - 2f'(x)y]p + h(x, y)$. The unknown functions $f(x), g(x)$ and $h(x, y)$ are determined from the remaining two equations. As a result, one obtains the following characteristic function:

$$\begin{aligned}W &= C_1 + C_2 x + C_3 x^2 + C_4 y + C_5 p + C_6 xp + C_7(x^2 p - 2xy) \\ &\quad + C_8 p^2 + C_9(xp^2 - 2yp) + C_{10}(x^2 p^2 - 4xyp + 4y^2).\end{aligned} \quad (9.53)$$

Substituting (9.53) in (9.48), one obtains a ten-dimensional Lie algebra spanned, in accordance with Theorem 8.3.8, by the operators (8.58), viz.

$$X_1 = \frac{\partial}{\partial x}, \quad X_2 = \frac{\partial}{\partial y}, \quad X_3 = x\frac{\partial}{\partial y}, \quad X_4 = x\frac{\partial}{\partial x},$$

$$X_5 = y\frac{\partial}{\partial y}, \quad X_6 = x^2\frac{\partial}{\partial y}, \quad X_7 = x^2\frac{\partial}{\partial x} + 2xy\frac{\partial}{\partial y} \qquad (9.54)$$

and

$$X_8 = 2p\frac{\partial}{\partial x} + p^2\frac{\partial}{\partial y}, \quad X_9 = (y - xp)\frac{\partial}{\partial x} - \frac{1}{2}xp^2\frac{\partial}{\partial y} - \frac{1}{2}p^2\frac{\partial}{\partial p},$$

$$X_{10} = \left(xy - \frac{1}{2}x^2 p\right)\frac{\partial}{\partial x} + \left(y^2 - \frac{1}{4}x^2 p^2\right)\frac{\partial}{\partial y} + \left(yp - \frac{1}{2}xp^2\right)\frac{\partial}{\partial p}. \qquad (9.55)$$

Operators (9.54) span a seven-dimensional Lie algebra of point transformations. According to Theorem 1 of Section 9.3.1, it is the maximal Lie algebra of point symmetries of equation (9.49). Here, their extension to $y' = p$ is, naturally, not quoted. The operators (9.55) generate proper contact transformations.

Remark. Recall that the method of determining equations is not efficient for calculating point symmetries of first-order ordinary differential equations as well as for systems of such equations. The reason is that, e.g. for a single equation $y' = f(x, y)$ the determining equation

$$X(y' - f)|_{y'=f} \equiv \eta_x + (\eta_y - \xi_x)f - \xi_y f^2 - \xi f_x - \eta f_y = 0$$

does not contain the variable y'. Consequently, the split into an overdetermined system does not occur here (cf. Section 9.3.1). That is why any first-order equation admits an infinite group (Theorem 1, Section 9.3.1).

Likewise, the method of determining equations for contact symmetries is efficient if the order of an ordinary differential equation is equal to or greater than three. For lower orders, the determining equation does not split into an overdetermined system (cf. Section 9.3.5).

9.4 Invariant solutions

9.4.1 Lie's theory of invariant solutions

In his fundamental paper [5.11], Lie developed, *inter alia*, a theory of group invariant solutions for arbitrary systems of partial differential equations (9.1).

Definition. Let a system S of differential equations (9.1) admit a group G, and let H be a subgroup of G. A solution of equations (9.1)

$$u^\alpha = h^\alpha(x), \quad \alpha = 1, \ldots, m, \qquad (9.56)$$

9.4. INVARIANT SOLUTIONS

is called an *invariant solution* (specifically, an H-invariant solution) of the system S if equations (9.56) determine an invariant manifold for H. This invariant manifold may be regular or singular with respect to H (Definition 7.4.4). Accordingly, (9.56) is said to be a *regular* or *singular* invariant solution.

The majority of exact solutions of nonlinear partial differential equations, widely encountered, e.g. in fluid mechanics, are invariant solutions. They describe steady-state gas flows, as well as one-dimensional, planar, axially-symmetric and other particular types of flows [5.33]. Numerous examples of invariant solutions are supplied by the general relativity where invariant solutions are identified with Riemannian spaces admitting groups of isometric motions. These are examples of regular invariant solutions. An example of a singular invariant solution of physical significance is provided by the Schwarzschild metric. It can be obtained by group methods as a singular invariant solution (with respect to the rotation group) for Einstein's equations in the general theory of relativity [9.9]. Central to the whole subject is the following theorem due to Lie [9.10].

Theorem. Let a system S of differential equations admit a continuous group G (finite or infinite), and let H be its r-parameter subgroup generated by

$$X_\nu = \xi^i_\nu(x,u)\frac{\partial}{\partial x^i} + \eta^\alpha_\nu(x,u)\frac{\partial}{\partial u^\alpha}, \quad \nu = 1,\ldots,r. \tag{9.57}$$

Let $r_* = \text{rank}\,\|\xi^i_\nu, \eta^\alpha_\nu\|$ (cf. (7.103)). Hence, H has $m+n-r_*$ functionally independent invariants (Theorem 2 of Section 7.4.3):

$$J_1(x,u),\ldots,J_{m+n-r_*}(x,u).$$

Suppose that the Jacobian of J with respect to u is of rank m, e.g.

$$\text{rank}\left\|\frac{\partial J_\beta(x,u)}{\partial u^\alpha}\right\| = m, \tag{9.58}$$

where $\beta = 1,\ldots,m$ and $\alpha = 1,\ldots,m$. Then, setting

$$v^\beta = J_\beta(x,u),\ \lambda^j = J_{m+j}(x,u), \quad \beta = 1,\ldots,m;\ j = 1,\ldots,n-r_*, \tag{9.59}$$

one can write the regular invariant solution (9.56) of the system S in the form

$$v^\beta = \Phi^\beta(\lambda^1,\ldots,\lambda^{n-r_*}), \quad \beta = 1,\ldots,m. \tag{9.60}$$

The functions Φ^β are determined by a system of differential equations, denoted by S/H, involving only invariants, viz. $n - r_*$ independent variables λ^j, the dependent variables v^β and partial derivatives of v^β with respect to λ^j.

Remark. In most applications, one can choose invariants (9.59) such that the variables u^α do not occur in the invariants λ^j. This facilitates the calculation of invariant solutions. The condition (9.58) guarantees that equations (9.60) can be solved for u^α. Substituting them into (9.1), one obtains the system S/H.

Example 1 ([5.11], §57). Consider a linear partial differential equation

$$\alpha(y,u)\frac{\partial u}{\partial x} + \beta(y,u)\frac{\partial u}{\partial y} - \gamma(y,u) = 0.$$

Since x does not occur in α, β, and γ, the equation admits the infinitesimal translation $X = \partial/\partial x$. Invariants of the latter are y and u. According to the above Remark, we take (9.59) with $v = u, \lambda = y$. Then (9.60) provides translational invariant solutions $u = \Phi(y)$ with Φ defined by the ordinary differential equation

$$\beta(y,\Phi)\frac{d\Phi}{dy} - \gamma(y,\Phi) = 0.$$

Example 2. Let us find invariant solutions of the Riccati equation (9.7), $y' + y^2 - 2/x^2 = 0$, using its infinitesimal symmetry (9.9), $X = x\partial/\partial x - y\partial/\partial y$. The only independent invariant is $\psi = xy$. Hence, (9.60) is written $xy = k$ with an arbitrary constant k. Substitution of the expression $y = k/x$ into equation (9.7) yields $k^2 - k - 2 = 0$, whence $k = 2$ and $k = -1$. We thus obtain two invariant solutions of the Riccati equation (9.7):

$$y_1 = \frac{2}{x} \quad \text{and} \quad y_2 = -\frac{1}{x}. \tag{9.61}$$

Example 3. The Black-Scholes equation (1.5)

$$u_t + \frac{1}{2}A^2x^2u_{xx} + Bxu_x - Cu = 0, \quad A, B, C = \text{const.},$$

admits an infinite continuous group (Problem 9.7). Let us find invariant solutions provided by

$$X_3 = 2t\frac{\partial}{\partial t} + (\ln x + Kt)x\frac{\partial}{\partial x} + 2Ctu\frac{\partial}{\partial u}, \quad K = B - A^2/2.$$

A basis of invariants is provided by

$$v = ue^{-Ct}, \quad \lambda = \frac{\ln x}{\sqrt{t}} - K\sqrt{t}.$$

Equation (9.60) $v = \Phi(\lambda)$ yields, upon solving for u:

$$u = e^{Ct}\Phi(\lambda).$$

Substitution into equation (1.5), invoking the expression for λ, yields the ordinary differential equation of the second order, $A^2\Phi'' - \lambda\Phi' = 0$, whence

$$\Phi(\lambda) = l_1\int_0^\lambda e^{s^2/(2A^2)}ds + l_2, \quad l_1, l_2 = \text{const.}$$

Hence, the invariant solution:

$$u(t,x) = l_1 e^{Ct}\int_0^\lambda e^{s^2/(2A^2)}ds + l_2 e^{Ct}, \quad \lambda = \frac{\ln x}{\sqrt{t}} - K\sqrt{t}.$$

9.4. INVARIANT SOLUTIONS

Example 4. The following nonlinear partial differential equation is used for modelling soil water infiltration and redistribution in irrigation systems [9.11]:

$$C(\psi)\psi_t = (K(\psi)\psi_x)_x + (K(\psi)(\psi_z - 1))_z - S(\psi). \quad (9.62)$$

Here ψ is the soil moisture pressure head, $C(\psi)$ is the specific water capacity, $K(\psi)$ is the unsaturated hydraulic conductivity, $S(\psi)$ is the sink or source term, t is the time, x is the horizontal axis and z is the vertical axis which is considered positive downward. Line source drip systems produce a continuous wetted band along the length of the lateral (the y-axis), and the phenomenon actually involves all three space coordinates, $x, y,$ and z [9.12].

The infinitesimal symmetries of equation (9.62) with arbitrary coefficients form a Lie algebra called the *principal Lie algebra* L_P for equation (9.62). It is a three-dimensional algebra spanned by (infinitesimal translations)

$$X_1 = \frac{\partial}{\partial t}, \quad X_2 = \frac{\partial}{\partial x}, \quad X_3 = \frac{\partial}{\partial z}.$$

There are 29 particular types of the coefficients $C(\psi), K(\psi), S(\psi)$ when an extension of the algebra L_P occurs. Let us consider here one of cases when L_P extends by three operators. Namely, consider the equation ($M = $ const.)

$$\frac{4}{Me^{4\psi}-1}\psi_t = \left(e^{-4\psi}\psi_x\right)_x + \left(e^{-4\psi}\psi_z\right)_z + 4e^{-4\psi}\psi_z + M - e^{-4\psi}. \quad (9.63)$$

Equation (9.63) admits a six-dimensional Lie algebra L_6 obtained by adding to the basis X_1, X_2, X_3 of L_P the following three operators:

$$X_4 = t\frac{\partial}{\partial t} - \frac{1}{4}(Me^{4\psi} - 1)\frac{\partial}{\partial \psi},$$

$$X_5 = \sin x\, e^{-z}\frac{\partial}{\partial x} - \cos x\, e^{-z}\frac{\partial}{\partial z} + \frac{1}{2}\cos x\, e^{-z}(Me^{4\psi} - 1)\frac{\partial}{\partial \psi},$$

$$X_6 = \cos x\, e^{-z}\frac{\partial}{\partial x} + \sin x\, e^{-z}\frac{\partial}{\partial z} - \frac{1}{2}\sin x\, e^{-z}(Me^{4\psi} - 1)\frac{\partial}{\partial \psi}.$$

According to the above theorem, the search for invariant solutions reduces the number of independent variables by r_*. Hence, if we are interested in those solution of equation (9.62) described by ordinary differential equations, we should use a subalgebra with $r_* = 2$. Let us find invariant solutions based on the subalgebra $L_2 \subset L_6$ spanned by X_4, X_5. Invariants $J(t, x, z, \psi)$ of L_2 are defined by the system of linear partial differential equations

$$X_4(J) = 0, \quad X_5(J) = 0. \quad (9.64)$$

Solution of this system yields the basis of invariants (see Problem 9.11):

$$v = te^{2z}(e^{-4\psi} - M), \quad \lambda = e^z \sin x.$$

9. SYMMETRY OF DIFFERENTIAL EQUATIONS

The representation (9.60), $te^{2z}(e^{-4\psi} - M) = \Phi(\lambda)$, yields upon solving for ψ:

$$\psi = -\frac{1}{4}\ln\left|M + \frac{e^{-2z}}{t}\Phi(\lambda)\right|.$$

Substitution into equation (9.63) yields $\Phi''(\lambda) = 4$, whence $\Phi(\lambda) = 2\lambda^2 + l_1\lambda + l_2$. Thus, the invariant solution is given by

$$\psi = -\frac{1}{4}\ln\left|M + \frac{e^{-2z}}{t}\left(2e^{2z}\sin^2 x + l_1 e^z \sin x + l_2\right)\right|, \quad l_1, l_2 = \text{const}. \quad (9.65)$$

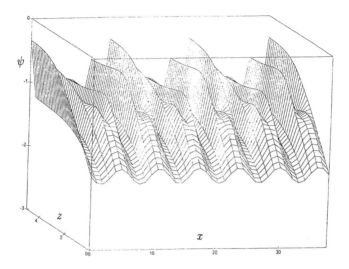

Figure 9.1 Plot of the invariant solution (9.65), $M = 4, l_1 = -2, l_2 = -4, t = 0,01$.

Example 5. We now illustrate the approach to systems by considering invariant solutions of the differential equations of the shallow-water theory,

$$\begin{aligned} \boldsymbol{v}_t + (\boldsymbol{v} \cdot \nabla)\boldsymbol{v} + g\nabla h &= 0, \\ h_t + \boldsymbol{v} \cdot \nabla h + h\,\text{div}\,\boldsymbol{v} &= 0, \end{aligned} \quad (9.66)$$

where $g =$const. Equations (9.66) admit the Lie algebra L_9 (see Problem 9.9). Let us find the invariant solutions based on its subalgebra L_2 spanned by [9.13]

$$X_4 = x^2\frac{\partial}{\partial x^1} - x^1\frac{\partial}{\partial x^2} + v^2\frac{\partial}{\partial v^1} - v^1\frac{\partial}{\partial v^2},$$

$$X_0 + X_3 = (1+t^2)\frac{\partial}{\partial t} + tx^i\frac{\partial}{\partial x^i} + (x^i - tv^i)\frac{\partial}{\partial v^i} - 2th\frac{\partial}{\partial h}.$$

Solution of the system of equations $X_4(J) = 0$, $(X_0 + X_3)(J) = 0$ yields four independent invariants (cf. Example 4 in Section 7.2.3):

$$J_1 = x^1 v^1 + x^2 v^2 - \frac{tr^2}{1+t^2}, \quad J_2 = x^1 v^2 - x^2 v^1, \quad J_3 = (1+t^2)h, \quad J_4 = \frac{r}{\sqrt{1+t^2}},$$

9.4. INVARIANT SOLUTIONS

where $r = \sqrt{(x^1)^2 + (x^2)^2}$. We denote $J_4 = \lambda$ and write the representation (9.60) in the form $J_1 = U(\lambda)$, $J_2 = V(\lambda)$, $J_3 = H(\lambda)$, whence

$$v^1 = \frac{tx^1}{1+t^2} + \frac{1}{r^2}\left[x^1 U(\lambda) - x^2 V(\lambda)\right], \quad h = \frac{H(\lambda)}{1+t^2},$$
$$v^2 = \frac{tx^2}{1+t^2} + \frac{1}{r^2}\left[x^2 U(\lambda) + x^1 V(\lambda)\right], \quad \lambda = \frac{r}{\sqrt{1+t^2}}.$$
(9.67)

The ansatz (9.67) reduces (9.66) to a system of ordinary differential equations:

$$\lambda U U' - U^2 - V^2 + g\lambda^3 H' + \lambda^4 = 0, \quad UV' = 0, \quad (UH)' = 0. \quad (9.68)$$

Equations (9.68) describe various rotationally symmetric flows of shallow water over a flat bottom. One of them is obtained by setting $U = 0$, $V = K\lambda$ and integrating (9.68). Then (9.67) yields:

$$v^1 = \frac{tx^1}{1+t^2} - K\frac{x^2}{r\sqrt{1+t^2}}, \quad v^2 = \frac{tx^2}{1+t^2} + K\frac{x^1}{r\sqrt{1+t^2}},$$
$$h = \frac{1}{(1+t^2)g}\left[C - \frac{r^2}{2(1+t^2)} + K^2 \ln \frac{r}{\sqrt{1+t^2}}\right], \quad K, C = \text{const}.$$
(9.69)

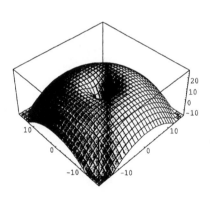

Figure 9.2 Invariant solution (9.69) describes the motion of a bounded annular mass of water with the initial ($t = 0$) radii $r_1(0) > 0$ and $r_2(0) > r_1(0)$ of the boundary circles and the initial angular velocity $\omega(0) = K/r$. As the water spreads under the action of gravity and rotation, the angular velocity ω decays as $\omega = K/(r\sqrt{1+t^2})$. The bottom boundary ($h = 0$) of the water consists of two concentric circles with the radii r_1 and r_2: $r_1 = r_1(0)\sqrt{1+t^2}$, $r_2 = r_2(0)\sqrt{1+t^2}$.

9.4.2 Inherited symmetries of equations for invariant solutions

Consider differential equations S/H for invariant solutions given by Theorem 9.4.1. The question naturally arises of whether these equations inherit any symmetries of the original system of differential equations (9.1). An answer is given by the following statement [9.14].

Theorem. Let a system of differential equations (9.1) admit a Lie algebra L, and let L_r be a subalgebra of L. Let $Q \subset L$ be the maximal subalgebra of L such that L_r is an ideal of Q. Then the differential equations for invariant solutions based on the subalgebra L_r admit the quotient algebra Q/L_r.

Consequently, operators from the quotient algebra Q/L_r are termed *inherited symmetries* of equations S/H for invariant solutions, where H is the group generated by L_r. In general, Q/L_r is not the maximal Lie algebra admitted by equations S/H. However, in particular cases the above theorem may provide the maximal symmetry algebra, e.g. in the following example.

Example. Consider again equation (9.40) and investigate inherited symmetries of the equation for its invariant solutions based on the rotation group. For this purpose, it is convenient to use the polar coordinates r and θ defined by $x = r\cos\theta$, $y = r\sin\theta$. Then equation (9.40) is written

$$\frac{\partial^2 u}{\partial r^2} + \frac{1}{r}\frac{\partial u}{\partial r} + \frac{1}{r^2}\frac{\partial^2 u}{\partial \theta^2} = e^u, \qquad (9.70)$$

while the admitted operator (9.42) and the Cauchy-Riemann system (9.41) become:

$$X = A(r,\theta)\frac{\partial}{\partial r} + B(r,\theta)\frac{\partial}{\partial \theta} - 2A_r(r,\theta)\frac{\partial}{\partial u}, \qquad (9.71)$$

$$B_r + \frac{1}{r^2}A_\theta = 0, \quad B_\theta - A_r + \frac{A}{r} = 0. \qquad (9.72)$$

The trivial solution $A = 0, B = $ const. of equations (9.72) yields the generator of translations in polar coordinates:

$$X_1 = \frac{\partial}{\partial \theta}. \qquad (9.73)$$

Thus, the rotational invariance means the θ-translational invariance of solutions for equation (9.70). Hence, the equation S/H is obtained by letting $u = u(r)$ in (9.70) (cf. Example 1 of the previous section):

$$\frac{\partial^2 u}{\partial r^2} + \frac{1}{r}\frac{\partial u}{\partial r} = e^u. \qquad (9.74)$$

The subalgebra Q introduced in the above theorem is determined by the relation $[X_1, X] \in L_1$ for the operators X_1 (9.73) and X (9.71), where L_1 is the one-dimensional algebra spanned by X_1. This relation is written

$$A_\theta \frac{\partial}{\partial r} + B_\theta \frac{\partial}{\partial \theta} - 2A_{r\theta}\frac{\partial}{\partial u} = C_1 \frac{\partial}{\partial \theta}, \quad C_1 = \text{const.},$$

whence $A_\theta = 0$, $B_\theta = C$. Invoking (9.72), we obtain $A = C_1 r \ln r + C_2 r$ and $B = C_1 \theta + C_3$. Hence, Q is the three-dimensional Lie algebra spanned by

$$X_1 = \frac{\partial}{\partial \theta}, \quad X_2 = r\frac{\partial}{\partial r} - 2\frac{\partial}{\partial u}, \quad X_3 = r\ln r \frac{\partial}{\partial r} + \theta \frac{\partial}{\partial \theta} - 2(1 + \ln r)\frac{\partial}{\partial u}.$$

The quotient algebra Q/L_1 has the dimension two and is spanned by the operators X_2 and X_3 (see Section 7.3.4). Restricting the action of the latter to r and u, we obtain the following operators admitted by equation (9.74):

$$Y_1 = r\frac{\partial}{\partial r} - 2\frac{\partial}{\partial u}, \quad Y_2 = r\ln r \frac{\partial}{\partial r} - 2(1+\ln r)\frac{\partial}{\partial u}. \qquad (9.75)$$

Comparison of (9.74) and (9.75) with (9.22) and (9.25), respectively, shows that the inherited symmetries (9.75) furnish the maximal Lie algebra admitted by equation (9.74).

9.5 Approximate symmetries of equations with a small parameter

The purpose of this section is to carry over the infinitesimal method of Sections 9.2 and 9.3 to approximate symmetries of differential equations with a small parameter ε. The notation is taken from Section 7.5. We will consider approximations in the first order of precision in ε. A detailed presentation with proofs of the basic statements and numerous applications is to be found in [5.38](ii); see also [5.35], vol. 3, Chap. 2, 9 and references therein.

9.5.1 Definition of an approximate symmetry group

Definition. Let G be a one-parameter approximate transformation group:

$$\bar{z}^i \approx f(z, a, \varepsilon) \equiv f_0^i(z, a) + \varepsilon f_1^i(z, a), \quad i = 1, \ldots, N. \tag{9.76}$$

An approximate equation

$$F(z, \varepsilon) \equiv F_0(z) + \varepsilon F_1(z) \approx 0 \tag{9.77}$$

is said to be *approximately invariant* with respect to G (alias (9.77) *admits* G) if

$$F(f(z, a, \varepsilon), \varepsilon) = o(\varepsilon)$$

whenever $z = (z^1, \ldots, z^N)$ satisfies equation (9.77).

If we identify z with $[z]$ (see Section 8.2.1), i.e. set $z = (x, u, u_{(1)}, \ldots, u_{(k)})$, then (9.77) becomes an approximate differential equation of order k, and G is an *approximate symmetry group* of the differential equation.

9.5.2 Determining equations

Theorem. Equation (9.77) is approximately invariant under the approximate transformation group G with the generator (7.128),

$$X = X^0 + \varepsilon X^1 \equiv \xi_0^i(z) \frac{\partial}{\partial z^i} + \varepsilon \xi_1^i(z) \frac{\partial}{\partial z^i}, \tag{9.78}$$

if and only if $[XF(z, \varepsilon)]_{F \approx 0} = o(\varepsilon)$, or

$$\left[X^0 F_0(z) + \varepsilon \left(X^1 F_0(z) + X^0 F_1(z) \right) \right]_{(9.77)} = o(\varepsilon). \tag{9.79}$$

Proof. Substitute the expressions (9.77) and (9.78) into the determining equation (9.5) and single out the principal part.

Equation (9.79) is the *determining equation*. X is called an *infinitesimal approximate symmetry*, or an *approximate operator admitted* by (9.77).

Corollary 1. Invoking (9.6), the determining equation (9.79) can be written

$$X^0 F_0(z) = \lambda(z) F_0(z), \qquad (9.80)$$

$$X^1 F_0(z) + X^0 F_1(z) = \lambda(z) F_1(z). \qquad (9.81)$$

The factor $\lambda(z)$ is determined by (9.80) and then substituted into (9.81). The latter equation must hold for all solutions of $F_0(z) = 0$.

Comparing (9.80) with the determining equation of exact symmetries (9.6), one concludes the following.

Corollary 2. Let equation (9.77) admit an approximate transformation group with the generator $X = X^0 + \varepsilon X^1$, where $X^0 \neq 0$. Then the operator

$$X^0 = \xi_0^i(z) \frac{\partial}{\partial z^i} \qquad (9.82)$$

generates a one-parameter exact symmetry group for the equation

$$F_0(z) = 0. \qquad (9.83)$$

Definition. Equation (9.83) is termed an *unperturbed equation*, and (9.77) a *perturbed equation*. Under the conditions of Corollary 2, X^0 is called a *stable symmetry* of the unperturbed equation (9.83). The corresponding approximate symmetry generator $X = X^0 + \varepsilon X^1$ for the perturbed equation (9.82) is called a *deformation of the infinitesimal symmetry* X^0 of equation (9.83) caused by the perturbation $\varepsilon F_1(z)$. In particular, if the most general symmetry Lie algebra of equation (9.83) is stable, we say that the perturbed equation (9.77) *inherits the symmetries of the unperturbed equation.*

9.5.3 Calculation of infinitesimal approximate symmetries

Corollaries 1 and 2 of the previous section furnish us with an algorithm for calculating infinitesimal approximate symmetries of differential equations (9.73) with a small parameter ε. The algorithm comprises three steps.

1st step. Find the exact infinitesimal symmetries X^0 of the unperturbed equation (9.83), e.g. by solving the determining equation (9.5):

$$\left. X^0 F_0(z) \right|_{F_0(z)=0} = 0. \qquad (9.84)$$

2nd step. Given X^0 and a perturbation $\varepsilon F_1(z)$, define the *auxiliary function* H by virtue of equations (9.80)–(9.81) and (9.77), viz.

$$H = \frac{1}{\varepsilon} \left[X^0 \big(F_0(z) + \varepsilon F_1(z) \big) \Big|_{F_0(z)+\varepsilon F_1(z)=0} \right]. \qquad (9.85)$$

10.5. APPROXIMATE SYMMETRIES

3rd step. Define the first-order deformations (9.78), i.e. find the operators X^1 from the *determining equation for deformations*:

$$X^1 F_0(z)\Big|_{F_0(z)=0} + H = 0. \tag{9.86}$$

Note that equation (9.86), unlike the determining equation (9.84) for exact symmetries, is *inhomogeneous*.

Example 1. Let us apply the algorithm to a nonlinear wave equation with a small dissipation εu_t:

$$u_{tt} + \varepsilon u_t = (u^\sigma u_x)_x, \tag{9.87}$$

where $\sigma \neq 0$ is an arbitrary constant. An approximate group generator is written

$$X = X^0 + \varepsilon X^1 \equiv (\tau_0 + \varepsilon \tau_1)\frac{\partial}{\partial t} + (\xi_0 + \varepsilon \xi_1)\frac{\partial}{\partial x} + (\eta_0 + \varepsilon \eta_1)\frac{\partial}{\partial u},$$

where τ_ν, ξ_ν, and η_ν ($\nu = 0, 1$) are unknown functions of t, x, and u.

1st step. This step requires the calculation of exact symmetries for the unperturbed equation

$$u_{tt} = (u^\sigma u_x)_x, \quad \sigma \neq 0. \tag{9.88}$$

The determining equation (9.84) yields (see Problem 9.17)

$$X^0 = (C_1 + C_3 t)\frac{\partial}{\partial t} + (C_2 + C_3 x + C_4 x)\frac{\partial}{\partial x} + C_4\frac{2u}{\sigma}\frac{\partial}{\partial u},$$

where $C_1, \ldots, C_4$ are arbitrary constants. Hence, equation (9.88) admits an L_4.

2nd step. Substitution of the above generator X^0 into (9.85) yields the auxiliary function $H = C_3 u_t$.

3rd step. Now the determining equation (9.86) for deformations is written:

$$X^1\left(u_{tt} - u^\sigma u_{xx} + \sigma u^{\sigma-1} u_x^2\right)\Big|_{(9.87)} + C_3 u_t = 0,$$

where X^1 denotes the extension of the operator

$$X^1 = \tau_1 \frac{\partial}{\partial t} + \xi_1 \frac{\partial}{\partial x} + \eta_1 \frac{\partial}{\partial u}$$

to the derivatives of u involved in equation (9.87). Upon setting $u_{tt} = (u^\sigma u_x)_x$, the left-hand side of the latter equation becomes a polynomial in the variables u_{tx}, u_{xx}, u_t, u_x. Equating to zero its coefficients, we obtain:

$$\tau_1 = \tau_1(t), \ \tau_1'''(t) = 0; \ \xi_1 = \xi_1(x), \ \xi_1'''(x) = 0; \ \eta_1 = 2\frac{u}{\sigma}[\xi_1'(x) - \tau_1'(t)];$$

$$\left(\sigma + \frac{4}{3}\right)\xi_1'' = 0, \quad \left(\frac{4}{\sigma} + 1\right)\tau_1'' = C_3. \tag{9.89}$$

The solution of this system can be written in the form:

$$\tau_1 = C_3 \frac{\sigma t^2}{2(\sigma+4)} + A_1 + A_3 t, \quad \xi_1 = A_2 + A_3 x + A_4 x, \quad \eta_1 = \frac{2u}{\sigma}\left(A_4 - \frac{\sigma}{\sigma+4}C_3 t\right).$$

Thus, equation (9.87) with an arbitrary constant $\sigma \neq 0$ has the following infinitesimal approximate symmetries:

$$X_1 = \frac{\partial}{\partial t}, \quad X_2 = \frac{\partial}{\partial x}, \quad X_3 = \left(t + \frac{\varepsilon \sigma t^2}{2(\sigma+4)}\right)\frac{\partial}{\partial t} + x\frac{\partial}{\partial x} - \frac{2\varepsilon t u}{\sigma+4}\frac{\partial}{\partial u},$$

$$X_4 = x\frac{\partial}{\partial x} + \frac{2u}{\sigma}\frac{\partial}{\partial t}, \quad X_5 = \varepsilon X_1, X_6 = \varepsilon X_2, X_7 = \varepsilon X_4, X_8 = \varepsilon\left(t\frac{\partial}{\partial t} + x\frac{\partial}{\partial x}\right) \approx \varepsilon X_3.$$

They generate an eight-parameter approximate transformations group. The operators X_1 to X_4 show that the symmetry Lie algebra L_4 of equation (9.88) is stable. Hence, the perturbed equation (8.87) inherits the symmetries of the unperturbed equation (9.88) with an arbitrary σ. The operators X_5 to X_8 are of minor significance and can be regarded as incidental approximate symmetries reflecting a general property of the determining equation (9.79); see Problem 9.19.

Example 2. It is transparent from the last two equations (9.89) that two numerical values of the exponent of nonlinearity, $\sigma = -4/3$ and $\sigma = -4$, have a particular significance for the approximate group analysis. Let us examine the case $\sigma = -4$.

1st step. Equation $u_{tt} = \left(u^{-4} u_x\right)_x$ admits L_5 consisting of the operators

$$X^0 = \left(C_1 + C_3 t + C_5 t^2\right)\frac{\partial}{\partial t} + \left(C_2 + C_3 x + C_4 x\right)\frac{\partial}{\partial x} + \left(-\frac{1}{2}C_4 + C_5 t\right)u\frac{\partial}{\partial u}.$$

2nd step. The auxiliary function has the form $H = C_3 u_t + 2C_5 t u_t + C_5 u$.

3rd step. The determining equation (9.86) for deformations yields, *inter alia*, that $C_3 = C_5 = 0$. Proceeding further as in the previous example, one obtains the following infinitesimal approximate symmetries of the perturbed equation $u_{tt} + \varepsilon u_t = \left(u^{-4} u_x\right)_x$:

$$X_1 = \frac{\partial}{\partial t}, \quad X_2 = \frac{\partial}{\partial x}, \quad X_3 = \varepsilon\left(t\frac{\partial}{\partial t} + x\frac{\partial}{\partial x}\right), \quad X_4 = x\frac{\partial}{\partial x} - \frac{1}{2}u\frac{\partial}{\partial t},$$

$$X_5 = \varepsilon X_1, \quad X_6 = \varepsilon X_2, \quad X_7 = \varepsilon X_4, \quad X_8 = \varepsilon\left(t^2\frac{\partial}{\partial t} + tu\frac{\partial}{\partial u}\right).$$

It follows that the operators

$$X^0 = t\frac{\partial}{\partial t} + x\frac{\partial}{\partial x} \quad \text{and} \quad X^0 = t^2\frac{\partial}{\partial t} + tu\frac{\partial}{\partial u}$$

of the symmetry Lie algebra L_5 of the unperturbed equation are unstable. Hence, in this example the perturbed equation does not inherit the symmetries of the unperturbed equation.

9.6 On discontinuous and non-local symmetries

Lie's infinitesimal method furnishes a general approach for constructing continuous (local) groups admitted by differential equations. Let us consider simple examples revealing the existence of other types of symmetries.

Example 1. The equation $(y'')^2 = [(y'/x) - e^y]^2$ (cf. Problem 9.3(xi) and equation (9.22)) is invariant under a mixed group comprising, as its continuous component, the one-parameter group $\bar{x} = xe^a$, $\bar{y} = y - 2a$, and the reflection of the x-axis, $\tilde{x} = -x$. The question of whether this is a maximal symmetry group remains open (see Problem 9.13**).

Example 2. The maximal symmetry group of the second-order equation

$$y'' = \frac{1}{x}y' + \frac{4}{x^3}y^2 \tag{9.90}$$

is a mixed group made up of the one-parameter subgroup of uniform dilations, $\bar{x} = ax$, $\bar{y} = ay$ (a continuous component), and the isolated transformation [9.15]

$$S: \tilde{x} = \frac{1}{x}, \quad \tilde{y} = \frac{y}{x^2}. \tag{9.91}$$

Example 3. The operator (7.39),

$$L = e^q \left(x^2 \frac{\partial}{\partial x} + (3xy + x^2) \frac{\partial}{\partial y} \right), \quad \text{where} \quad q = \int \frac{dx}{y},$$

provides a non-local infinitesimal symmetry for equation (7.38),

$$\frac{dy}{dx} = 3\frac{y}{x} + \frac{x^2}{2y} + x + 1.$$

Indeed, the coefficient ζ_1 (8.17) of the extended operator L is given by

$$\zeta_1 = D_x[e^q(3xy + x^2)] - y' D_x e^q x^2].$$

Using the definition of the non-local variable q in the form $D_x(q) = 1/y$, one obtains

$$\zeta_1 = e^q \left(\left(x - \frac{x^2}{y} \right) y' + \frac{x^2}{y} + 3y + 5x \right).$$

Thus, the extended operator is of the form

$$L = e^q \left(x^2 \frac{\partial}{\partial x} + (3xy + x^2) \frac{\partial}{\partial y} + \left[\left(x - \frac{x^2}{y} \right) y' + \frac{x^2}{y} + 3y + 5x \right] \frac{\partial}{\partial y'} \right).$$

The reckoning shows that $L(y' - 3y/x - x^2/(2y) - x - 1)|_{(7.38)} = 0$.

9.7 Symmetry and conservation laws

Noether's theorem, proved in this section, gives a constructive way of determining conservation laws for Euler-Lagrange equations once their symmetries are known.

9.7.1 Principle of least action. Euler-Lagrange equations

Hamilton's principle of least action states: A mechanical system is characterized by its *Lagrangian*, $L = L(t, q, v)$. The motion of the system concerned is determined by the requirement that the trajectories of the particles of the system provide an extremum of the *action*

$$S = \int_{t_1}^{t_2} L(t, q, v) dt.$$

Here t is time, and $q = (q^1, \ldots, q^m)$ and $v = q' \equiv dq/dt$ denote coordinates and velocities of particles of the sytem. The action is defined on the set of functions $q^\alpha = q^\alpha(t)$ such that the integral exists in an arbitrary interval of time $t_1 \leq t \leq t_2$.

Consider a variation of q when it is replaced by $q + \delta q$. It is assumed that the increment is a function $\delta q = \delta q(t)$ such that it is small everywhere in the interval $t_1 \leq t \leq t_2$ and vanishes at the boundary, $\delta q(t_1) = \delta q(t_2) = 0$. Differentiation yields $\delta v = d[\delta q(t)]/dt$. This causes the variation of the action integral by

$$\delta S = \int_{t_1}^{t_2} [L(t, q + \delta q, v + \delta v) - L(t, q, v)] dt.$$

Expansion of the integrand in powers of the increments δq and δv yields the linear principal part of δS (summation in $\alpha = 1, \ldots, m$):

$$\delta S = \int_{t_1}^{t_2} \left(\frac{\partial L}{\partial q^\alpha} \delta q^\alpha + \frac{\partial L}{\partial v^\alpha} \delta v^\alpha \right) dt.$$

On integrating the second term by parts, it is written:

$$\delta S = \int_{t_1}^{t_2} \left(\frac{\partial L}{\partial q^\alpha} - \frac{d}{dt} \frac{\partial L}{\partial v^\alpha} \right) \delta q^\alpha dt + \left[\frac{\partial L}{\partial v^\alpha} \delta q^\alpha \right]_{t_1}^{t_2}$$

or, invoking the boundary conditions $\delta q(t_1) = \delta q(t_2) = 0$:

$$\delta S = \int_{t_1}^{t_2} \left(\frac{\partial L}{\partial q^\alpha} - \frac{d}{dt} \frac{\partial L}{\partial v^\alpha} \right) \delta q^\alpha dt.$$

The necessary condition for S to have an extremum is that $\delta S = 0$. Since the time interval $t_1 \leq t \leq t_2$ and the increment δx are arbitrary, it follows:

$$\frac{d}{dt} \frac{\partial L}{\partial v^\alpha} - \frac{\partial L}{\partial q^\alpha} = 0, \quad \alpha = 1, \ldots, m. \tag{9.92}$$

Differential equations (9.92) are known as the *Euler–Lagrange equations*. Thus, the trajectory $q = q(t)$ of a mechanical system with the Lagrangian $L(t, q, v)$ solves the Euler-Lagrange equations (9.92).

9.7. SYMMETRY AND CONSERVATION LAWS

Likewise, one can treat a multi-dimensional case with independent variables $x = (x^1, \ldots, x^n)$ and differential variables $u = (u^1, \ldots, u^m)$. Let us use the differential algebraic notation and terminology of Chapter 8.

Let $L \in \mathcal{A}$ be a differential function of the first order. Let $V \subset \mathbb{R}^n$ be an arbitrary n-dimensional volume in the space of the independent variables x with the boundary ∂V. An *action* is the integral (also termed a *variational integral*)

$$l[u] = \int_V L(x, u, u_{(1)}) dx \qquad (9.93)$$

defined on the set of functions $u = u(x)$ such that the integral (9.93) exists. The variation $\delta l[u]$ of the integral (9.93), caused by the variation $u + h(x)$ of u, is the principal linear part (in h) of the integral

$$\int_V [L(x, u+h, u_{(1)} + h_{(1)}) - L(x, u, u_{(1)})] dx,$$

i.e.

$$\delta l[u] = \int_V \left[\frac{\partial L}{\partial u^\alpha} h^\alpha + \frac{\partial L}{\partial u_i^\alpha} h_i^\alpha \right] dx.$$

On integrating the second term by parts, it is written:

$$\delta l[u] = \int_V \left[\frac{\partial L}{\partial u^\alpha} - D_i\left(\frac{\partial L}{\partial u_i^\alpha}\right) \right] h^\alpha dx + \int_V D_i\left(\frac{\partial L}{\partial u_i^\alpha} h^\alpha \right) dx$$

or, using the Gauss divergence formula,

$$\delta l[u] = \int_V \left[\frac{\partial L}{\partial u^\alpha} - D_i\left(\frac{\partial L}{\partial u_i^\alpha}\right) \right] h^\alpha dx + \int_{\partial V} \frac{\partial L}{\partial u_i^\alpha} h^\alpha \nu^i dx,$$

where $\nu = (\nu^1, \ldots, \nu^n)$ is the unit outer normal to ∂V. Provided that the functions $h^\alpha(x)$ vanish on the boundary ∂V, we arrive at the following:

$$\delta l[u] = \int_V \left[\frac{\partial L}{\partial u^\alpha} - D_i\left(\frac{\partial L}{\partial u_i^\alpha}\right) \right] h^\alpha dx.$$

A function $u = u(x)$ is called an *extremum* of the variational integral (9.93) if $\delta l[u(x)] = 0$ for any volume V and any increment $h = h(x)$ vanishing on ∂V. It follows from the above expression for $\delta l[u]$ that a necessary condition for u to be an extremum is given by the *Euler-Lagrange equations* (cf. Section 8.4.1):

$$\frac{\delta L}{\delta u^\alpha} \equiv \frac{\partial L}{\partial u^\alpha} - D_i\left(\frac{\partial L}{\partial u_i^\alpha}\right) = 0, \quad \alpha = 1, \ldots, m. \qquad (9.94)$$

In the generic case, (9.94) is a system of m partial differential equations of the second order.

If $L \in \mathcal{A}$ is a differential function of an arbitrary order, the left-hand side of (9.94) is replaced by the general Euler-Lagrange operator (8.59). If, e.g., $L = L(x, u, u_{(1)}, u_{(2)})$ the Euler–Lagrange equations have the form

$$\frac{\delta L}{\delta u^\alpha} \equiv \frac{\partial L}{\partial u^\alpha} - D_i\left(\frac{\partial L}{\partial u_i^\alpha}\right) + D_i D_j\left(\frac{\partial L}{\partial u_{ij}^\alpha}\right) = 0, \quad \alpha = 1, \ldots, m.$$

9.7.2 Differential algebraic proof of Noether's theorem

Definition 1. Let G be a one-parameter group of transformations

$$\bar{x}^i = f^i(x, u, a), \quad \bar{u}^\alpha = \varphi^\alpha(x, u, a), \quad i = 1, \ldots, n; \ \alpha = 1, \ldots, m, \tag{9.95}$$

with the generator

$$X = \xi^i(x, u)\frac{\partial}{\partial x^i} + \eta^\alpha(x, u)\frac{\partial}{\partial u^\alpha}. \tag{9.96}$$

A variational integral (9.93) is said to be invariant under G if

$$\int_{\bar{V}} L(\bar{x}, \bar{u}, \bar{u}_{(1)}) d\bar{x} = \int_V L(x, u, u_{(1)}) dx, \tag{9.97}$$

where $\bar{V} \subset \mathbb{R}^n$ is a volume obtained from V by transformation (9.95).

Lemma. An integral (9.93) is invariant under the group G if and only if

$$X(L) + LD_i(\xi^i) = 0, \tag{9.98}$$

where X denotes the extended action (8.31) of the generator (9.96), i.e.

$$X(L) = \xi^i\frac{\partial L}{\partial x^i} + \eta^\alpha\frac{\partial L}{\partial u^\alpha} + \zeta^\alpha_i\frac{\partial L}{\partial u^\alpha_i}, \quad \zeta^\alpha_i = D_i(\eta^\alpha) - u^\alpha_j D_i(\xi^j).$$

Proof. Since V is an arbitrary volume, the integral equation (9.97) is equivalent to the invariance of the *elementary action* Ldx [5.28]. Therefore, one can apply the infinitesimal invariance test (Section 7.1.7) with an appropriate modification. To expose formal changes required by this modification, let us deduce (9.98) from (9.97). We first rewrite the left-hand side of (9.97) in the form

$$\int_{\bar{V}} L(\bar{x}, \bar{u}, \bar{u}_{(1)}) d\bar{x} = \int_V L(\bar{x}, \bar{u}, \bar{u}_{(1)}) J dx,$$

where $J = \det\|D_j(\bar{x}^i)\|$ is the Jacobian of the first transformation (9.95). Since $J = 1$ at $a = 0$, the Jacobian is positive for sufficiently small a. Furthermore, the rule of differentiation of determinants yields $(\partial J/\partial a)|_{a=0} = D_i(\xi^i)$. Now both integrals in (9.97) are taken over the same volume V. Since this volume is arbitrary, the integral equation is equivalent to $L(\bar{x}, \bar{u}, \bar{u}_{(1)})J = L(x, u, u_{(1)})$. Upon substituting $\bar{x}^i \approx x^i + a\xi^i$, $\bar{u}^\alpha \approx u^\alpha + a\eta^\alpha$, $\bar{u}^\alpha_i \approx u^\alpha_i + a\zeta^\alpha_i$, $J \approx 1 + aD_i(\xi^i)$, it follows $L(\bar{x}, \bar{u}, \bar{u}_{(1)})J \approx L(x, u, u_{(1)}) + a[X(L) + LD_i(\xi^i)]$, hence, (9.98).

Theorem 1. Let the variational integral (9.93) be invariant under the group with the generator (9.96). Then the vector $T = (T^1, \ldots, T^n)$, $T^i \in \mathcal{A}$, defined by

$$T^i = L\xi^i + (\eta^\alpha - \xi^j u^\alpha_j)\frac{\partial L}{\partial u^\alpha_i}, \quad i = 1, \ldots, n, \tag{9.99}$$

is a conserved vector of the Euler-Lagrange equations (9.94), i.e. $D_i(T^i) = 0$ on the solutions of (9.94) [9.16].

9.7. SYMMETRY AND CONSERVATION LAWS

Proof. We use the above Lemma and write the operator (9.96) in the form (8.74):

$$X = \xi^i D_i + W^\alpha \frac{\partial}{\partial u^\alpha} + D_i(W^\alpha)\frac{\partial}{\partial u_i^\alpha} + \cdots, \quad \text{where} \quad W^\alpha = \eta^\alpha - \xi^j u_j^\alpha.$$

The left-hand side of equation (9.98) can be written as follows:

$$\begin{aligned}
X(L) + LD_i(\xi^i) &= \xi^i D_i(L) + W^\alpha \frac{\partial L}{\partial u^\alpha} + D_i(W^\alpha)\frac{\partial L}{\partial u_i^\alpha} + LD_i(\xi^i) \\
&= D_i(L\xi^i) + W^\alpha \frac{\partial L}{\partial u^\alpha} + D_i\left(W^\alpha \frac{\partial L}{\partial u_i^\alpha}\right) - W^\alpha D_i\left(\frac{\partial L}{\partial u_i^\alpha}\right) \\
&= D_i\left(L\xi^i + W^\alpha \frac{\partial L}{\partial u_i^\alpha}\right) + W^\alpha \left[\frac{\partial L}{\partial u^\alpha} - D_i\left(\frac{\partial L}{\partial u_i^\alpha}\right)\right].
\end{aligned}$$

Thus, the infinitesimal invariance test (9.98) is written

$$X(L) + LD_i(\xi^i) \equiv D_i(T^i) + W^\alpha \frac{\delta L}{\delta u^\alpha} = 0,$$

and hence $D_i(T^i) = 0$ on the solutions of (9.94).

Remark. Comparing with the derivation of equations (9.94), one might conclude that the integral equation (9.97) holds even when $L(\overline{x}, \overline{u}, \overline{u}_{(1)})J = L(x, u, u_{(1)}) + D_i(F^i)$, provided that functions $F^i \in \mathcal{A}$ vanish on the boundary ∂V of the volume V. However, since F^i (unlike h^α) are not arbitrary but determined by the Lagrangian L and the transformations (9.95), their vanishing on the boundary of an *arbitrary* volume V implies that $F^i = 0$ identically. But, on the other hand, our main concern is not so much with the variational integrals. Rather we wish to study differential equations represented in the form of Euler-Lagrange equations. Therefore, one can adopt Theorem 1 when the infinitesimal invariance condition (9.98) is replaced by a *divergence relation*, $X(L) + LD_i(\xi^i) = D_i(B^i)$, $B^i \in \mathcal{A}$. Then the conserved vector (9.99) is replaced by $C^i = T^i - B^i$, i.e. $D_i(C^i) = 0$ on the solutions of (9.94).

Theorem 2. Let $L(x, u, u_{(1)}, \ldots, u_{(s)}) \in \mathcal{A}$ be a differential function of any order. Consider, using notation (8.59), the Euler-Lagrange equations:

$$\frac{\delta L}{\delta u^\alpha} = 0, \quad \alpha = 1, \ldots, m. \tag{9.100}$$

Let X be a symmetry (point or Lie-Bäcklund (8.72)) of (9.100) such that

$$X(L) + LD_i(\xi^i) = D_i(B^i), \quad B^i \in \mathcal{A}. \tag{9.101}$$

Then equations (9.100) admit a conservation law, $D_i(C^i) = 0$, defined by

$$C^i = \mathsf{N}^i(L) - B^i, \quad i = 1, \ldots, n, \tag{9.102}$$

where N^i are the operators (8.78) associated with X.

Proof. The statement follows from the operator identity (8.82).

Definition 2. An X admitted by equations (9.100) is termed a *Noether symmetry* for (9.100) if (9.101) holds identically or on the solutions of (9.100).

9.7.3 Examples from classical mechanics

Consider the well-known conservation laws of mechanics from the group standpoint. The position of a particle in space will be defined by its Cartesian coordinates x^1, x^2, x^3 regarded as the components of the position vector $\boldsymbol{x} = (x^1, x^2, x^3)$. Accordingly, the velocity of the particle is a vector $\boldsymbol{v} = (v^1, v^2, v^3)$ defined as the derivative $\boldsymbol{v} = d\boldsymbol{x}/dt$ of $\boldsymbol{x}$ with respect to the time t. The motion of a single particle in a potential field $U(t, \boldsymbol{x})$ is described by the Lagrangian

$$L = \frac{m}{2} \sum_{k=1}^{3} (v^k)^2 - U(t, \boldsymbol{x}), \tag{9.103}$$

where $m =$ const. is the mass of the particle. Substitution of (9.103) into the Euler-Lagrange equations (9.92) gives the equations of motion:

$$m \frac{d^2 x^k}{dt^2} = -\frac{\partial U}{\partial x^k}, \quad k = 1, 2, 3. \tag{9.104}$$

Generators of point symmetry groups will be written in the form

$$X = \xi(t, \boldsymbol{x}) \frac{\partial}{\partial t} + \eta^k(t, \boldsymbol{x}) \frac{\partial}{\partial x^k}. \tag{9.105}$$

Denoting $\delta t = a\xi$, $\delta x^k = a\eta^k$, where a is a group parameter, we obtain another common form for infinitesimal transformations:

$$\bar{t} = t + \delta t, \quad \bar{\boldsymbol{x}} = \boldsymbol{x} + \delta \boldsymbol{x}. \tag{9.106}$$

In our case, formula (9.99) gives the quantity

$$T = \xi L + (\eta^k - \xi v^k) \frac{\partial L}{\partial v^k}, \tag{9.107}$$

which is a constant of motion, i.e. $dT/dt = 0$ on the solutions of equations (9.104).

Example 1. *Free motion of a particle* is obtained by letting $U = 0$ in (9.103). Hence the Lagrangian of a free particle has the form

$$L = \frac{m}{2} |\boldsymbol{v}|^2, \quad \text{where} \quad |\boldsymbol{v}|^2 = \sum_{k=1}^{3} (v^k)^2. \tag{9.108}$$

The action integral is invariant under the Galilean group (cf. Example 8 in Section 6.3.3) with the basic infinitesimal generators

$$X_0 = \frac{\partial}{\partial t}, \quad X_i = \frac{\partial}{\partial x^i}, \quad X_{ij} = x^j \frac{\partial}{\partial x^i} - x^i \frac{\partial}{\partial x^j}, \quad Y_i = t \frac{\partial}{\partial x^i}, \tag{9.109}$$

where $i, j = 1, 2, 3$. Let us apply Noether's theorem to each of these generators.

9.7. SYMMETRY AND CONSERVATION LAWS

(i) *Time translation.* Its generator X_0 has the coordinates $\xi = 1$ and $\eta^1 = \eta^2 = \eta^3 = 0$. Substitution into (9.107) yields, upon setting $E = -T$, the conservation of energy $E = m|\boldsymbol{v}|^2/2$.

(ii) *Space translations.* Consider the x^1-translation generated by X_1 with the coordinates $\xi = 0$, $\eta^1 = 1$, $\eta^2 = \eta^3 = 0$. Substitution into (9.107) yields, upon setting $T = p^1$, the conserved quantity $p^1 = mv^1$. The use of all space translations with the generators X_i furnishes us with the vector valued conservation quantity, viz. the linear momentum $\boldsymbol{p} = m\boldsymbol{v}$.

(iii) *Rotations.* Consider the rotation around the x^3-axis. Its generator X_{12} has the coordinates $\xi = 0$, $\eta^1 = x^2$, $\eta^2 = -x^1$, $\eta^3 = 0$. Substitution into (9.107) yields, upon setting $T = M_3$, the conserved quantity $M_3 = m(x^2 v^1 - x^1 v^2)$. The use of all rotations with the generators X_{12}, X_{13}, X_{23} furnishes us with the conservation of the angular momentum $\boldsymbol{M} = m(\boldsymbol{x} \times \boldsymbol{v})$.

(iv) *Galilean transformations.* Their generators Y_i, unlike X_0, X_i, X_{ij}, do not satisfy the invariance condition (9.98). Indeed, the extended action of Y_1,

$$Y_1 = t\frac{\partial}{\partial x^1} + \frac{\partial}{\partial v^1},$$

yields $Y_1(L) + LD_t(\xi) = mv^1 \equiv D_t(mx^1)$. Hence, one can apply Remark 9.7.2 with $B = mx^1$, while (9.107) gives $T = mtv^1$. This yields the conserved quantity $C = m(tv^1 - x^1)$. Using all operators Y_i and denoting C by Q, one obtains the vector valued conserved quantity $\boldsymbol{Q} = m(t\boldsymbol{v} - \boldsymbol{x})$. Conservation of the vector $\boldsymbol{Q}$ is known, in the case of a system of particles, as the center-of-mass theorem.

Remark. Conservation of the angular momentum follows, in general, from the rotational symmetry and is valid for Lagrangians (9.103) with a central potential field $U = U(r)$, $r = |\boldsymbol{x}|$. Note also that the three-parameter family of infinitesimal rotations is written in the form (9.106) with the increments $\delta t = 0$, $\delta \boldsymbol{x} = \boldsymbol{x} \times \boldsymbol{a}$, where $\boldsymbol{a} = (a^1, a^2, a^3)$.

9.7.4 Derivation of Kepler's laws from symmetries

The two-body problem (e.g. the sun and a planet), known also as Kepler's problem, is described by Newton's gravitation force (1.33), i.e. by the Lagrangian

$$L = \frac{m}{2}\sum_{k=1}^{3}(v^k)^2 - \frac{\mu}{r}, \quad \mu = \text{const.} \tag{9.110}$$

The equations of motion (cf. equations (1.34))

$$m\frac{d^2 x^k}{dt^2} = \mu\frac{x^k}{r^3}, \quad k = 1, 2, 3, \tag{9.111}$$

possess five infinitesimal point symmetries:

$$X_0 = \frac{\partial}{\partial t}, \quad X_{ij} = x^j\frac{\partial}{\partial x^i} - x^i\frac{\partial}{\partial x^j}, \quad Z = 3t\frac{\partial}{\partial x^i} + 2x^i\frac{\partial}{\partial x^i}. \tag{9.112}$$

The operators X_0 and X_{ij} provide, via Noether's theorem, the well-known conservation laws of energy and angular momentum, respectively:

$$E = \frac{m}{2}|\boldsymbol{v}|^2 + \frac{\mu}{r}, \quad \boldsymbol{M} = m(\boldsymbol{x} \times \boldsymbol{v}).$$

It follows from the constancy of the vector product $\boldsymbol{x} \times \boldsymbol{v}$ that the orbit of a planet lies in a fixed plane perpendicular to the constant vector $\boldsymbol{M}$. If we choose the rectangular coordinate frame (x, y, z) with the sun at the origin and the z-axis directed along $\boldsymbol{M}$, then $z = 0$ on the orbit, i.e. the path of the planet lies on the (x, y) plane, while its velocity is $\boldsymbol{v} = (v^1, v^2, 0)$. Hence, $\boldsymbol{M} = (0, 0, M)$ with $M = m(xv^2 - yv^1)$ which is a conserved quantity, i.e. $M = $ const. on the orbit. In the polar coordinates, $x = r\cos\theta$, $y = r\sin\theta$, the conservation of the angular momentum is written $M = mr^2 d\theta/dt = $ const. Since $dS = (r^2/2)d\theta$ defines the area of the infinitesimal sector bounded by two neighboring position vectors and an element of the orbit, the conservation of the angular momentum means that the sectorial velocity dS/dt is constant. It follows after integration that the position vector $\boldsymbol{x}$ of the planet sweeps out equal areas in equal times. Hence, *Kepler's second law* is merely a geometrical formulation of the conservation of the angular momentum.

The generator Z (9.12) is not a Noether symmetry and hence it does not furnish a conservation law. However, *Kepler's third law* follows directly from the scaling invariance of equations (9.111). Indeed, the operator Z provides the invariant $J = t^2/r^3$, the constancy of which is a mathematical expression of Kepler's third law (cf. Section 1.3.4).

It remains to discuss a group theoretic meaning of *Kepler's first law*. Considering canonical Lie-Bäcklund operators (see Definition 2, Section 8.4.2) which are at most linear in the velocity components v^k, one can find [5.28] the following generators admitted by the equations of motion (9.111):

$$X_l = \left(2x^l v^k - x^k v^l - (\boldsymbol{x}\cdot\boldsymbol{v})\delta_l^k\right)\frac{\partial}{\partial x^k}, \quad l = 1, 2, 3. \tag{9.113}$$

They can be written in the form (5.39), $\delta\boldsymbol{x} = \boldsymbol{x}\times(\boldsymbol{v}\times\boldsymbol{a})+(\boldsymbol{x}\times\boldsymbol{v})\times\boldsymbol{a}$ (cf. Remark 9.7.3). We have $X_l(L) = D_t(-2\mu x^l/r)$ on the solutions of (9.111). Hence, (9.102) yields three conserved quantities, viz. the components of Laplace's vector (5.38):

$$\boldsymbol{A} = \boldsymbol{v}\times\boldsymbol{M} + \mu\frac{\boldsymbol{x}}{r}. \tag{9.114}$$

On the orbit, (9.114) yields $\boldsymbol{A} = M(v^2, -v^1, 0) + (\mu/r)(x, y, 0)$, provided that we use the special coordinates (x, y, z) introduced above. Hence, $\boldsymbol{A} = (A_1, A_2, 0)$, where $A_1, A_2 = $ const. on the orbit. Now we take the scalar product of both sides of equation (9.114) with the position vector $\boldsymbol{x} = (x, y, 0)$. The resulting equation $A_1 x + A_2 y = M(xv^2 - yv^1) + \mu r$, after substituting $xv^2 - yv^1 = M/m$, defines an ellipse in polar coordinates:

$$r(A_1\cos\theta + A_2\sin\theta) = \frac{M^2}{m} + \mu r. \tag{9.115}$$

9.7. SYMMETRY AND CONSERVATION LAWS

Table 9.3 Group theoretic background of Kepler's laws

Kepler's law	Conserved quantity	Underlying symmetry
The first law Elliptic orbits	Laplace's vector $A = v \times M + \mu x/r$	Infinitesimal Lie-Bäcklund $\delta x = x \times (v \times a) + (x \times v) \times a$
The second law Equality of areas	Angular momentum $M = m(x \times v)$	Infinitesimal rotation $\delta x = x \times a$
The third law Invariance of t^2/r^3	None	Scaling $\bar{t} = a^3 t, \;\; \bar{x} = a^2 x$

9.7.5 Conservation laws of relativistic mechanics

An event occurring in a material particle is described by the place where it occurred and the time when it occurred. Thus, events are represented by points in a four-dimensional space of three space coordinates and the time. Points in this four-dimensional space are called *world points*. The totality of world points corresponding to each particle describe a line called a *world line*. This line determines the positions of the particle in all moments of time. The world line of a particle in uniform motion with a constant velocity is a straight line.

In what follows, world points will be defined by three space coordinates (x, y, z) referred to the rectangular cartesian frame and by time t.

An *inertial system of reference* comprises space coordinates and time such that the velocity of a freely moving body referred to this coordinates is constant.

The *principle of relativity* asserts that the laws of nature are identical in all inertial systems of reference. It means that the equations describing the laws of natural phenomena are invariant under the invertible transformations of space-time coordinates (x, y, z, t) from one inertial system to another. The commonly known example is the *Galilean principle of relativity* in classical mechanics.

The *fundamental assumption,* based on experiments with the propagation of light, is that the velocity of propagation of interactions is a universal constant. This constant velocity is the velocity c of light in empty space. Its numerical value is $c = 2.99793 \times 10^{10}$ cm/s.

Special relativity, formulated by A. Einstein in 1905, is the combination of the principle of relativity with the assumption on the finiteness of the velocity of propagation of interactions.

9. SYMMETRY OF DIFFERENTIAL EQUATIONS

Let a signal propagate, in an inertial system of reference, with the light velocity c. We suppose that one sends the signal at time t_1 from a point (x_1, y_1, z_1, t_1). Let the signal arrive at point (x_2, y_2, z_2) at time t_2. Since the signal propagates with velocity c, the distance $\sqrt{(x_2 - x_1)^2 + (y_2 - y_1)^2 + (z_2 - z_1)^2}$ covered in time $t_2 - t_1$ is equal to $c(t_2 - t_1)$. Hence, the world points with coordinates (x_1, y_1, z_1, t_1) and (x_1, y_1, z_1, t_1) are related by

$$c(t_2 - t_1) = \sqrt{(x_2 - x_1)^2 + (y_2 - y_1)^2 + (z_2 - z_1)^2}.$$

Let us write this relation in the symmetric form:

$$c^2(t_2 - t_1)^2 - (x_2 - x_1)^2 - (y_2 - y_1)^2 - (z_2 - z_1)^2 = 0.$$

The quantity

$$s_{12} = \sqrt{c^2(t_2 - t_1)^2 - (x_2 - x_1)^2 - (y_2 - y_1)^2 - (z_2 - z_1)^2}$$

is called the *interval* between the world points (or events) (x_1, y_1, z_1, t_1) and (x_1, y_1, z_1, t_1). The infinitesimal interval ds is defined then by

$$ds^2 = c^2 dt^2 - dx^2 - dy^2 - dz^2. \tag{9.116}$$

The assumption that light velocity c is universally constant implies that the differential form (9.116) is invariant with respect to any change of coordinates of world points. The four-dimensional Riemannian space of signature $(+---)$ with the metric form (9.116) is known as the *Minkowski space*, or space-time.

The group of isometric motions in the Minkowski space-time is the *Lorentz group* with the basic infinitesimal generators (cf. operators (9.109) and Example 9 in Section 6.3.3)

$$X_0 = \frac{\partial}{\partial t}, \quad X_i = \frac{\partial}{\partial x^i}, \quad X_{ij} = x^j \frac{\partial}{\partial x^i} - x^i \frac{\partial}{\partial x^j}, \quad X_{0i} = t \frac{\partial}{\partial x^i} + \frac{1}{c^2} x^i \frac{\partial}{\partial t}, \tag{9.117}$$

where $i, j = 1, 2, 3$.

One can develop a theory of relativity in any four-dimensional Riemannian space V_4 with a metric form

$$ds^2 = g_{ij}(x) dx^i dx^j \tag{9.118}$$

of the space-time signature $(+---)$. We discuss here the relativistic mechanics of particles, more specifically the motion of a free material particle. The motion of a particle with mass m is determined by the following rule.

Principle of least action: a free particle moves so that its world line $x^i = x^i(s)$, $i = 1, \ldots, 4$, is a geodesic. In other words, the motion of a particle in space with the metric (9.118) is determined by the Lagrangian

$$L = -mc\sqrt{g_{ij}(x) \dot{x}^i \dot{x}^j}. \tag{9.119}$$

9.7. SYMMETRY AND CONSERVATION LAWS

Here $x = (x^1, \ldots, x^4)$, and $\dot{x} = dx/ds$ is the derivative of the four-vector x with respect to the arc length s measured from a fixed point x_0. Thus, the equations of free motion are the Euler-Lagrange equations (9.92) with the Lagrangian (9.119). These equations have the form

$$\frac{d^2 x^i}{ds^2} + \Gamma^i_{jk}(x) \frac{dx^j}{ds} \frac{dx^k}{ds} = 0, \quad i = 1, \ldots, 4,$$

where the coefficients Γ^i_{jk} are known as the *Christoffel symbols* and are given by

$$\Gamma^i_{jk} = \frac{1}{2} g^{il} \left(\frac{\partial g_{lj}}{\partial x^k} + \frac{\partial g_{lk}}{\partial x^j} - \frac{\partial g_{jk}}{\partial x^l} \right).$$

A subsequent development of the relativistic mechanics is based on the group of isometric motions. The generators

$$X = \eta^i(x) \frac{\partial}{\partial x^i} \tag{9.120}$$

of the isometric motions in the space V_4 with the metric form (9.118) are determined by the *Killing equations*:

$$\eta^k \frac{\partial g_{ij}}{\partial x^k} + g_{ik} \frac{\partial \eta^k}{\partial x^j} + g_{jk} \frac{\partial \eta^k}{\partial x^i} = 0, \quad i \leq j, \ i, j = 1, \ldots, 4. \tag{9.121}$$

We now turn to the *special relativity* based on the Minkowski space-time with the metric form (9.116). In what follows, the vector of space variables is denoted by $\boldsymbol{x} = (x^1, x^2, x^3)$. Consequently, the physical velocity $\boldsymbol{v} = d\boldsymbol{x}/dt$ is a three-dimensional vector $\boldsymbol{v} = (v^1, v^2, v^3)$. Their scalar and vector products are denoted by $(\boldsymbol{x} \cdot \boldsymbol{v})$ and $\boldsymbol{x} \times \boldsymbol{v}$, respectively. Equation (9.116), written in the form

$$ds = c\sqrt{1 - \beta^2}\, dt \quad \text{with } \beta^2 = |\boldsymbol{v}|^2/c^2, \tag{9.122}$$

yields:

$$\frac{d\boldsymbol{x}}{ds} = \frac{\boldsymbol{v}}{c\sqrt{1-\beta^2}}, \quad \frac{dt}{ds} = \frac{1}{c\sqrt{1-\beta^2}}.$$

The action integral is defined by $S = \mu \int_a^b ds$, where $\int_a^b$ is an integral along the world line of the particle between two arbitrary world points a and b and α is a normalizing factor. One represents the action as an integral with respect to the time by substituting the expression (9.122) for ds:

$$S = \mu c \int_{t_1}^{t_2} \sqrt{1 - \beta^2}\, dt, \quad \mu = \text{const}.$$

Thus, we arrive at the action integral of Section 9.7.1 with the Lagrangian

$$L = \mu c \sqrt{1 - \beta^2}. \tag{9.123}$$

The constant factor μ is to be found from the requirement that in classical mechanics where $|\boldsymbol{v}| \ll c$, the function (9.123) must go over into the free particle Lagrangian (9.108) $L = m|\boldsymbol{v}|^2/2$. Expanding the function (9.123) in powers of β^2 and neglecting terms of higher order, one obtains:

$$L = \mu c \sqrt{1 - \beta^2} \approx \mu c \left(1 - \frac{\beta^2}{2}\right) = \mu c - \frac{\mu |\boldsymbol{v}|^2}{2c}.$$

Dropping the constant term μc and comparing with the classical Lagrangian (9.108), one finds $\mu = -mc$. Hence, the relativistic Lagrangian for a free particle with mass m has the form

$$L = -mc^2 \sqrt{1 - \beta^2}, \qquad (9.124)$$

where $\beta^2 = |\boldsymbol{v}|^2/c^2$, $|\boldsymbol{v}|^2 = \sum_{i=1}^{3}(v^i)^2$.

Now we can find the conservation laws in special relativity. We write the generators (9.117) of the Lorentz group in the form (9.105),

$$X = \xi(t, x)\frac{\partial}{\partial t} + \eta^i(t, x)\frac{\partial}{\partial x^i},$$

and employ formula (9.107) for conserved quantities,

$$T = \xi L + (\eta^i - \xi v^i)\frac{\partial L}{\partial v^i}.$$

(i) *Time translation.* Its generator X_0 has the coordinates $\xi = 1$ and $\eta^1 = \eta^2 = \eta^3 = 0$. Formula (9.107) yields, upon setting $E = -T$, the *relativistic energy*:

$$E = \frac{mc^2}{\sqrt{1 - \beta^2}}.$$

(ii) *Space translations.* Similarly, the operators X_i yield the *relativistic momentum*

$$\boldsymbol{p} = \frac{m\boldsymbol{v}}{\sqrt{1 - \beta^2}}.$$

(iii) *Rotations.* Using the operators X_{ij}, one obtains the *relativistic angular momentum* $\boldsymbol{M} = \boldsymbol{x} \times \boldsymbol{p}$.

(iv) *Lorentz transformations.* Their generators X_{0i} give rise to the vector

$$\boldsymbol{Q} = \frac{m(\boldsymbol{x} - t\boldsymbol{v})}{\sqrt{1 - \beta^2}}.$$

Conservation of the vector $\boldsymbol{Q}$ provides the *relativistic center-of-mass theorem*.

Remark. In the classical limit as $\beta^2 = |\boldsymbol{v}|^2/c^2 \to 0$, the above conservation laws go over into the conservation laws of Section 9.7.3.

Problems

9.1. The differential equation of the system of conics, $C_1 x^2 + C_2 y^2 = 1$, has the form:

$$y'' + \frac{y'^2}{y} - \frac{y'}{x} = 0.$$

Deduce this equation by eliminating the parameters and find its scaling symmetries.

9.2. Prove that a differential equation $y^{(n)} = f(x, y, y', \ldots, y^{(n-1)})$ admits the translation group $\bar{x} = x + a$ if and only if the variable x does not occur in f, i.e. the equation has the form $y^{(n)} = f(y, y', \ldots, y^{(n-1)})$. Show also that the invariance under y-translations, $\bar{y} = y + a$, implies $y^{(n)} = f(x, y', \ldots, y^{(n-1)})$. Cf. Example 1 in Section 5.2.2.

9.3. Find the point symmetries of the following second-order equations ($k = \text{const.} \neq 0$):

(i) $y'' + ky = 0$, (ii) $y'' + 3yy' + y'^3 = 0$, (iii) $y'' - \dfrac{y' + y'^3}{x} = 0$, (iv) $y'' + \dfrac{y' + y'^3}{2x} = 0$,

(v) $y'' + \dfrac{y'}{x} - \dfrac{y}{4x^2} - \dfrac{k}{x^2 y^3} = 0$, (vi) $y'' - 2\dfrac{y' + ky'^{3/2} + y'^2}{y - x} = 0$, (vii) $y'' + e^{3y} y'^4 + y'^2 = 0$,

(viii) $(x^2 + y^2)y'' + (y - xy')(1 + y'^2) = 0$, (ix) $2y'' + 3x^{-5/2} y^{-1/2} = 0$,

(x) $y'' + y - k(y' - y'^3) = 0$, (see (5.46)), (xi) $y'' - \dfrac{y'}{x} + e^y = 0$, (xii) $3y'' + xy^{-5/3} = 0$,

(xiii) $y'' = [(x + x^2)y' + (1 + 2x)]e^y$, (xiv) $y'' + y'^2 + xy = 0$, (xv) $y'' - 6y^2 - x = 0$.

9.4. Find the point symmetries of third-order equations (for (i)–(iii), see (1.17)–(1.19)):

(i) $y''' = 0$, (ii) $y''' = 3\dfrac{y' y''^2}{1 + y'^2}$, (iii) $y''' = \dfrac{3}{2}\dfrac{y''^2}{y'}$, (iv) $y''' = e^{-y''}$, (v) $y''' = y^{-2}$ [9.17].

9.5. Find the contact symmetries of the third-order equations (ii)–(v) given above.

9.6. Calculate the point symmetries of the heat conduction equations:

(i) $u_t = u_{xx}$, (ii) $u_t = u_{xx} + u_{yy}$, (iii) $u_t = u_{xx} + u_{yy} + u_{zz}$.

9.7. Find the Lie algebra of point symmetries of the Black-Scholes equation (1.5), $u_t + (A^2/2)x^2 u_{xx} + Bx u_x - Cu = 0$, provided that $A \neq 0$.

9.8. Solve the determining equation in Example 2 of Section 9.3.2.

9.9. Find the symmetries of the shallow-water equations (see Section 5.4.4):

$$v_t + (v \cdot \nabla)v + g\nabla h = 0, \quad h_t + v \cdot \nabla h + h \operatorname{div} v = 0,$$

where $v = (v^1, v^2)$ is the two-dimensional velocity vector, h the depth of the water over a flat bottom, and g the gravitational constant (see Section 1.3.1). The two-dimensional vector-operator ∇ acts with respect to the position vector $x = (x^1, x^2)$.

9.10. Find the symmetries of the system $\dfrac{d^2 x}{dt^2} + k_1 x = 0$, $\dfrac{d^2 y}{dt^2} + k_2 y = 0$, $\dfrac{d^2 z}{dt^2} + k_3 z = 0$ with constant coefficients k_i.

9.11. Solve the system of two linear partial differential equations (9.64).

9.12. Deduce the system of ordinary differential equations (9.68) for invariant solutions by substituting (9.67) into equations (9.66). Integrate the system (9.68) and study shallow-water flows different from that described in Example 5, Section 9.4.1.

9.13.** Study symmetries of ordinary differential equations $F(x, y, y', \ldots, y^{(n)}) = 0$ not solved for $y^{(n)}$. As an example, investigate the maximal group admitted by the second-order equation $(y'')^2 = \left[(y'/x) - e^y\right]^2$ not solved for y'' (cf. Example 1 of Section 9.6).

9.14.** Investigate ordinary differential equations of order $n \geq 3$ admitting irreducible contact transformation groups. In particular, find all third-order equations of this type and clarify if they are reducible to $y''' = 0$ by suitable contact transformations.

9.15*. Find all transformations of variables x, y that carry every second-order equation of the form $y'' = F(x, y)$ into an equation of the same form. In other words, find the equivalence transformations for the class of equations $y'' = F(x, y)$.

9.16*. Show that (7.111) is an approximate symmetry group of the neutrinos' equations (5.45) and (5.45), provided that the wave functions ϕ and φ undergo suitable approximate linear transformations. Find the latter transformations.

9.17. Find exact infinitesimal symmetries,

$$X^0 = \tau_0(t, x, u)\frac{\partial}{\partial t} + \xi_0(t, x, u)\frac{\partial}{\partial x} + \eta_0(t, x, u)\frac{\partial}{\partial u},$$

of the unperturbed nonlinear wave equation (9.88), $u_{tt} = \left(u^\sigma u_x\right)_x$, $\sigma \neq 0$.

9.18. Prove that if X^0 is admitted by equation (9.83) then εX^0 is an infinitesimal approximate symmetry of the perturbed equation (9.77).

9.19. Examine for stability the symmetries of equation (9.88) in the case $\sigma = -4/3$ by calculating approximate symmetries of the perturbed equation (9.87).

9.20. The second-order ordinary differential equation $y'' - x - \varepsilon y^2 = 0$ has no exact point symmetries if ε is treated as a constant coefficient (cf. Problem 9.3(xv)). On the other hand, the linear equation $y'' - x = 0$ admits a Lie algebra L_8. Find the algebra L_8 and investigate its stability under the perturbation $-\varepsilon y^2$. In other words, find all approximate symmetries of the perturbed equation $y'' - x - \varepsilon y^2 = 0$.

9.21. Check that equation (9.90) is invariant under transformation (9.91).

9.22. Check that the Lagrangian $L = m|v|^2/2 + \beta/r^2$ of a particle in the Newton-Cotes field (5.47) and the generators of dilation and projective transformation groups,

$$X_1 = 2t\frac{\partial}{\partial t} + x^k\frac{\partial}{\partial x^k}, \quad X_1 = t^2\frac{\partial}{\partial t} + tx^k\frac{\partial}{\partial x^k},$$

satisfy the conditions of Noether's theorem and deduce the integrals of motion (5.49).

9.23. Transform equation (9.115) of an ellipse to its canonical form $r = p/(1 + e\cos\phi)$.

9.24*. Prove that (9.113) are Noether symmetries for equations of motion (9.104) in a central field $U(r)$ if and only if U is Newton's potential, $U = \mu/r$. See [5.39](i).

9.25.** Investigate approximate symmetries of the system consisting of the sun and two planets. Treat the interaction "planet-planet" as a small perturbation of the interaction "sun-planet". In particular, examine the stability of the Lie-Bäcklund symmetries (9.113) and the existence of an approximate version of the Laplace vector (9.114).

Chapter 10

Invariants of algebraic and differential equations

The above title is an abbreviation from *Invariants of groups of equivalence transformations of algebraic equations and differential invariants of groups of equivalence transformations of families of differential equations*. Following the classical literature, I will also use the nomenclature *algebraic and differential invariants*.

The invention of both algebraic and differential invariants can be dated back to 1773 when J.L. Lagrange noted the invariance of the discriminant of the general binary quadratic form under special linear transformations of the variables (with the determinant 1), and simultaneously P.S. Laplace introduced [10.1] his renowned invariants of linear partial differential equations of the second order.

It took another 70 years before G. Boole in 1841-42 generalized Lagrange's incidental observation to rational homogeneous functions (of many variables and of an arbitrary order) with the discovery of so-called *covariants* with respect to arbitrary linear transformations. Boole's success provided A. Cayley [10.2] with an incentive to begin in 1845 the systematic development of a new approach to linear transformations. His general theory of algebraic forms and their invariants became one of the dominating fields of pure mathematics in the 19th century.

Differential invariants were seriously considered in the 1880s [10.3]. They were found in the problem of the practical solution of differential equations by reducing them to equivalent but readily integrable forms (e.g. in [10.3](1), (3), (6)), or by using relations between solutions of a given equation (e.g. in [10.3](2)). Modern group analysis reveals that differential invariants furnish a powerful tool for tackling initial value problems (Riemann's method), qualitative analysis of differential equations (e.g. Huygens' principle), etc.

Lie noticed ([10.3](5)) that the background of the theory of differential invariants of linear differential equations is an infinite group but that, however, his contemporaries had not noticed this substantial fact. The purpose of this chapter is to outline the use of infinite groups and Lie's infinitesimal method in the theory of invariants of algebraic and differential equations [10.4].

10.1 Invariants of algebraic equations

10.1.1 Preliminaries

It is advantageous, for successive calculation of invariants, to write algebraic and linear ordinary differential equations in a *standard form* involving the binomial coefficients. The standard form of algebraic equations of the nth degree is

$$P_n(x) \equiv C_0 x^n + n C_1 x^{n-1} + \frac{n(n-1)}{2!} C_2 x^{n-2} + \cdots + n C_{n-1} x + C_n = 0. \quad (10.1)$$

Definition 1. An *equivalence transformation* of equations (10.1) is an invertible transformation $\bar{x} = f(x)$ such that the substitution of $x = f^{-1}(\bar{x})$ into $P_n(x)$ converts every equation (10.1) of the nth degree into an algebraic equation $\overline{P}_n(\bar{x}) = 0$ of the same degree n but, in general, with new coefficients $\overline{C}_i$.

Proposition. The most general group of equivalence transformations of equations (10.1) is provided by the linear fractional transformations (6.22),

$$\bar{x} = \frac{ax + \varepsilon}{b + \delta x}, \quad (10.2)$$

subject to the invertibility condition:

$$ab - \varepsilon \delta \neq 0. \quad (10.3)$$

Definition 2. An *invariant of equations* (10.1) is a function F of the coefficients[1] C_i unalterable under the equivalence transformations (10.2):

$$F(C_0, C_1, \ldots, C_n) = F(\overline{C}_0, \overline{C}_1, \ldots, \overline{C}_n). \quad (10.4)$$

Here $\overline{C}_i$ are given by Definition 1 applied to equivalence transformations (10.2).

In what follows we will encounter invariants of subgroups of equivalence groups. Following Laguerre's suggestion ([10.3](1)(ii)) and in accordance with Cayley's theory, this type of invariants will be termed *seminvariants*. Furthermore, certain invariance properties are represented by invariant equations,

$$H_\nu(C_0, C_1, \ldots, C_n) = 0, \quad \nu = 1, 2, \ldots, \quad (10.5)$$

where the functions H_ν are not necessarily invariants.

One encounters invariants of algebraic equations, e.g. when one applies a linear transformation of the variable x to the equation (10.1) so as to annul the term next to the highest (cf. Section 6.1.1). After this transformation, the coefficients of the transformed polynomial $\overline{P}_n(\bar{x})$ are given as rational functions of C_i. For example, the quadratic equation

$$P_2(x) \equiv C_0 x^2 + 2 C_1 x + C_2 = 0 \quad (10.6)$$

---
[1] Boole's *covariants*, unlike the invariants, are functions of both C_i and x.

10.1. INVARIANTS OF ALGEBRAIC EQUATIONS

is transformed to an equation lacking the second term,

$$\overline{P}_2(\bar{x}) \equiv \bar{x}^2 + H_1 = 0,$$

e.g. by the substitution $\bar{x} = C_0 x + C_1$. Then H_1 is the *discriminant*,

$$H_1 = C_0 C_2 - C_1^2. \tag{10.7}$$

Likewise, the substitution $\bar{x} = C_0 x + C_1$ transforms the cubic equation

$$P_3(x) \equiv C_0 x^3 + 3 C_1 x^2 + 3 C_2 x + C_3 = 0 \tag{10.8}$$

to an equation lacking the second term,

$$\overline{P}_3(\bar{x}) \equiv \bar{x}^3 + 3 H_1 \bar{x} + H_2 = 0,$$

where the first coefficient H_1 is again given by (10.7), and the second one is

$$H_2 = C_0^2 C_3 - 3 C_0 C_1 C_2 + 2 C_1^3. \tag{10.9}$$

The vanishing of both H_1 and H_2 defines a system of invariant equations (10.5) *and provides the necessary and sufficient condition for equation* (10.8) *to have three equal roots.*

Remark. The equation of fourth degree has three functions $H_1, H_2,$ and H_3 that provide invariant equations, the first two functions, H_1 and H_2, being (10.7) and (10.9). For equations (10.1) of an arbitrary degree, invariant equations are obtained recursively by adding to $H_1, H_2, H_3, \ldots$ only one more H at each step. This is an advantage gained from using the binomial coefficients in (10.1).

10.1.2 Tschirnhausen's transformation. An approach to the Galois group

E.W. Tschirnhausen in 1683 applied to equation (10.1) more general, than (10.2), transformations to eliminate the terms of the degrees $n-1, n-2, n-3$ [5.1]. Namely, he considered transformations given by the rational fractions:

$$\bar{x} = \frac{A_0 x^r + A_1 x^{r-1} + \cdots + A_r}{B_0 x^s + B_1 x^{s-1} + \cdots + B_s}. \tag{10.10}$$

It was noticed by G.W. Leibnitz that the application of Tschirnhausen's transformation trying to get rid also of the $(n-4)$th term requires the solution of equations more complicated than the original one.

Tschirnhausen's transformation (10.10) is not an equivalence transformation because its inverse is not single-valued. Nevertheless, it is useful in the transformation theory of equations. For example, it furnishes a tool for a conceptually simple approach to the Galois group suggested in [5.36](ii) to draw a parallel between Galois groups of algebraic equations and Lie symmetries of differential equations. Unlike the algebraic Galois theory where the Galois group is defined by means of extensions of fields, this approach is based on the notion of invariance of equations. Consider the following illustrative examples.

Example 1. Let us find all linear fractional transformations (10.2) with complex coefficients $a, b, \varepsilon, \delta$, carrying the quadratic equation

$$P_2(x) \equiv x^2 + 1 = 0 \tag{10.11}$$

into itself, $P_2(\bar{x}) = 0$ whenever $P_2(x) = 0$. In other words, let us find the coefficients $a, b, \varepsilon, \delta$ of (10.2) by solving the *determining equation*:

$$(\bar{x}^2 + 1)|_{x^2 = -1} = 0. \tag{10.12}$$

The substitution of (10.2) into the polynomial $P(\bar{x}) = \bar{x}^1 + 1$ yields:

$$P(\bar{x}) = (\delta x + b)^{-2}[(a^2 + \delta^2)x^2 + 2(a\varepsilon + b\delta)x + b^2 + \varepsilon^2].$$

Hence, the determining equation (10.12) is written

$$2(a\varepsilon + b\delta)x + b^2 + \varepsilon^2 - a^2 - \delta^2 = 0$$

and splits into the following two equations:

$$a\varepsilon + b\delta = 0, \tag{10.13}$$

$$b^2 + \varepsilon^2 - a^2 - \delta^2 = 0. \tag{10.14}$$

If $\delta = 0$, equation (10.13) and the condition (10.3) yield $\varepsilon = 0$. Then (10.14) reduces to $b = \pm a$. Thus, equation (10.11) is invariant under the following two transformations (10.2) with $\delta = 0$:

$$\bar{x} = x \quad \text{and} \quad \bar{x} = -x. \tag{10.15}$$

In the case $\delta \neq 0$, equations (10.13) and (10.14) are written:

$$b = -\frac{a\varepsilon}{\delta}, \quad (a^2 + \delta^2)(\varepsilon^2 - \delta^2) = 0.$$

These equations together with the condition (10.3) yield $\varepsilon^2 - \delta^2 = 0$. Hence,

$$\delta = \varepsilon, \ b = -a, \quad \text{or} \quad \delta = -\varepsilon, \ b = a,$$

where $\varepsilon \neq 0$ and a are arbitrary constants. Thus, if $\delta \neq 0$ we obtain the following two types of transformations:

$$\bar{x} = \frac{ax + \varepsilon}{a - \varepsilon x} \quad \text{and} \quad \bar{x} = \frac{ax + \varepsilon}{\varepsilon x - a}. \tag{10.16}$$

If $a = 0$, these transformations reduce to

$$\bar{x} = -1/x \quad \text{and} \quad \bar{x} = 1/x. \tag{10.17}$$

10.1. INVARIANTS OF ALGEBRAIC EQUATIONS

If $a \neq 0$, the first transformation (10.16) provides a one-parameter local group:

$$T_\alpha : \bar{x} = \frac{x + \alpha}{1 - \alpha x}, \tag{10.18}$$

while the second transformation (10.16) can be written (here the parameter is denoted by β to distinguish the two one-parameter families of transformations):

$$S_\beta : \bar{x} = \frac{x + \beta}{\beta x - 1}. \tag{10.19}$$

The transformations (10.18)–(10.19) reduce to (10.15) when $\alpha = 0$ and $\beta = 0$, and to (10.17) when $\alpha = \infty$ and $\beta = \infty$. Thus, equation (10.11) is invariant with respect to the one-parameter families of transformations (10.18)–(10.19), where the parameters α and β range over the *extended complex plane*.

The transformations (10.18) and (10.19) form a group (see Problem 10.3). It is a group of mixed type, in the terminology of Section 6.3.4.

Since equation (10.11) is invariant under the group G of transformations (10.18)–(10.19), the roots $x_1 = i$, $x_2 = -i$ of (10.11) are merely permuted among themselves (or are individually unaltered) by the transformations (10.18)–(10.19). Consequently, one can consider the action of the group G restricted on the set $\{x_1, x_2\}$. This restriction is a group (an *induced group* by Definition 1 of Section 7.2.2). This induced group comprises permutations of the roots x_1, x_2 and is the *Galois group* of equation (10.11). It is denoted here by $\mathcal{G}$.

To find the elements of the Galois group $\mathcal{G}$, let us consider the action of transformations (10.18) and (10.19) on the roots:

$$T_\alpha(x_1) = \frac{i + \alpha}{1 - \alpha i} = i \equiv x_1, \quad T_\alpha(x_2) = \frac{-i + \alpha}{1 + \alpha i} = -i \equiv x_2,$$

$$S_\beta(x_1) = \frac{i + \beta}{\beta i - 1} = -i \equiv x_2, \quad S_\beta(x_2) = \frac{-i + \beta}{-\beta i - 1} = i \equiv x_1.$$

Hence, the restriction of T_α on the roots is the identical transformation which is designated by 1 (the unit), while S_β permutes the roots and is denoted by (x_1, x_2). Thus, the Galois group $\mathcal{G}$ of equation (10.11) comprises two elements:

$$\mathcal{G} = \{1, (x_1, x_2)\}.$$

Here, the Galois group consists of all possible permutations of x_1, x_2. In this sense, the example is trivial. Therefore, let us consider examples of equations of the fourth degree when the Galois group contains four elements of all possible 24 permutations of four roots of the equations under consideration.

Example 2. Consider the transformations (10.2) leaving invariant the equation

$$P_4(x) \equiv x^4 - x^2 + 1 = 0. \tag{10.20}$$

Substituting (10.2) into the determining equation

$$(\bar{x}^4 - \bar{x}^2 + 1)|_{x^4 = x^2 - 1} = 0, \qquad (10.21)$$

one can find the following four linear fractional transformations (10.2) that leave invariant equation (10.20):

$$I: \bar{x} = x; \quad S: \bar{x} = -x; \quad R: \bar{x} = \frac{1}{x}; \quad T: \bar{x} = -\frac{1}{x}.$$

Denoting by $\tilde{I}, \tilde{S}, \tilde{R}, \tilde{T}$ the restriction of I, S, R, T on the roots

$$x_1 = \sqrt{(1+i\sqrt{3})/2}, \quad x_2 = -x_1, \quad x_3 = \sqrt{(1-i\sqrt{3})/2}, \quad x_4 = -x_3$$

of equation (10.20), we obtain $\tilde{I} = 1$ (the unit) and the following permutations:

$$\tilde{S} = \begin{pmatrix} x_1 & x_2 & x_3 & x_4 \\ x_2 & x_1 & x_4 & x_3 \end{pmatrix}, \quad \tilde{R} = \begin{pmatrix} x_1 & x_2 & x_3 & x_4 \\ x_3 & x_4 & x_1 & x_2 \end{pmatrix}, \quad \tilde{T} = \begin{pmatrix} x_1 & x_2 & x_3 & x_4 \\ x_4 & x_3 & x_2 & x_1 \end{pmatrix}.$$

Hence, the Galois group of equation (10.20) comprises four elements:

$$\mathcal{G} = \{1, (x_1, x_2)(x_3, x_4), (x_1, x_3)(x_2, x_4), (x_1, x_4)(x_2, x_3)\}.$$

Example 3. Consider the equation

$$P_4(x) \equiv x^4 + x^3 + x^2 + x + 1 = 0. \qquad (10.22)$$

It is invariant, along with the identical transformation $\bar{x} = x$, under the following Tschirnhausen's transformations:

$$\bar{x} = x^2, \quad \bar{x} = x^3, \quad \bar{x} = \frac{1}{x}, \quad \bar{x} = -(x^3 + x^2 + x + 1). \qquad (10.23)$$

Furthermore, equation (10.22) admits the transformation $\bar{x} = x^n$ with any integer n indivisible by 5.

Restriction of these transformations to the roots

$$x_1 = \epsilon, \quad x_2 = \epsilon^2, \quad x_3 = \epsilon^3, \quad x_4 = \epsilon^4$$

of equation (10.22), where $\epsilon = e^{2\pi i/5}$, yields precisely four different permutations, namely the unit 1 and the following permutations:

$$\begin{pmatrix} x_1 & x_2 & x_3 & x_4 \\ x_2 & x_4 & x_1 & x_3 \end{pmatrix}, \quad \begin{pmatrix} x_1 & x_2 & x_3 & x_4 \\ x_3 & x_1 & x_4 & x_2 \end{pmatrix}, \quad \begin{pmatrix} x_1 & x_2 & x_3 & x_4 \\ x_4 & x_3 & x_2 & x_1 \end{pmatrix}.$$

Hence, the Galois group of equation (10.22) comprises four elements:

$$\mathcal{G} = \{1, (x_1, x_2, x_4, x_3), (x_1, x_3, x_4, x_2), (x_1, x_4)(x_2, x_3)\}.$$

10.1. INVARIANTS OF ALGEBRAIC EQUATIONS

10.1.3 Infinitesimal method

Equation (10.1) is rewritten, by setting $x = u/v$ and multiplying the resulting equation by v^n, in the homogeneous form:

$$Q_n(u,v) \equiv C_0 u^n + nC_1 u^{n-1} v + \frac{n(n-1)}{2!} C_2 u^{n-2} v^2 + \cdots + nC_{n-1} uv^{n-1} + C_n v^n = 0.$$

Then the group of linear fractional transformations (10.2) is represented as the linear homogeneous group:

$$\bar{u} = au + \varepsilon v, \quad \bar{v} = bv + \delta u.$$

This simplifies calculations, e.g. makes Proposition 10.1.1 self-evident.

I further set $a = e^\alpha, b = e^\beta$ and consider the *infinitesimal equivalence transformations*,

$$\bar{u} \approx u + (\alpha u + \varepsilon v), \quad \bar{v} \approx v + (\beta v + \delta u).$$

Its inverse, written in the first order of precision with respect to the small parameters $\alpha, \beta, \varepsilon, \delta$, has the form:

$$u \approx \bar{u} - (\alpha \bar{u} + \varepsilon \bar{v}), \quad v \approx \bar{v} - (\beta \bar{v} + \delta \bar{u}). \tag{10.24}$$

The infinitesimal approach to the theory of algebraic invariants will be exhibited here by applying the method to the quadratic and the cubic equations.

THE QUADRATIC EQUATION

Substitution of (10.24) into the quadratic equation

$$Q_2(u,v) \equiv C_0 u^2 + 2C_1 uv + C_2 v^2 = 0 \tag{10.25}$$

converts it into $\overline{Q}_2(\bar{u}, \bar{v}) \equiv \overline{C}_0 \bar{u}^2 + 2\overline{C}_1 \bar{u}\bar{v} + \overline{C}_2 \bar{v}^2 = 0$, where

$$\overline{C}_0 \approx C_0 - 2(\alpha C_0 + \delta C_1), \quad \overline{C}_1 \approx C_1 - (\alpha C_1 + \beta C_1 + \varepsilon C_0 + \delta C_2),$$

$$\overline{C}_2 \approx C_2 - 2(\beta C_2 + \varepsilon C_1). \tag{10.26}$$

Formulas (10.26) provide the infinitesimal transformations of the four-parameter group of transformations of coefficients C_i of the quadratic equation (10.25). Let us denote this group by G_4^2. According to (10.26), its generators are:

$$X_1 = 2C_0 \frac{\partial}{\partial C_0} + C_1 \frac{\partial}{\partial C_1}, \quad X_2 = C_1 \frac{\partial}{\partial C_1} + 2C_2 \frac{\partial}{\partial C_2},$$

$$X_3 = C_0 \frac{\partial}{\partial C_1} + 2C_1 \frac{\partial}{\partial C_2}, \quad X_4 = 2C_1 \frac{\partial}{\partial C_0} + C_2 \frac{\partial}{\partial C_1}.$$

The invariants $F(C_0, C_1, C_2)$ of the group G_4^2 are determined by the equations $X_s(F) = 0, \ s = 1, \ldots, 4$. We know from Section 4.5.1 that a necessary condition for the existence of non-trivial ($F \neq$ const.) solutions of these equations is $r_* < 3$, where r_* is the *generic rank* of the matrix A associated with the operators X_s:

$$A = \begin{Vmatrix} 2C_0 & C_1 & 0 \\ 0 & C_1 & 2C_2 \\ 0 & C_0 & 2C_1 \\ 2C_1 & C_2 & 0 \end{Vmatrix}. \tag{10.27}$$

But here $r_* = 3$, and hence the group G_4^2 has no invariants. Furthermore, it follows from $r_* = 3$ that G_4^2 has no *regular* invariant equations (10.5) either. However, it may have *singular* invariant equations $H(C_0, C_1, C_2) = 0$. The latter are found, according to Section 7.4.4, by imposing the condition $rank A|_{H=0} < 3$ on the elements of the matrix (10.27) and testing the infinitesimal invariance criterion $X_s(H)|_{H=0} = 0$, $s = 1, \ldots, 4$. Implementation of this algorithm shows that there is only one singular invariant equation, viz. $H_1 \equiv C_0 C_2 - C_1^2 = 0$, where H_1 is the discriminant (10.7).

THE CUBIC EQUATION

Likewise, by substituting (10.24) into the cubic form $Q_3(u,v) = C_0 u^3 + 3C_1 u^2 v + 3C_2 u v^2 + C_3 v^3$, one arrives at the 4-parameter group G_4^3 generated by

$$X_1 = 3C_0 \frac{\partial}{\partial C_0} + 2C_1 \frac{\partial}{\partial C_1} + C_2 \frac{\partial}{\partial C_2}, \quad X_2 = C_1 \frac{\partial}{\partial C_1} + 2C_2 \frac{\partial}{\partial C_2} + 3C_3 \frac{\partial}{\partial C_3},$$

$$X_3 = C_0 \frac{\partial}{\partial C_1} + 2C_1 \frac{\partial}{\partial C_2} + 3C_2 \frac{\partial}{\partial C_3}, \quad X_4 = 3C_1 \frac{\partial}{\partial C_0} + 2C_2 \frac{\partial}{\partial C_1} + C_3 \frac{\partial}{\partial C_2},$$

and at the associated matrix

$$A = \begin{Vmatrix} 3C_0 & 2C_1 & C_2 & 0 \\ 0 & C_1 & 2C_2 & 3C_3 \\ 0 & C_0 & 2C_1 & 3C_2 \\ 3C_1 & 2C_2 & C_3 & 0 \end{Vmatrix}.$$

Here again the rank of A is equal to the number of the transformed quantities C_i, i.e. $r_* = 4$. Hence G_4^3 has neither invariants nor regular invariant equations.

But it has singular invariant equations of two types. One of them is obtained by letting rank $A = 3$, i.e. det $A = 0$. It is given by

$$\Delta \equiv (C_0 C_3)^2 - 6 C_0 C_1 C_2 C_3 + 4 C_0 C_2^3 - 3(C_1 C_2)^2 + 4 C_1^3 C_3 = 0, \tag{10.28}$$

where Δ is known as the *discriminant* of the cubic equation (cf. Section 6.1.1). The second type of singular invariant equations is obtained by letting rank $A = 2$, i.e. by annulling the minors of all elements of A, and applying the invariance test. It follows that:

$$H \equiv C_0 C_2 - C_1^2 = 0, \quad F \equiv C_0^2 C_3 - C_1^3 = 0. \tag{10.29}$$

Noting that here $H = H_1$ and $F = H_2 + 3C_1 H_1$ with H_1 and H_2 given by (10.7) and (10.9), one can rewrite (10.29) in the equivalent form $H_1 = 0$, $H_2 = 0$.

10.2 Linear ordinary differential equations

10.2.1 Preliminaries

The standard form of the general linear homogeneous ordinary differential equation of the nth order with (regular) variable coefficients $c_i(x)$ is

$$y^{(n)} + nc_1(x)y^{(n-1)} + \frac{n!c_2(x)}{(n-2)!2!}y^{(n-2)} + \cdots + nc_{n-1}(x)y' + c_n(x)y = 0. \quad (10.30)$$

Definition. An equivalence transformation of equations (10.30) is an invertible transformation of the independent and dependent variables, x and y, preserving the order n of equations (10.30) as well as their linearity and homogeneity.

Proposition The most general group of equivalence transformations of the equations (10.30) is an infinite group composed of linear transformations of the dependent variable:

$$y = \sigma(x)z, \quad \sigma(x) \neq 0, \quad (10.31)$$

and invertible changes of the independent variable:

$$\bar{x} = \phi(x), \quad \phi'(x) \neq 0, \quad (10.32)$$

where $\sigma(x)$ and $\phi(x)$ are arbitrary n times continuously differentiable functions.

Papers cited in [10.3] deal with differential invariants of the subgroup (10.31) of the general group of equivalence transformations. These differential invariants are therefore termed in what follows *seminvariants* of equations (10.30). For example, the second-order equations

$$y'' + 2c_1(x)y' + c_2(x)y = 0 \quad (10.33)$$

have one differential seminvariant:

$$h_1 = c_2 - c_1^2 - c_1'; \quad (10.34)$$

the third-order equations have two seminvariants, namely (10.34) and

$$h_2 = c_3 - 3c_1c_2 + 2c_1^3 - c_1''. \quad (10.35)$$

An analog of Tschirnhausen's transformation for differential equations (10.30) was also discovered in the nineteenth century. Namely, Cockle [10.3](6)(ii) (for $n = 3$) and Laguerre [10.3](1)(ii) (for the general equation) independently showed that the two terms of orders next below the highest can be simultaneously removed in any equation (10.30) by the transformations (10.31) and (10.32).

Exercise. Find the equivalence transformations (10.31)–(10.32) reducing equation (10.30) to that with $c_1 = c_2 = 0$, i.e. to *Laguerre's canonical form*:

$$y^{(n)} + \frac{n!c_3(x)}{(n-3)!3!}y^{(n-3)} + \cdots + nc_{n-1}(x)y' + c_n(x)y = 0. \quad (10.36)$$

Hint: Use results of Exercise 2 in Section 8.2.3.

Solution. Let us remove the $(n-1)$th term by transforming the differential variable only, i.e. using the substitution (10.31), $y = \sigma(x)z$. Since

$$y^{(n)} + nc_1 y^{(n-1)} = \sigma z^{(n)} + n(\sigma' + c_1\sigma)z^{(n-1)} + \cdots,$$

we get rid of the $(n-1)$th term by letting $\sigma' + c_1\sigma = 0$. Hence, $y = z\,\mathrm{e}^{-\int c_1(x)\mathrm{d}x}$.

Thus, without loss of generality, we may assume that $c_1 = 0$ in equation (10.30). Let us apply to this equation the general equivalence transformations (10.31)–(10.32). Invoking (8.12)–(8.14), we transform the principal part (namely, the terms of the orders n and $n-2$) of the equation in question as follows:

$$y^{(n)} + \frac{n!c_2(x)}{(n-2)!2!}y^{(n-2)} = \sigma D_x^n(z) + n\sigma' D_x^{n-1}(z) + \frac{n!}{2(n-1)!}(\sigma'' + c_2)D_x^{n-2}(z)$$

$$= \sigma\left[\phi'^n z^{(n)} + \frac{n!\phi'^{n-2}\phi''}{2(n-2)!}z^{(n-1)} + \left(\frac{n!\phi'^{n-4}\phi''^2}{(n-4)!2^3} + \frac{n!\phi'^{n-3}\phi'''}{(n-3)!3!}\right)z^{(n-2)}\right]$$

$$+ n\sigma'\left[\phi'^{n-1}z^{(n-1)} + \frac{(n-1)!\phi'^{n-3}\phi''}{(n-3)!2}z^{(n-2)}\right] + \frac{n!}{(n-2)!2}(\sigma'' + c_2\sigma)\phi'^{n-2}z^{(n-2)}.$$

Thus, to remove the terms with $z^{(n-1)}$ and $z^{(n-2)}$, we have to solve the equations

$$(n-1)\sigma\phi'' + 2\sigma'\phi' = 0,$$

$$\frac{(n-2)(n-3)}{4}\sigma\phi''^2 + \frac{n-2}{3}\sigma\phi'\phi''' + (n-2)\sigma'\phi'\phi'' + (\sigma'' + c_2\sigma)\phi'^2 = 0.$$

We can satisfy the first equation by setting $\phi' = h^{-2}(x)$, $\sigma = h^{n-1}(x)$ with an arbitrary function $h(x)$. Then the second equation becomes:

$$(n+1)h'' + 3c_2(x)h = 0. \tag{10.37}$$

Thus, given an equation (10.30) with $c_1 = 0$, we remove the terms of the orders $n-1$ and $n-2$ by first solving the differential equation (10.37), then passing to the new independent variable $\bar{x}$ and the new differential variable z defined by

$$\bar{x} = \int h^{-2}(x)\mathrm{d}x, \quad y = z\,h^{n-1}(x). \tag{10.38}$$

10.2.2 Infinitesimal method

Here, the method is illustrated by the third-order equation (10.30):

$$y''' + 3c_1(x)y'' + 3c_2(x)y' + c_3(x)y = 0. \tag{10.39}$$

Let us implement the transformation (10.31) by letting $\sigma(x) = 1 - \varepsilon\eta(x)$ with a small parameter ε. Then equation (10.39) becomes

$$z''' + 3\bar{c}_1(x)z'' + 3\bar{c}_2(x)z' + \bar{c}_3(x)z = 0,$$

where

$$\bar{c}_1 \approx c_1 - \varepsilon\eta', \quad \bar{c}_2 \approx c_2 - \varepsilon(\eta'' + 2c_1\eta'), \quad \bar{c}_3 \approx c_3 - \varepsilon(\eta''' + 3c_1\eta'' + 3c_2\eta'). \tag{10.40}$$

10.2. LINEAR ORDINARY DIFFERENTIAL EQUATIONS

Formulae (10.40) provide the group generator (prolonged to derivatives of $c_i(x)$):

$$X_\eta = \eta'\frac{\partial}{\partial c_1} + (\eta'' + 2c_1\eta')\frac{\partial}{\partial c_2} + (\eta''' + 3c_1\eta'' + 3c_2\eta')\frac{\partial}{\partial c_3} + \eta''\frac{\partial}{\partial c_1'}$$

$$+ (\eta''' + 2c_1\eta'' + 2c_1'\eta')\frac{\partial}{\partial c_2'} + (\eta^{(iv)} + 3c_1\eta''' + 3c_2\eta'' + 3c_1'\eta'' + 3c_2'\eta')\frac{\partial}{\partial c_3'} + \ldots$$

Definition 1. *Seminvariants* [10.5] of equation (10.39) are differential invariants of the infinitesimal transformation (10.40). They are functions $h(c, c', c'', \ldots)$ of the coefficients $c = (c_1, c_2, c_3)$ and their derivatives $c', c'', \ldots$ of a finite order satisfying the invariance test $X_\eta(h) = 0$.

Lemma. Equation (10.39) has two independent seminvariants,

$$h = c_2 - c_1^2 - c_1', \quad f = c_3 - 3c_1 c_2 + 2c_1^3 + 2c_1 c_1' - c_2'. \tag{10.41}$$

Any seminvariant is a function of h and f and their derivatives.

Note. The basic seminvariants h and f can be replaced by h_1 and h_2 given in (10.34) and (10.35), respectively. Indeed, $h_1 = h$ and $h_2 = f + h'$.

Proof of Lemma. Let $h = h(c)$ and

$$X_\eta(h) \equiv \eta'\frac{\partial h}{\partial c_1} + (\eta'' + 2c_1\eta')\frac{\partial h}{\partial c_2} + (\eta''' + 3c_1\eta'' + 3c_2\eta')\frac{\partial h}{\partial c_3} = 0. \tag{10.42}$$

Since the function $\eta(x)$ is arbitrary, there are no relations between its derivatives. Therefore equation (10.42) splits into the following three equations obtained by annulling separately the terms with η''', η'' and η':

$$\frac{\partial h}{\partial c_3} = 0, \quad \frac{\partial h}{\partial c_2} + 3c_1\frac{\partial h}{\partial c_3} = 0, \quad \frac{\partial h}{\partial c_1} + 2c_1\frac{\partial h}{\partial c_2} + 3c_2\frac{\partial h}{\partial c_3} = 0.$$

Whence $h = \text{const.}$, i.e. there are no differential invariants of the order 0.

Likewise, substitution of $h = h(c, c')$ into the equation $X_\eta(h) = 0$ yields, after annulling the term with $\eta^{(iv)}$, that $\partial h/\partial c_3' = 0$. The terms with η''', η'', η' give three linear partial differential equations for the function h of five variables, c_1, c_2, c_3, c_1', and c_2'. These equations have precisely two functionally independent solutions, e.g. (10.41).

It can be easily verified that the equation $X_\eta(h) = 0$ has precisely four functionally independent solutions involving c, c', and c''. Since h and f together with their first derivatives h' and f' provide four functionally independent solutions of this type, the Lemma is proved for differential invariants of the second order. The iteration completes the proof.

Definition 2. *Invariants* are seminvariants $J = J(h, f, h', f', \ldots)$ satisfying the additional invariance condition with respect to the transformation (10.32).

Let us find J by implementing the infinitesimal transformation (10.32), $\bar{x} \approx x + \varepsilon \xi(x)$. In the notation of Section 8.3.1, we have:
$$\bar{y}' \approx (1+\varepsilon \xi')\bar{y}', \quad \bar{y}'' \approx (1+2\varepsilon \xi')\bar{y}''+\varepsilon \bar{y}'\xi'', \quad \bar{y}''' \approx (1+3\varepsilon \xi')\bar{y}'''+3\varepsilon \bar{y}''\xi''+\varepsilon \bar{y}'\xi'''.$$
Consequently equation (10.39) becomes $\bar{y}''' + 3\bar{c}_1 \bar{y}'' + 3\bar{c}_2 \bar{y}' + \bar{c}_3 \bar{y} = 0$, where
$$\bar{c}_1 \approx c_1 + \varepsilon(\xi'' - c_1 \xi'), \quad \bar{c}_2 \approx c_2 + \varepsilon\Big(\frac{1}{3}\xi''' + c_1\xi'' - 2c_2\xi'\Big), \quad \bar{c}_3 \approx c_3 - 3\varepsilon c_3 \xi'. \quad (10.43)$$
The corresponding group generator (extended to derivatives of $c(x)$) is
$$X_\xi = \xi\frac{\partial}{\partial x} + (\xi'' - c_1 \xi')\frac{\partial}{\partial c_1} + \Big(\frac{1}{3}\xi''' + c_1 \xi'' - 2c_2 \xi'\Big)\frac{\partial}{\partial c_2} - 3c_3 \xi'\frac{\partial}{\partial c_3}$$
$$+(\xi''' - c_1 \xi'' - 2c_1' \xi')\frac{\partial}{\partial c_1'} + \Big(\frac{1}{3}\xi^{(iv)} + c_1 \xi''' + c_1' \xi'' - 2c_2 \xi'' - 3c_2' \xi'\Big)\frac{\partial}{\partial c_2'} + \cdots.$$
It is written in the space of the seminvariants (10.41) as follows:
$$X_\xi = \xi\frac{\partial}{\partial x} - \Big(\frac{2}{3}\xi''' + 2h\xi'\Big)\frac{\partial}{\partial h} - \Big(\frac{1}{3}\xi^{(iv)} + h\xi'' + 3f\xi'\Big)\frac{\partial}{\partial f} - \Big(\frac{2}{3}\xi^{(iv)} + 2h\xi'' + 3h'\xi'\Big)\frac{\partial}{\partial h'}$$
$$-\Big(\frac{1}{3}\xi^{(v)} + h\xi''' + h'\xi'' + 3f\xi'' + 4f'\xi'\Big)\frac{\partial}{\partial f'} - \Big(\frac{2}{3}\xi^{(v)} + 2h\xi''' + 5h'\xi'' + 4h''\xi'\Big)\frac{\partial}{\partial h''} + \cdots.$$

By applying the philosophy of the proof of the above Lemma to the equation $X_\xi(J) = 0$, one can prove the following statements [5.4].

Theorem 1. The third-order differential equation (10.39) has a singular invariant equation with respect to the group of general equivalence transformations (10.31)–(10.32). Namely, the equation:
$$\lambda \equiv h' - 2f = 0, \quad (10.44)$$
where h and f are the seminvariants (10.41).

Theorem 2. The *least invariant* of equation (10.39), i.e. a solution to $X_\xi(J) = 0$ involving the derivatives of h and f of the lowest order is
$$\theta = \frac{1}{\lambda^2}\bigg[7\Big(\frac{\lambda'}{\lambda}\Big)^2 - 6\frac{\lambda''}{\lambda} + 27h\bigg]^3, \quad (10.45)$$
where $\lambda = h' - 2f$. The higher-order invariants are obtained from θ by means of invariant differentiation (see Section 8.3.5). Any invariant J of an arbitrary order is a function of the least invariant θ and its invariant derivatives.

Example 1. An equation (10.39) is equivalent to $y''' = 0$ if and only if its coefficients $c_1(x), c_2(x), c_3(x)$ satisfy equation (10.44), $\lambda = 0$.

Example 2. The necessary and sufficient condition for an equation (10.39) to be equivalent to $y''' + y = 0$ is that $\lambda \equiv h' - 2f \neq 0$ and that the invariant (10.45) vanishes, $\theta = 0$. For example, $y''' + c(x)y = 0$ can be transformed to the form $z''' + z = 0$ only in the case $c(x) = (kx + l)^{-6}$, where the constants k and l do not vanish simultaneously.

10.3 Nonlinear ordinary differential equations

The infinitesimal method can be used for investigation of invariants of nonlinear differential equations as well. Consider an example due to S. Lie mentioned in Section 5.2.4. The set of nonlinear equations of the form (5.7):

$$y'' + a(x,y)y'^3 + b(x,y)y'^2 + c(x,y)y' + d(x,y) = 0, \qquad (10.46)$$

contains all second-order equations obtained from the linear equation (10.33) by arbitrary changes of the variables,

$$\bar{x} = f(x,y), \quad \bar{y} = g(x,y). \qquad (10.47)$$

The variety of equations (10.46) with arbitrary $a(x,y), b(x,y), c(x,y), d(x,y)$ is invariant under the transformations (10.47). Consequently, (10.47) form an infinite group G of equivalence transformations of the equations (10.46).

Theorem. The following system of equations:

$$H \equiv 3a_{xx} - 2b_{xy} + c_{yy} - 3ac_x - 3ca_x + 2bb_x + 3da_y + 6ad_y - bc_y = 0, \qquad (10.48)$$

$$K \equiv 3d_{yy} - 2c_{xy} + b_{xx} - 3ad_x - 6da_x + 3bd_y + 3db_x - 2cc_y + cb_x = 0, \qquad (10.49)$$

is invariant under the group G of transformations (10.47) and specifies, among the nonlinear equations (10.46), all linearizable ones.

10.4 Linear partial differential equations

10.4.1 The Laplace invariants

Some 100 years prior to the aforementioned historical events Laplace, in his fundamental memoir dedicated to the integration of linear partial differential equations, discovered *inter alia* two invariants [10.1]:

$$h = a_\xi + ab - c, \quad k = b_\eta + ab - c, \qquad (10.50)$$

for the general hyperbolic second-order equations with two independent variables,

$$u_{\xi\eta} + a(\xi,\eta)u_\xi + b(\xi,\eta)u_\eta + c(\xi,\eta)u = 0. \qquad (10.51)$$

Here, as usual, u_ξ etc. denote partial derivatives.

Setting $\xi = \eta = x$, $u(x,x) = y(x)$, $a(x,x) = b(x,x) = c_1(x)$, $c(x,x) = c_2(x)$ in (10.51) and (10.50), one obtains the second-order ordinary differential equation (10.33) and its invariant (10.34):

$$h = k = c_1' + c_1^2 - c_2 \equiv -h_1.$$

Proposition. The most general group of equivalence transformations of the equations (10.51) is an infinite group composed of linear transformations of the dependent variable:
$$u = \sigma(\xi,\eta)v, \quad \sigma(\xi,\eta) \neq 0, \tag{10.52}$$
and invertible changes of the independent variables of the form:
$$\bar{\xi} = \phi(\xi), \quad \bar{\eta} = \psi(\eta), \tag{10.53}$$
where σ, ϕ, ψ are arbitrary regular functions.

The Laplace invariants are invariant only with respect to the transformations (10.52). Hence h and k are *differential seminvariants* of (10.51).

10.4.2 The Ovsyannikov invariants

It took almost another 200 years before L.V. Ovsyannikov [5.34] discovered two proper *differential invariants*:
$$p = \frac{k}{h} \quad \text{and} \quad q = \frac{1}{h}\frac{d^2 \ln|h|}{d\xi d\eta}, \tag{10.54}$$
that are invariant under the general equivalence transformations (10.52)-(10.53).

The infinitesimal method of calculation of invariants presented above can naturally be extended to all algebraic and linear ordinary differential equations, (10.1) and (10.30). Moreover, it can be used also in the case of partial differential equations (10.51) with the following result [10.4].

Theorem. Ovsyannikov's invariants (10.54) provide a complete set of invariants of equations (10.51). Any other invariant is a function of p and q and their invariant derivatives.

10.5 The Maxwell equations

The concept of invariants of differential equations can be useful for (linear or nonlinear) evolutionary equations. For illustration purposes, let us consider the Maxwell equations. Their evolutionary part,
$$\boldsymbol{E}_t = \nabla \times \boldsymbol{B}, \quad \boldsymbol{B}_t = -\nabla \times \boldsymbol{E}, \tag{10.55}$$
defines a particular Lie-Bäcklund group (t is a group parameter) with the infinitesimal generator
$$X = \sum_{i=1}^{3}\left((\nabla \times \boldsymbol{B})^i \frac{\partial}{\partial E^i} - (\nabla \times \boldsymbol{E})^i \frac{\partial}{\partial B^i}\right). \tag{10.56}$$

Theorem. The Lie-Bäcklund transformation group defined by equations (10.55) has the following basis of invariants ([5.28], p. 242):
$$\operatorname{div} \boldsymbol{E}, \quad \operatorname{div} \boldsymbol{B}. \tag{10.57}$$

All other differential invariants of an arbitrary order are functions of the basic invariants (10.57) and their successive derivatives with respect to the spatial variables $x = (x^1, x^2, x^3)$.

This result guarantees the solvability of an arbitrary initial value problem for the *overdetermined* system of generalized Maxwell equations comprising (10.55) and additional differential equations of the form

$$\text{div } \boldsymbol{E} = 4\pi\rho(x), \quad \text{div } \boldsymbol{B} = g(x), \tag{10.58}$$

where $g(x) = 0$ corresponds to Maxwell's equations. The above theorem states that the additional equations (10.58) are satisfied identically for initial data

$$\boldsymbol{E}|_{t=0} = \boldsymbol{E}_0(x), \quad \boldsymbol{B}|_{t=0} = \boldsymbol{B}_0(x) \tag{10.59}$$

subject to the constraints (10.58):

$$\text{div } \boldsymbol{E}_0 = 4\pi\rho(x), \quad \text{div } \boldsymbol{B}_0 = g(x). \tag{10.60}$$

Problems

10.1. Transform the equation $y''' + 4y' = 0$ to Laguerre's canonical form (10.36).

10.2*. Prove Proposition 10.1.1 on equivalence transformations of algebraic equations.

10.3. Show that the transformations (10.18)–(10.19) form a group.

10.4. Find the transformations S, R, T of Example 2 of Section 10.1.2 by solving the determining equation (10.21).

10.5. Show that equation (10.22) admits Tschirnhausen's transformations (10.23).

10.6**. Investigate whether or not the Galois group of any algebraic equation (10.1) can be obtained by restricting Tschirnhausen's transformations, admitted by equation (10.1), to the roots of (10.1).

10.7*. Apply the infinitesimal method of Section 10.1.3 to the equation

$$P_4(x) \equiv C_0 x^4 + 4C_1 x^3 + 6C_2 x^2 + 4C_3 x + C_4 = 0.$$

Use the corresponding homogeneous form $Q_4(u, v)$ and find the transformation of its coefficients under the infinitesimal equivalence transformation (10.11). Investigate all invariants and invariant equations; in particular, find the discriminant of $P_4(x)$.

10.8*. Prove Proposition 10.2.1.

10.9. Prove statements of Examples 1 and 2 of Section 10.2.2. Find an equivalence transformation (10.31)-(10.32) mapping the equation $y''' + x^{-6}y = 0$ into $z''' + z = 0$. Use this transformation to integrate $y''' + x^{-6}y = 0$.

10.10**. Find all invariants of the nonlinear equations (10.46),

$$y'' + a(x,y)y'^3 + b(x,y)y'^2 + c(x,y)y' + d(x,y) = 0,$$

with respect to the group of equivalence transformations (10.47), $\bar{x} = f(x,y), \bar{y} = g(x,y)$.

10.11*. Prove Proposition 10.4.1.

Part III

BASIC INTEGRATION METHODS

> *I noticed that the majority of ordinary differential equations which were integrable by the old methods were left invariant under certain transformations, and that these integration methods consisted in using that property. Once I had thus represented many old integration methods from a common viewpoint, I set myself the natural problem: to develop a general theory of integration for all ordinary differential equations admitting finite or infinitesimal transformations.*
>
> **S. Lie, 1888**

Group theory sheds light on a whole range of approaches and results concerning integration of particular types of ordinary differential equations that are widely used in practice. It allows one to understand better the interconnection between various *ad hoc* methods, presented in Chapter 2, and provides a universal tool for tackling considerable numbers of nonlinear equations when other means of integration fail.

This part contains basic methods of integration of ordinary differential equations admitting Lie groups. The group methods, unlike the *ad hoc* methods of Chapter 2, do not depend upon the particular appearance of equations. This property makes them especially well suited for applications.

Chapter 11

First-order equations and systems

This chapter presents the simplest methods of integration provided by group theory. It also contains a discussion of superposition formulas for nonlinear equations and provides an introduction to the Vessiot-Guldberg-Lie algebra.

11.1 Integration by using an infinitesimal symmetry

Group theory furnishes two approaches for the integration of first-order ordinary differential equations with a known infinitesimal symmetry. The first provides a method for finding an integrating factor (see Section 2.1.5). The second, the method of canonical variables (see Section 7.1.8), provides a method for finding a suitable change of variables that simplifies the frame. The latter method is also applicable for reducing the order of higher-order equations.

11.1.1 Lie's integrating factor

The following theorem (Lie, 1874, see, e.g. [5.14]) establishes the relationship between integrating factors and infinitesimal symmetries of differential equations of the first order. The equations will be written in the symmetric form (2.2).

Theorem. A first-order ordinary differential equation

$$M(x,y)\mathrm{d}x + N(x,y)\mathrm{d}y = 0 \tag{11.1}$$

admits a one-parameter group G with an infinitesimal generator

$$X = \xi(x,y)\frac{\partial}{\partial x} + \eta(x,y)\frac{\partial}{\partial y} \tag{11.2}$$

if and only if the function

$$\mu = \frac{1}{\xi M + \eta N} \tag{11.3}$$

is an integrating factor for equation (11.1), provided that $\xi M + \eta N \neq 0$ [11.1].

Proof. Let us treat (11.1) as the characteristic system for the homogeneous linear partial differential equation (see Section 4.2.2)

$$N(x,y)\frac{\partial F}{\partial x} - M(x,y)\frac{\partial F}{\partial y} = 0. \qquad (11.4)$$

Let G be a group admitted by equation (11.1). Then G maps every solution of (11.1) into a solution of the same equation. Hence, a group transformation $(x,y) \mapsto (\bar{x}, \bar{y})$ converts any first integral $F(x,y) = C$ of equation (11.1) into a first integral $F(\bar{x}, \bar{y}) = C'$. In particular, the infinitesimal transformation $F(\bar{x}, \bar{y}) \approx F(x,y) + aXF(x,y)$ yields a first integral $XF(x,y) = \text{const.}$ Hence, by Theorem 4.2.1, $XF = \Phi(F)$. Since the first integrals of the characteristic equation (11.1) solve equation (11.4), we conclude that

$$N(x,y)\frac{\partial F}{\partial x} - M(x,y)\frac{\partial F}{\partial y} = 0, \quad \xi(x,y)\frac{\partial F}{\partial x} + \eta(x,y)\frac{\partial F}{\partial y} = \Phi(F).$$

Invoking the condition $\xi M + \eta N \neq 0$, one can solve this system with respect to the partial derivatives of F:

$$\frac{\partial F}{\partial x} = \frac{M\Phi}{\xi M + \eta N}, \quad \frac{\partial F}{\partial y} = \frac{N\Phi}{\xi M + \eta N},$$

whence

$$\frac{M\,dx + N\,dy}{\xi M + \eta N} = \frac{dF}{\Phi(F)}.$$

Since $dF/\Phi(F)$ is a total differential, the left-hand side of the latter equation is an exact 1-form. Hence, (11.3) is an integrating factor known as *Lie's integrating factor*. Finally, we notice that the above calculations are invertible, thus concluding the proof.

Example. Consider the Riccati equation (9.7):

$$y' + y^2 = \frac{2}{x^2}.$$

Writing it in the form (11.1),

$$dy + (y^2 - 2/x^2)dx = 0,$$

and using the known infinitesimal symmetry (9.9),

$$X = x\frac{\partial}{\partial x} - y\frac{\partial}{\partial y},$$

one obtains by formula (11.3) the integrating factor

$$\mu = \frac{x}{x^2 y^2 - xy - 2}.$$

11.1. INTEGRATION BY USING AN INFINITESIMAL SYMMETRY

After multiplication by this factor, the Riccati equation is written:

$$\frac{xdy + (xy^2 - 2/x)dx}{x^2y^2 - xy - 2} = \frac{xdy + ydx}{x^2y^2 - xy - 2} + \frac{dx}{x} = d\left(\ln|x| + \frac{1}{3}\ln\left|\frac{xy-2}{xy+1}\right|\right) = 0.$$

Hence, upon integration:

$$x^3 \frac{xy-2}{xy+1} = C \quad \text{or} \quad y = \frac{2x^3 + C}{x(x^3 - C)}, \quad C = \text{const}.$$

The integration presumes that the expressions $xy - 2$ and $xy + 1$ do not vanish. In the case when they do, one arrives at the invariant solutions (9.61).

Exercise. Deduce Lie's integrating factor for the non-homogeneous linear equation (2.3) $y' + P(x)y = Q(x)$ from the linear superposition principle and solve the equation.

Solution. Let us take a particular solution, e.g. $y_0(x) = \exp[-\int P(x)dx]$ of the homogeneous equation (2.4) $y' + P(x)y = 0$. The linear superposition $\bar{y} = y + ay_0(x)$ converts any solution of the non-homogeneous equation into a solution of the same equation and forms a one-parameter group with infinitesimals $\xi = 0, \eta = y_0(x)$. We rewrite our equation in the form (11.1), $dy + (Py - Q)dx = 0$, and obtain by virtue of (11.3) Lie's integrating factor $\mu = y_0^{-1}(x) \equiv \exp[\int P(x)dx]$. Using Theorem 11.1.1, we write

$$[dy + (Py - Q)dx]e^{\int P(x)dx} = dF$$

with an unknown function $F(x, y)$. It follows that

$$\frac{\partial F}{\partial y} = e^{\int P(x)dx}, \quad \frac{\partial F}{\partial x} = [P(x)y - Q(x)]e^{\int P(x)dx}.$$

The first equation yields $F = y \exp[\int P(x)dx] + f(x)$. Upon substituting this expression into the second equation, one obtains $f'(x) = -Q(x)\exp[\int P(x)dx]$, whence

$$f(x) = -\int Q(x)e^{\int P(x)dx}dx.$$

Now substitute this back into the expression for F and set $dF = 0$, i.e. $F = C$, to obtain

$$ye^{\int P(x)dx} - \int Q(x)e^{\int P(x)dx}dx = C.$$

Hence, the solution formula (2.6): $y = \left(C + \int Q(x)e^{\int P(x)dx}dx\right)e^{-\int P(x)dx}$.

11.1.2 Method of canonical variables

The simplest way to integrate a first-order equation

$$y' = f(x, y) \tag{11.5}$$

with a known infinitesimal symmetry

$$X = \xi(x, y)\frac{\partial}{\partial x} + \eta(x, y)\frac{\partial}{\partial y}$$

is furnished by the method of canonical variables. By introducing canonical variables (see Section 7.1.8), we convert the frame of equation (11.5) into a cylinder, thus eliminating the explicit dependence of (11.5) on one of the variables x or y. Consequently, our equation can be integrated by quadrature. In the case of higher-order equations, this method leads to a reduction of order. The following example illustrates the method.

Example. Consider again the Riccati equation (9.7)

$$y' + y^2 = \frac{2}{x^2}$$

with a known infinitesimal symmetry

$$X = x\frac{\partial}{\partial x} - y\frac{\partial}{\partial y}.$$

Canonical variables are obtained by solving the equations $X(u) = 0$, $X(t) = 1$ and have the form

$$t = \ln|x|, \quad u = xy.$$

In these variables the Riccati equation takes the integrable form (cf. Fig. 8.3 in Section 8.5):

$$u' + u^2 - u - 2 = 0.$$

Hence, upon integration:

$$\ln\left|\frac{u+1}{u-2}\right| - 3t = \text{const.},$$

provided that $u + 1 \neq 0$ and $u - 2 \neq 0$ (cf. Example 11.1.1). Substituting here the expressions for t and u in terms of x and y, one arrives at the solution given in the previous section:

$$y = \frac{2x^3 + C}{x(x^3 - C)}, \quad C = \text{const.}$$

11.1.3 Group interpretation of variation of parameters

Consider the general non-homogeneous linear equation (2.3):

$$y' + P(x)y = Q(x).$$

The linear superposition principle asserts that the equation admits the one-parameter group $\bar{x} = x$, $\bar{y} = y + ay_0(x)$ with the generator

$$X = y_0(x)\frac{\partial}{\partial y},$$

where $y_0(x)$ is any solution of the homogeneous equation, $y_0' + P(x)y_0 = 0$. The independent variable x is an invariant of this group. Furthermore, the equation

11.3. SYSTEMS ADMITTING NONLINEAR SUPERPOSITION

$X(u) \equiv y_0(x)\partial u/\partial y = 1$ yields $u = y/y_0(x)$. Accordingly, the canonical variables t, u are given by $t = x$, $u = y/y_0(x)$. In these variables, the non-homogeneous linear equation is written

$$y_0(x)\frac{du}{dx} = Q(x), \quad \text{whence} \quad u = \int \frac{Q(x)}{y_0(x)} dx + C.$$

Taking, e.g. $y_0(x) = \exp(-\int P(x)dx)$ and substituting the above expression for u into $y = u\, y_0(x)$, one arrives at the solution formula (2.6):

$$y = \left(C + \int Q\, e^{\int P dx} dx\right) e^{-\int P dx}.$$

Remark. Introduction of canonical variables is equivalent to the substitution

$$y = u(x)\, e^{-\int P(x)dx}$$

obtained in Section 2.1.1 by variation of the parameter C in (2.5). Thus, group analysis provides a theoretical background to the method of variation of parameters. This approach can be easily extended to higher-order equations. It will be discussed, in the case of second-order equations, in Section 12.2.5.

11.2 Systems admitting nonlinear superposition

A nonlinear superposition of solutions was discussed in Section 2.1.1 for a single first-order equation with separated variables. This section presents a generalization to systems due to E. Vessiot, A. Guldberg and S. Lie [11.2]. The main result is that a nonlinear superposition requires a generalized separation of variables complying to a Lie algebra structure. For convenience of citation, I called the corresponding algebra a *Vessiot-Guldberg-Lie algebra* ([5.35], vol. 2, Chap. 6).

11.2.1 Superposition formulas for nonlinear systems

Definition. A system of ordinary differential equations

$$\frac{dx^i}{dt} = F_i(t, x^1, \ldots, x^n), \quad i = 1, \ldots, n, \tag{11.6}$$

possesses a (nonlinear) superposition if its general solution $x = (x^1, \ldots, x^n)$ can be expressed via a finite number m of particular solutions of (11.6),

$$x_1 = (x_1^1, \ldots, x_1^n), \quad \ldots, \quad x_m = (x_m^1, \ldots, x_m^n), \tag{11.7}$$

by the formulae

$$x^i = \varphi^i(x_1, \ldots, x_m, C_1, \ldots, C_n), \quad i = 1, \ldots, n, \tag{11.8}$$

involving n parameters $C_1, \ldots, C_n$.

The particular solutions (11.7) are referred to as a *fundamental system of solutions* for (11.6). It is required that the form of the superposition functions φ^i does not depend upon the choice of particular solutions (11.7). This, however, does not exclude, for a given system of equations (11.6), the possibility to have several distinct representations (11.8) of a general solution as well as different numbers m of necessary particular solutions (see Example 6 in Section 11.2.3).

11.2.2 Main theorem. The Vessiot-Guldberg-Lie algebra

The question of which of equations (11.6) have a fundamental system of solutions is solved by the following statement proved in [5.15], Chap. 24, pp. 793–804.

Theorem. Equations (11.6) possess a superposition of solutions if and only if they admit a *generalized separation of variables* (cf. (2.9)):

$$\frac{dx^i}{dt} = T_1(t)\xi_1^i(x) + \cdots + T_r(t)\xi_r^i(x), \tag{11.9}$$

whose coefficients $\xi_\alpha^i(x)$ satisfy the condition that the operators

$$X_\alpha = \xi_\alpha^i \frac{\partial}{\partial x^i}, \quad \alpha = 1, \ldots, r, \tag{11.10}$$

span an r-dimensional Lie algebra L_r. The number m of necessary particular solutions (11.7) is estimated by

$$nm \geq r. \tag{11.11}$$

Definition. The Lie algebra L_r spanned by (11.10) will be referred to as the Vessiot-Guldberg-Lie algebra for equations (11.9).

11.2.3 Examples

Example 1. Consider a single linear homogeneous equation

$$\frac{dx}{dt} = A(t)x.$$

Here $n = 1$, $m = 1$, $r = 1$, $X = xd/dx$, and (11.7) is the linear superposition (2.7), $x = Cx_1$. The condition (11.11) is satisfied as an equality.

Example 2. For the linear non-homogeneous equation

$$\frac{dx}{dt} = A(t)x + B(t)$$

we have $n = 1$, $m = 2$, $r = 2$. The Vessiot-Guldberg-Lie algebra is spanned by the operators (11.10), $X_1 = d/dx$, $X_2 = xd/dx$. It is a two-dimensional Lie algebra since $[X_1, X_2] = X_1$. Formula (11.8) reduces to the linear superposition (2.8), $x = (1 - C)x_1 + Cx_2$.

11.3. SYSTEMS ADMITTING NONLINEAR SUPERPOSITION

Example 3. An example of a non-linear equation with a fundamental system of solutions is the Riccati equation

$$\frac{dx}{dt} = P(t) + Q(t)x + R(t)x^2. \tag{11.12}$$

It has the special form (11.9) with $r = 3$ and the operators (11.10) have the form

$$X_1 = \frac{d}{dx}, \quad X_2 = x\frac{d}{dx}, \quad X_3 = x^2\frac{d}{dx}, \tag{11.13}$$

which generate a Lie algebra L_3. Formula (11.11) yields $m \geq 3$. Hence, a representation of the general solution of the Riccati equation (11.12) requires at least three particular solutions. We know from Section 2.1.8 that, in fact, it suffices to know three solutions due to the property (2.32); see also Section 11.2.4.

Example 4. A system of two homogeneous linear equations

$$\frac{dx}{dt} = a_{11}(t)x + a_{12}(t)y, \quad \frac{dy}{dt} = a_{21}(t)x + a_{22}(t)y$$

has the special form (11.9) with coefficients

$$T_1 = a_{11}(t), \quad T_2 = a_{12}(t), \quad T_3 = a_{21}(t), \quad T_4 = a_{22}(t),$$
$$\xi_1 = (x, 0), \quad \xi_2 = (y, 0), \quad \xi_3 = (0, x), \quad \xi_4 = (0, y).$$

Consequently, the Vessiot-Guldberg-Lie algebra is a four-dimensional algebra L_4 spanned by the operators

$$X_1 = x\frac{\partial}{\partial x}, \quad X_2 = y\frac{\partial}{\partial x}, \quad X_3 = x\frac{\partial}{\partial y}, \quad X_4 = y\frac{\partial}{\partial y}.$$

Formula (11.8) is provided by the linear superposition representing the general solution (x, y) via two particular solutions (x_1, y_1) and (x_2, y_2) :

$$x = C_1 x_1 + C_2 x_2, \quad y = C_1 y_1 + C_2 y_2.$$

Here, $n = 2, m = 2, r = 4$, and the condition (11.11) is an equality.

Example 5. In the case of the general non-homogeneous system

$$\frac{dx}{dt} = a_{11}(t)x + a_{12}(t)y + b_1(t), \quad \frac{dy}{dt} = a_{21}(t)x + a_{22}(t)y + b_2(t),$$

one has to add to the coefficients T_α and ξ_α of Example 4 the following:

$$T_5 = b_1(t), \quad \xi_5 = (1, 0); \quad T_6 = b_2(t), \quad \xi_6 = (0, 1).$$

Consequently, the algebra L_4 of Example 4 extends to an algebra L_6 spanned by

$$X_1 = x\frac{\partial}{\partial x}, \quad X_2 = y\frac{\partial}{\partial x}, \quad X_3 = x\frac{\partial}{\partial y}, \quad X_4 = y\frac{\partial}{\partial y}, \quad X_5 = \frac{\partial}{\partial x}, \quad X_6 = \frac{\partial}{\partial y}.$$

The superposition formula (11.8) is well known:
$$x = x_1 + C_1(x_2 - x_1) + C_2(x_3 - x_1), \quad y = y_1 + C_1(y_2 - y_1) + C_2(y_3 - y_1).$$

Example 6. Lie [5.15] gave an example of a non-homogeneous system of two linear equations, viz.
$$\frac{dx}{dt} = a_{12}(t)y + b_1(t), \quad \frac{dy}{dt} = -a_{12}(t)x + b_2(t), \tag{11.14}$$

which, unlike the general system treated in Example 5, requires only two particular solutions. Indeed, the Vessiot-Guldberg-Lie algebra of system (11.14) is spanned by the operators
$$X_1 = y\frac{\partial}{\partial x} - x\frac{\partial}{\partial y}, \quad X_2 = \frac{\partial}{\partial x}, \quad X_3 = \frac{\partial}{\partial y}$$

and generates the group of motions in the plane (rotation and translations, cf. Example 3 in Section 6.3.2). Since the motions conserve all distances, any three solutions (x_1, y_1), (x_2, y_2), (x, y) are connected by the relations
$$(x - x_1)^2 + (y - y_1)^2 = C_1, \quad (x - x_2)^2 + (y - y_2)^2 = C_2. \tag{11.15}$$

Consequently, system (11.14) admits two representations of the general solution, viz. as a linear superposition of *three* solutions (Example 5) or as a nonlinear superposition (11.15) of *two* solutions.

11.2.4 Projective interpretation of the Riccati equation

Theorem 11.2.2 reduces the problem of identifying all ordinary differential equations (11.6) with linear or nonlinear superposition of solutions to that of enumerating all possible finite-dimensional Lie algebras of operators (11.10) in the n-dimensional space of variables $x = (x^1, \ldots, x^n)$.

Recall (Theorem 6.3.1) that for $n = 1$, i.e. in the case of a single variable x, any finite-dimensional Lie algebra is similar to the three-dimensional algebra (11.13) of the projective group or to its subalgebra. It follows that any equation $x' = f(t, x)$ admitting a superposition of solutions can be transformed, by a suitable change $\bar{x} = \phi(x)$ of the dependent variable, to the Riccati equation (11.12). On the other hand, there are many different types of non-similar finite-dimensional Lie algebras with $n \geq 2$ variables x^i (see, e.g. Section 7.3.8). Therefore, the variety of distinctly different systems of equations with nonlinear superposition is much greater than in the case of a single equation [11.3].

Thus, *the Riccati equation is in a sense a realization of the group of projective transformations.* This fact can be used for the proof of the constancy of the cross-ratio of any four solutions of a Riccati equation. Namely, let us introduce, following Lie [5.15], Chap. 24, homogeneous projective coordinates u and v by setting $x = u/v$. Then the Riccati equation (11.12) assumes the form
$$v\left(\frac{du}{dt} - \frac{1}{2}Qu - Pv\right) - u\left(\frac{dv}{dt} + Ru + \frac{1}{2}Qv\right) = 0. \tag{11.16}$$

11.3. SYSTEMS ADMITTING NONLINEAR SUPERPOSITION

Since the definition $x = u/v$ contains two functions, one can constrain them, e.g., by equating to zero the first left-hand term in (11.16) to obtain a representation of the Riccati equation in projective space as a system of two linear equations:

$$\frac{du}{dt} = \frac{1}{2}Q(t)u + P(t)v, \quad \frac{dv}{dt} = -R(t)u - \frac{1}{2}Q(t)v.$$

Let (u_1, v_1) and (u_2, v_2) be two particular solutions of this system, chosen so that the ratios u_1/v_1 and u_2/v_2 are not equal to the same constant. Then

$$u = C_1 u_1 + C_2 u_2, \quad v = C_1 v_1 + C_2 v_2,$$

and the general solution of the Riccati equation (11.12) is given by

$$x = \frac{C_1 u_1 + C_2 u_2}{C_1 v_1 + C_2 v_2} = \frac{u_1 + K u_2}{v_1 + K v_2}.$$

Let $x_1, \ldots, x_4$ be four solutions of the Riccati equation corresponding to particular numerical values $K_1, \ldots, K_4$ of the parameter K. Then (cf. (2.33))

$$\frac{x_1 - x_2}{x_3 - x_2} : \frac{x_1 - x_4}{x_3 - x_4} = \frac{K_1 - K_2}{K_3 - K_2} : \frac{K_1 - K_4}{K_3 - K_4} = C. \quad (11.17)$$

Hence, the cross-ratio of any four solutions of the Riccati equation is constant.

11.2.5 Linearizable Riccati equations

The Vessiot-Guldberg-Lie algebra (11.10) of the general Riccati equation (11.12) is a three-dimensional algebra L_3 spanned by (11.13). However, for particular coefficients P, Q, R, operators (11.10) may span a subalgebra of L_3, so that $r = 2$ or $r = 1$. Then the corresponding Riccati equation can be linearized by a suitable change of the dependent variable x. Hence, the following theorem [11.4].

Theorem 1. A first-order ordinary differential equation possessing a superposition formula can be linearized by a change of the dependent variable if and only if its Vessiot-Guldberg-Lie algebra has the dimension ≤ 2. In other words, the equation admits a generalized separation of variables in a two-term form:

$$\frac{dx}{dt} = T_1(t)\xi_1(x) + T_2(t)\xi_2(x) \quad (11.18)$$

such that the operators (11.10),

$$X_1 = \xi_1(x)\frac{d}{dx}, \quad X_2 = \xi_2(x)\frac{d}{dx}, \quad (11.19)$$

span a two-dimensional (or one-dimensional if $\xi_2(x) = c\xi_1(x)$) Lie algebra, i.e.

$$[X_1, X_2] = aX_1 + bX_2, \quad a, b = \text{const}. \quad (11.20)$$

Remark. The form of equation (11.18) and the Lie algebra structure (11.20) are unalterable under any change of the dependent variable x.

Proof of Theorem 1. We know from Examples 1 and 2 of Section 11.2.3 that the Vessiot-Guldberg-Lie algebra of linear equations has dimension one or two. Hence, by virtue of the Remark, this is true also for any linearizable equation possessing a superposition formula.

Conversely, let the operators (11.19) span a Lie algebra L_r of dimension $r \leq 2$. The case $r = 1$ yields $\xi_2(x) = c\xi_1(x)$, $c = \text{const}$. Then equation (11.18) has the form (2.9) with separated variables,

$$\frac{dx}{dt} = [T_1(t) + cT_2(t)]\xi_1(x).$$

It is linearized by introducing a canonical variable for $X_1 = \xi_1(x)d/dx$. Indeed, after defining the canonical variable y by $X_1(y) \equiv \xi_1(x)dy/dx = 1$, the above equation becomes a linear one:

$$\frac{dy}{dt} = T_1(t) + cT_2(t).$$

Suppose now that the operators (11.19) span a two-dimensional algebra L_2. It is clear that equation (11.18) will be linearized if one transforms the operators (11.19) to the form (cf. Example 2 in Section 11.2.3)

$$X_1 = \frac{d}{dx}, \quad X_2 = x\frac{d}{dx}. \tag{11.21}$$

One can assume that the first operator (11.19) is already written in the canonical variable x, i.e. it has the form given in (11.19), $X_1 = d/dx$. Let $X_2 = f(x)d/dx$ be the second operator (11.19). We have $[X_1, X_2] = f'(x)d/dx$, and (11.20) yields

$$\frac{df}{dx} = a + bf. \tag{11.22}$$

It follows that at least one of the coefficients a or b is not zero since the operators X_1 and X_2 are linearly independent, hence $f'(x) \neq 0$. Equation (11.22) yields:

$$f = ax + C \quad\Longrightarrow\quad X_2 = ax\frac{d}{dx} + CX_1, \quad \text{if} \quad b = 0,$$

$$f = Ce^{bx} - \frac{a}{b} \quad\Longrightarrow\quad X_2 = Ce^{bx}\frac{d}{dx} - \frac{a}{b}X_1, \quad \text{if} \quad b \neq 0.$$

In the first case, a basis of L_2 is provided by (11.21). While in the second case, one can take basic operators in the form $X_1 = d/dx, X_2 = e^x d/dx$ (by assigning bx as new x) and then transform them to (11.21) by the substitution $\bar{x} = e^{-x}$.

Example 1. Consider equation (vi) of Problem 11.2:

$$\frac{dx}{dt} = P(t) + Q(t)x + [Q(t) - P(t)]x^2. \tag{11.23}$$

11.3. SYSTEMS ADMITTING NONLINEAR SUPERPOSITION

It can be written in the form (11.18) with $T_1 = P, T_2 = Q, \xi_1 = 1-x^2, \xi_2 = x+x^2$. Hence the operators (11.19):

$$X_1 = (1-x^2)\frac{d}{dx}, \quad X_2 = (x+x^2)\frac{d}{dx}. \tag{11.24}$$

Their commutator $[X_1, X_2] = X_1 + 2X_2$ shows that (11.24) span an L_2. Hence, equation (11.23) is linearizable. To implement the linearization algorithm furnished by Theorem 1, let us choose, instead of (11.24), a new basis,

$$X = X_1 + 2X_2 = (1+x)^2\frac{d}{dx}, \quad Y = X_2 = (x+x^2)\frac{d}{dx},$$

satisfying the same commutator relation $[X, Y] = X$ as the operators (11.21). Now we find a canonical variable for X from equation $X(y) \equiv (1+x)^2 dy/dx = 1$:

$$y = -(1+x)^{-1}. \tag{11.25}$$

Transformation (11.25) converts the algebra spanned by (11.24) into that spanned by (11.21). Accordingly, (11.23) becomes a linear equation:

$$\frac{dy}{dt} = Q(t) - P(t) + [Q(t) - 2P(t)]y.$$

Example 2. The equation

$$\frac{dx}{dt} = t + x^2$$

has the form (11.18) with $T_1 = t, T_2 = 1, \xi_1 = 1, \xi_2 = x^2$. Hence (11.19):

$$X_1 = \frac{d}{dx}, \quad X_2 = x^2\frac{d}{dx}.$$

Their commutator $[X_1, X_2] = 2x d/dx$ cannot be expressed as a linear combination of X_1 and X_2 with constant coefficients. Therefore, to obtain a Lie algebra, one should add $X_3 = x d/dx$ to X_1, X_2. Hence, the Vessiot-Gulberg-Lie algebra is L_3 spanned by (11.13). Consequently, the equation in question is not linearizable.

Example 3. The equation

$$\frac{dx}{dt} = t + (\sqrt{2}+t)^2 x + 2\sqrt{2}(2+t^2)x^2 \tag{11.26}$$

has the form (11.9) with

$$T_1 = t, \quad T_2 = (\sqrt{2}+t)^2, \quad T_3 = 2\sqrt{2}(2+t^2); \quad \xi_1 = 1, \quad \xi_2 = x, \quad \xi_3 = x^2.$$

Hence, the corresponding operators (11.10) coincide with (11.13). Since they span an L_3, one might conclude that equation (11.26) is not linearizable. However, (11.26) can be represented in a two-term form (11.18):

$$\frac{dx}{dt} = t(1 + 2\sqrt{2}x) + (2+t^2)(x + 2\sqrt{2}x^2),$$

such that the operators (11.19),

$$X_1 = \left(1 + 2\sqrt{2}\,x\right)\frac{\mathrm{d}}{\mathrm{d}x}, \quad X_2 = \left(x + 2\sqrt{2}\,x^2\right)\frac{\mathrm{d}}{\mathrm{d}x},$$

span an L_2. Hence, one can apply Theorem 1 and find a linearization.

The difficulty encountered in the last example is due to the fact that the form of a generalized separation of variables (11.9) is not unique. To overcome this inconvenience, the following simple modification of Theorem 1 can be used for purposes of practical linearization.

Theorem 2. If a Riccati equation (11.12),

$$\frac{\mathrm{d}x}{\mathrm{d}t} = P(t) + Q(t)x + R(t)x^2,$$

has any one of the following four properties, it also has the other three:

(1) equation (11.12) is linearizable by a change of the dependent variable x;
(2) equation (11.12) can be written in a two-term form (11.18),

$$\frac{\mathrm{d}x}{\mathrm{d}t} = T_1(t)\xi_1(x) + T_2(t)\xi_2(x),$$

such that the operators (11.19), $X_1 = \xi_1(x)\mathrm{d}/\mathrm{d}x$, $X_2 = \xi_2(x)\mathrm{d}/\mathrm{d}x$, span a two-dimensional (or one-dimensional if $\xi_2(x) = c\xi_1(x)$) Lie algebra;
(3) equation (11.12) has either the form

$$\frac{\mathrm{d}x}{\mathrm{d}t} = Q(t)x + R(t)x^2 \qquad (11.27)$$

or the form

$$\frac{\mathrm{d}x}{\mathrm{d}t} = P(t) + Q(t)x + k[Q(t) - kP(t)]x^2 \qquad (11.28)$$

with a constant (in general complex) coefficient k;
(4) equation (11.12) has a constant (in general complex) solution.

Note. Equation (11.28) has a constant solution $x = -1/k$. The linear equation $x' = P(t) + Q(t)x$, which is a particular case of (11.28) for $k = 0$, may be regarded as a Riccati equation having $x = \infty$ as its particular solution [11.5].

Example 4. Let us test equation (11.23) for property (4) of Theorem 2. Upon letting $x = c = $ const., equation (11.23) yields $(1 - c^2)P(t) + c(1 + c)Q(t) = 0$. Hence, $c = -1$, i.e. equation (11.23) has a constant solution $x = -1$ and therefore it is linearizable.

Example 5. It is evident that the equation $\mathrm{d}x/\mathrm{d}t = t + x^2$ from Example 2 has no constant solutions. Indeed, $x = c$ does not solve the equation since $t + c^2 = 0$ is not satisfied identically in t. Hence, the equation is not linearizable.

Example 6. Consider equation (11.26). Substituting $x = c$ and equating to zero coefficients of all powers of t, one readily obtains $c = -1/(2\sqrt{2})$. Thus, equation (11.26) has a constant solution and hence is linearizable.

11.3. SYSTEMS ADMITTING NONLINEAR SUPERPOSITION

11.2.6 Decoupling and integration of systems using Vessiot-Guldberg-Lie algebras

Vessiot-Guldberg-Lie algebras combined with the idea of decoupling may furnish a theoretical basis for a new general integration theory for systems of ordinary differential equations admitting a superposition of solutions [11.6]. Let us consider here the simple case of systems of two coupled equations:

$$\frac{dx}{dt} = T_1(t)\xi_1^1(x,y) + T_2(t)\xi_2^1(x,y),$$
$$\frac{dy}{dt} = T_1(t)\xi_1^2(x,y) + T_2(t)\xi_2^2(x,y), \quad (11.29)$$

with a two-dimensional Vessiot-Guldberg-Lie algebra L_2 spanned by

$$X_1 = \xi_1^1(x,y)\frac{\partial}{\partial x} + \xi_1^2(x,y)\frac{\partial}{\partial x}, \quad X_2 = \xi_2^1(x,y)\frac{\partial}{\partial x} + \xi_2^2(x,y)\frac{\partial}{\partial x}. \quad (11.30)$$

To solve the system (11.29), it suffices to transform the basic operators (11.30) to the canonical forms (see Section 7.3.8) by a suitable change of variables x, y. Then, according to Remark 11.2.5, a system (11.29) reduces to one of the following *canonical forms* associated with types I, II, III and IV of Table 7.1, Section 7.3.8.

Table 11.1 Canonical forms of operators (11.30) and systems (11.29)

	Vessiot-Guldberg-Lie algebra	Canonical forms of (11.29)
I	$X_1 = \dfrac{\partial}{\partial x},\ X_2 = \dfrac{\partial}{\partial y}$	$\dfrac{dx}{dt} = T_1(t),\ \dfrac{dy}{dt} = T_2(t)$
II	$X_1 = \dfrac{\partial}{\partial y},\ X_2 = x\dfrac{\partial}{\partial y}$	$\dfrac{dx}{dt} = 0,\ \dfrac{dy}{dt} = T_1(t) + T_2(t)x$
III	$X_1 = \dfrac{\partial}{\partial y},\ X_2 = x\dfrac{\partial}{\partial x} + y\dfrac{\partial}{\partial y}$	$\dfrac{dx}{dt} = T_2(t)x,\ \dfrac{dy}{dt} = T_1(t) + T_2(t)y$
IV	$X_1 = \dfrac{\partial}{\partial y},\ X_2 = y\dfrac{\partial}{\partial y}$	$\dfrac{dx}{dt} = 0,\ \dfrac{dy}{dt} = T_1(t) + T_2(t)y$

11.2.7 Application: An invariant solution of nonlinear equations modelling laser systems

The phenomena of the wave front correction for optical radiations in laser systems are simulated by nonlinear equations called the system of phase-conjugated reflection equations, known also as wave front reversal [11.7]. A simplified model obtained by considering steady-state waves is described, upon choosing particular parameters of a medium, by the following system:

$$\left(\frac{\partial}{\partial z} - i\Delta\right)E_1 = |E_2|^2 E_1, \quad \left(\frac{\partial}{\partial z} + i\Delta\right)E_2 = |E_1|^2 E_2, \quad (11.31)$$

where Δ is the Laplace operator in the (x,y) plane, E_1 and E_2 are complex amplitudes of incident and phase conjugated (amplified) light waves, respectively.

Equations (11.31) are evidently invariant under the translations of x, y, z, rotations in the (x,y) plane and appropriate dilations of the dependent and independent variables. One can find an additional symmetry group using an analogy of the left-hand sides of (11.31) and the heat equation. Namely, we shall look for an infinitesimal symmetry of system (11.31) in the following particular form (obtained as an analog of the generator of the Galilean transformation for the heat equation, cf. Problem 9.6):

$$X = 2z\frac{\partial}{\partial x} + \sum_{\alpha=1}^{2}\left(f_\alpha(x,z)E_\alpha\frac{\partial}{\partial E_\alpha} + g_\alpha(x,z)E_\alpha^*\frac{\partial}{\partial E_\alpha^*}\right),$$

where E_α and E_α^* are complex conjugate quantities. The choice of the operator in this specific form simplifies the solution of the determining equations and yields

$$X = 2z\frac{\partial}{\partial x} + ix\left(E_1\frac{\partial}{\partial E_1} - E_2\frac{\partial}{\partial E_2} - E_1^*\frac{\partial}{\partial E_1^*} + E_2^*\frac{\partial}{\partial E_2^*}\right). \qquad (11.32)$$

Consider the two-dimensional Abelian Lie algebra spanned by the operator (11.32) and the operator $\partial/\partial y$. The basis of invariants is given by

$$z, \quad u_1 = E_1 e^{-ix^2/(4z)}, \quad u_2 = E_2 e^{ix^2/(4z)},$$

and the complex conjugates for u_1 and u_2. Hence, we have the following general form of the invariant solutions:

$$E_1 = u_1(z)e^{ix^2/(4z)}, \quad E_2 = u_2(z)e^{-ix^2/(4z)}. \qquad (11.33)$$

For simplicity sake, we consider the case of real functions u_1 and u_2. Then the substitution of expressions (11.33) in equations (11.31) yields

$$\frac{du_1}{dz} = u_2^2 u_1 - \frac{u_1}{2z}, \quad \frac{du_2}{dz} = u_1^2 u_2 - \frac{u_2}{2z}. \qquad (11.34)$$

Equations (11.34) have the form (11.29) with the coefficients

$$T_1(z) = 1, \quad T_2(z) = -\frac{1}{2z}$$

and with the Vessiot-Guldberg-Lie algebra spanned by

$$X_1 = u_2^2 u_1 \frac{\partial}{\partial u_1} + u_1^2 u_2 \frac{\partial}{\partial u_2}, \quad X_2 = u_1 \frac{\partial}{\partial u_1} + u_2 \frac{\partial}{\partial u_2}. \qquad (11.35)$$

We have $[X_1, X_2] = -2X_1$, $\xi_1\eta_2 - \eta_1\xi_2 = u_1 u_2(u_2^2 - u_1^2) \neq 0$. Hence, the operators (11.35) span an L_2 of type III in the classification of Table 7.1, Section 7.3.8.

11.3. SYSTEMS ADMITTING NONLINEAR SUPERPOSITION

Consequently, we can transform (11.35) and hence the system (11.34) to the canonical form III of Table 11.1. We first find canonical variables for the first operator (11.35) from the equations $X_1(v_1) = 0$, $X_1(v_2) = 1$, whence

$$v_1 = u_1^2 - u_2^2, \quad v_2 = \frac{\ln u_2 - \ln u_1}{u_1^2 - u_2^2}. \tag{11.36}$$

One can verify that the variables (11.36) are, in fact, the canonical variables required for our algebra L_2. Indeed, the operators (11.35) are written in the form of type III of Table 11.1 (up to the nonessential constant factor of X_2):

$$X_1 = \frac{\partial}{\partial v_2}, \quad X_2 = 2\left(v_1 \frac{\partial}{\partial v_1} + v_2 \frac{\partial}{\partial v_2}\right).$$

Hence, in variables (11.36), equations (11.34) reduce to the following:

$$\frac{dv_1}{dz} = -\frac{v_1}{z}, \quad \frac{dv_2}{dz} = 1 + \frac{v_2}{z}. \tag{11.37}$$

The solution of equations (11.37) is given by

$$v_1 = -\frac{C_1}{z}, \quad v_2 = C_2 z + z \ln z. \tag{11.38}$$

Now we reverse formulas (11.36):

$$u_1 = \sqrt{\frac{v_1}{1 - e^{2v_1 v_2}}}, \quad u_2 = \sqrt{\frac{v_1}{e^{-2v_1 v_2} - 1}} \tag{11.39}$$

and substitute herein solutions (11.38). It follows that the general solution of equations (11.34) is given by

$$u_1 = \sqrt{\frac{k}{z(1 - \zeta^2)}}, \quad u_2 = \zeta \sqrt{\frac{k}{z(1 - \zeta^2)}}, \tag{11.40}$$

where $\zeta = Cz^k$, $C, k =$ const. Substituting (11.40) in (11.33), one obtains the invariant solution of the system (11.31):

$$E_1 = \sqrt{\frac{k}{z(1 - \zeta^2)}} e^{ix^2/(4z)}, \quad E_2 = \zeta \sqrt{\frac{k}{z(1 - \zeta^2)}} e^{ix^2/(4z)},$$

where $\zeta = Cz^k$ with arbitrary constants C and k.

Problems

11.1. Solve the first-order equations with known symmetries given in Table 9.1 of Section 9.3.4 by using Lie's integrating factor or canonical variables.

11.2. Find Vessiot-Guldberg-Lie algebras of minimal dimensions for the following Riccati equations (where $P(t)$ and $Q(t)$ are arbitrary functions):

(i) $\dfrac{dx}{dt} = x^2 + 1,$ (ii) $\dfrac{dx}{dt} = x^2 + P(t),$ (iii) $\dfrac{dx}{dt} = x^2 + [1 + P(t)]x + P(t),$

(iv) $\dfrac{dx}{dt} = x^2 + [1 + 2P(t)]x + P(t),$ (v) $\dfrac{dx}{dt} = P(t) + Q(t)x + 2[Q(t) - 2P(t)]x^2,$

(vi) $\dfrac{dx}{dt} = P(t) + Q(t)x + [Q(t) - P(t)]x^2,$ (vii) $\dfrac{dx}{dt} = P(t) + Q(t)x + [Q(t) - 2P(t)]x^2.$

11.3. Investigate which of the above Riccati equations are linearizable by a change of x.

11.4. Show, using the linerization (11.25), that any three solutions x_1, x_2, x_3 of the nonlinear equation (11.23) are related by (cf. (11.17))

$$\frac{x_3 - x_1}{x_2 - x_1} : \frac{x_3 + 1}{x_2 + 1} = C.$$

11.5*. Prove Theorem 2 of Section 11.2.5.

Chapter 12

Second-order equations

This chapter contains practical devices, provided by group theory, for lowering the order, integration and/or linearization of second-order ordinary differential equations. The methods of reduction of order by one and successive integration are applicable to higher-order equations as well. The restriction to second order is essential, however, for the method of integration using canonical forms of L_2 and for Lie's linearization test.

12.1 Reduction of order

If an ordinary differential equation of the second (or higher) order admits a one-parameter group, then its order can be lowered by one. Namely, an equation

$$y'' = f(x, y, y') \qquad (12.1)$$

with a known infinitesimal symmetry

$$X = \xi(x,y)\frac{\partial}{\partial x} + \eta(x,y)\frac{\partial}{\partial y} \qquad (12.2)$$

can be integrated once either by introducing canonical variables or using the method of invariant differentiation [12.1].

12.1.1 Canonical variables

The introduction of canonical variables eliminates the explicit dependence of (12.1) on one of the variables x or y (cf. Section 11.1.2). Thus, an equation of the second order with a known symmetry can be converted to any of the forms $y'' = f(y, y')$ and $y'' = f(x, y')$ readily reducible to first-order equations by the elementary methods of Sections 2.2.2 and 2.2.6.

Example. The homogeneous linear equation

$$y'' + a_1(x)y' + a_2(x)y = 0 \qquad (12.3)$$

is evidently invariant under the transformation $\bar{x} = x, \bar{y} = cy$. Hence, (12.3) admits the dilation group with the generator

$$X = y\frac{\partial}{\partial y}. \tag{12.4}$$

The canonical variables are given by $t = x$ and $u = \ln|y|$. In the new variables, the generator (12.4) reduces to $X = \partial/\partial u$, and equation (12.3) is written

$$u'' + u'^2 + a_1(t)u' + a_2(t) = 0.$$

The latter is an equation discussed in Section 2.2.6 and reduces to a first-order equation upon setting $z = y'$. Namely, the resulting equation is a Riccati equation:

$$z' + z^2 + a_1(t)z + a_2(t) = 0. \tag{12.5}$$

Provided that the integral $z = \phi(t, C_1)$ of (12.5) is known, the general solution of the original equation (12.3) is obtained by quadrature, $y = \int \phi(t, C_1)dx + C_2$.

12.1.2 Invariant differentiation

The method, based on the following theorem [12.2], provides a generalization and a *covariant* (i.e. independent of the change of variables x and y) formulation of ad hoc methods of lowering the order of equations that do not explicitly involve either the independent or the dependent variable.

Theorem. The general ordinary differential equation (12.1) of the second order admitting a one-parameter group G with a generator (12.2) can be written in the form of a first-order equation

$$\frac{dv}{du} = F(u, v) \tag{12.6}$$

with an arbitrary function $F(u, v)$. Here $u = u(x, y)$ denotes an invariant of the group G, and $v = v(x, y, y')$ its differential invariant of the first order.

Proof. By definition of a symmetry group (Definition 2 of Section 9.2.1), the frame of equation (12.1) is an invariant manifold with respect to the twice-extended group G. Therefore, invoking Theorem 7.2.2 on a representation of invariant manifolds via invariants, one can rewrite the equation (12.1) in the form $w = F(u, v)$, where u is an invariant of the group G, and v and w are its differential invariants of the first and second order, respectively (cf. Sections 8.3.5 and 9.3.4). On the other hand, Theorem 1 of Section 8.3.5 asserts that the second-order differential invariant w can be chosen to be of the form $w = dv/du$. Substituting this into $w = F(u, v)$, one arrives at a representation of equation (12.1) in the form (12.6), thus completing the proof.

12.2. INTEGRATION BY MEANS OF TWO SYMMETRIES

This theorem lowers the order of the equation in question, namely it reduces the problem of integration of the second-order equation (12.1) to that of the first-order equation (12.6) and a quadrature. Indeed, if the integral

$$\Phi(u, v, C) = 0 \qquad (12.7)$$

of equation (12.6) is known, the solution of the original second-order equation can be reduced to quadratures. Indeed, substituting known expressions $u(x, y)$ and $v(x, y, y')$ into (12.7) leads to a first-order differential equation, which admits the group G in view of the invariance of u and v, and which can therefore be integrated by quadrature.

Example [12.3]. Consider again equation (12.3), $y'' + a_1(x)y' + a_2(x)y = 0$. Its infinitesimal symmetry (12.4) extended to y' by formula (8.18) is written, after dropping the subscript (1) according to [8.5], in the form $X = y\partial/\partial y + y'\partial/\partial y'$. For this operator, one readily finds an invariant $u = x$ and a first-order differential invariant $v = y'/y$. Hence, a second-order differential invariant is

$$\frac{dv}{du} = \frac{y''}{y} - \frac{y'^2}{y^2} = \frac{y''}{y} - v^2.$$

The first-order equation (12.6) is obtained by expressing y' and y'' in terms of the differential invariants, namely by substituting $y'/y = v$ and $y''/y = dv/du + v^2$ into (12.3). As a result, one arrives at the Riccati equation (12.5):

$$\frac{dv}{du} + v^2 + a_1(u)v + a_2(u) = 0. \qquad (12.8)$$

12.2 Integration by means of two symmetries

According to Lie's group classification of ordinary differential equations [5.20](i), the maximal Lie algebra L_r admitted by a second-order equation (12.1) may have only the dimensions $r = 1, 2, 3$, or 8. The dimensionality $r = 8$ indicates that equation (12.1) is linear or linearizable (Remark 9.3.1). If (12.1) admits L_r, $r \geq 2$, we can single out a two-dimensional subalgebra $L_2 \subset L_r$ by using Theorem 7.3.3. Therefore, we will assume in what follows that equation (12.1) admits an L_2.

12.2.1 Consecutive integration

The method of consecutive integration of equations (12.1), admitting an L_2, is based on the following statement due to Lie [12.4] (cf. also Theorem 9.4.2).

Theorem. Let a second-order equation (12.1) admit a Lie algebra L_2 with a basis X_1, X_2. Let the one-dimensional algebra L_1 spanned by X_1 be an ideal in L_2. Then, upon reducing the order of (12.1) by means of X_1, the resulting first-order equation admits the quotient algebra L_2/L_1.

Note. To implement the method, choose a basis X_1, X_2 so that $[X_1, X_2] = cX_1$, $c = \text{const}$. Then X_1 will span an ideal L_1 required by the theorem. Now the quotient algebra L_2/L_1 can be identified with the one-dimensional algebra spanned by X_2 (Section 7.3.4). Let the first-order equation obtained by reducing the order of (12.1) using X_1 be taken in the form (12.6). Then the latter equation admits X_2 rewritten in terms of the variables u and v, and hence can be integrated by any of the methods discussed in Section 11.1.

Example. Consider a linear equation (12.3) of a particular form:

$$y'' - \frac{2}{x^2} y = 0. \tag{12.9}$$

Along with (12.3), $X_1 = y\partial/\partial y$, it obviously admits the dilation of x, viz.

$$X_2 = x \frac{\partial}{\partial x}. \tag{12.10}$$

After reducing the order of (12.9) using X_1, we obtain the first-order equation (12.8) of the form

$$\frac{dv}{du} + v^2 - \frac{2}{u^2} = 0. \tag{12.11}$$

To obtain the expression of the operator (12.10) in terms of u and V, one should use X_2 in the once-extended form and rewrite it in the variables $u = x, v = y'/y$:

$$X_2 = x \frac{\partial}{\partial x} - y' \frac{\partial}{\partial y'} = X_2(u) \frac{\partial}{\partial u} + X_2(v) \frac{\partial}{\partial v} = u \frac{\partial}{\partial u} - v \frac{\partial}{\partial v}.$$

Hence, equation (12.11) admits the operator

$$X_2 = u \frac{\partial}{\partial u} - v \frac{\partial}{\partial v}. \tag{12.12}$$

Thus, we obtained the special Riccati equation (9.7) and its infinitesimal symmetry (9.9) discussed above from various points of view.

12.2.2 Canonical forms of two-dimensional Lie algebras

Lie's canonical forms of two-dimensional algebras furnish the simplest method of integration of second-order equations by reducing them to four integrable types.

Lemma [12.5]. Every two-dimensional Lie algebra L_2 can be reduced, by choosing an appropriate basis $X_1 = \xi_1 \partial/\partial x + \eta_1 \partial/\partial y$, $X_2 = \xi_2 \partial/\partial x + \eta_2 \partial/\partial y$, to one of the following four distinctly different types:

I. $[X_1, X_2] = 0$, $\quad \xi_1 \eta_2 - \eta_1 \xi_2 \neq 0 \quad$ (or $\lambda_1(x,y) X_1 + \lambda_2(x,y) X_2 \not\equiv 0$),

II. $[X_1, X_2] = 0$, $\quad \xi_1 \eta_2 - \eta_1 \xi_2 = 0 \quad$ (or $\lambda_1(x,y) X_1 + \lambda_2(x,y) X_2 \equiv 0$),

III. $[X_1, X_2] = X_1$, $\quad \xi_1 \eta_2 - \eta_1 \xi_2 \neq 0 \quad$ (or $\lambda_1(x,y) X_1 + \lambda_2(x,y) X_2 \not\equiv 0$),

IV. $[X_1, X_2] = X_1$, $\quad \xi_1 \eta_2 - \eta_1 \xi_2 = 0 \quad$ (or $\lambda_1(x,y) X_1 + \lambda_1(x,y) X_2 \equiv 0$).

12.2. INTEGRATION BY MEANS OF TWO SYMMETRIES

Remark. The identity $\lambda_1(x,y)X_1 + \lambda_2(x,y)X_2 \equiv 0$ means that the operators X_1 and X_2 are connected; see Definition 1 of Section 4.5.1.

The structure relations I–IV are invariant with respect any change of variables x and y. Using this invariance, one can simplify the form of basis operators of the two-dimensional algebras of all the above types and arrive at the following result presented in Table 7.1 of Section 7.3.8.

Theorem [12.6] Any two-dimensional Lie algebra L_2 can be transformed, by a suitable change of variables x, y, to one of the following *canonical forms*:

$$\text{I.} \quad X_1 = \frac{\partial}{\partial x}, \quad X_2 = \frac{\partial}{\partial y};$$

$$\text{II.} \quad X_1 = \frac{\partial}{\partial y}, \quad X_2 = x\frac{\partial}{\partial y};$$

$$\text{III.} \quad X_1 = \frac{\partial}{\partial y}, \quad X_2 = x\frac{\partial}{\partial x} + y\frac{\partial}{\partial y};$$

$$\text{IV.} \quad X_1 = \frac{\partial}{\partial y}, \quad X_2 = y\frac{\partial}{\partial y}.$$

Proof. To prove the statement, let us consider all types I–IV of the above Lemma.

Type I. By Theorem 1 of Section 7.1.8, one can choose canonical variables for X_1 such that it becomes $X_1 = \partial/\partial x$. Let

$$X_2 = \xi(x,y)\frac{\partial}{\partial x} + \eta(x,y)\frac{\partial}{\partial y}$$

be the second basic operator. Then $[X_1, X_2] = 0$ yields $\xi_x = \eta_x = 0$, while the condition $\xi_1\eta_2 - \eta_1\xi_2 \neq 0$ implies that $\eta(y) \neq 0$. Thus, in any canonical variables for X_1, a basis of L_2 of type I has the form

$$X_1 = \frac{\partial}{\partial x}, \quad X_2 = \xi(y)\frac{\partial}{\partial x} + \eta(y)\frac{\partial}{\partial y}, \quad \eta(y) \neq 0. \tag{12.13}$$

Canonical variables for X_1 can be determined in many different ways, which fact can be used for transforming X_2 into the required form. Let us find the general form of the transformations $t = t(x,y), u = u(x,y)$ preserving the canonical coordinates, i.e. leaving unaltered the form of the first operator (12.13):

$$X_1 \equiv \frac{\partial t}{\partial x}\frac{\partial}{\partial t} + \frac{\partial u}{\partial x}\frac{\partial}{\partial u} = \frac{\partial}{\partial t}.$$

Hence, $t_x = 1, u_x = 0$. Thus, the most general transformation preserving the canonical variables for X_1 contains two arbitrary functions of y and has the form:

$$t = x + f(y), \quad u = g(y). \tag{12.14}$$

After this transformation, the second operator (12.13) becomes

$$X_2 = \left(\xi(y) + \eta(y)f'(y)\right)\frac{\partial}{\partial t} + \eta(y)g'(y)\frac{\partial}{\partial u}.$$

It takes the form $X_2 = \partial/\partial u$ if the functions f and g solve the equations $\xi(y) + \eta(y)f'(y) = 0$ and $\eta(y)g'(y) = 1$, respectively. We ultimately arrive at the change of coordinates

$$t = x - \int \frac{\xi(y)}{\eta(y)} dy, \quad u = \int \frac{dy}{\eta(y)} \qquad (12.15)$$

transforming the operators (12.13) to the canonical form I:

$$X_1 = \frac{\partial}{\partial t}, \quad X_2 = \frac{\partial}{\partial u}.$$

Type II. Here we start by choosing the canonical variables x, y such that the first operator has the form $X_1 = \partial/\partial y$. Then we adopt the reasoning used in the previous case. The conditions $[X_1, X_2] = 0$, $\xi_1 \eta_2 - \eta_1 \xi_2 = 0$ yield (cf. (12.13)):

$$X_1 = \frac{\partial}{\partial y}, \quad X_2 = \eta(x)\frac{\partial}{\partial y}. \qquad (12.16)$$

The most general transformation preserving the canonical variables for X_1 is obtained from (12.14) merely by interchanging x and y as well as t and u:

$$t = f(x), \quad u = y + g(x). \qquad (12.17)$$

We set here $f(x) = \eta(x)$, $g(x) = 0$ to obtain the change of variables

$$t = \eta(x), \quad u = y \qquad (12.18)$$

transforming the operators (12.16) to the canonical form II:

$$X_1 = \frac{\partial}{\partial u}, \quad X_2 = t\frac{\partial}{\partial u}.$$

Type III. We take again $X_1 = \partial/\partial y$ and employ transformations (12.17). The conditions $[X_1, X_2] = X_1$, $\xi_1\eta_2 - \eta_1\xi_2 \neq 0$ yield:

$$X_1 = \frac{\partial}{\partial y}, \quad X_2 = \xi(x)\frac{\partial}{\partial x} + \left(y + \eta(x)\right)\frac{\partial}{\partial y} \quad \xi(x) \neq 0. \qquad (12.19)$$

After transformation (12.17), the second operator (12.19) becomes

$$X_2 = \xi(x)f'(x)\frac{\partial}{\partial t} + \left(y + \eta(x) + \xi(x)g'(x)\right)\frac{\partial}{\partial u}.$$

It takes the desired form $X_2 = t\partial/\partial t + u\partial/\partial u \equiv f(x)\partial/\partial t + ((y + g(x))\partial/\partial u$ if

$$\xi(x)f'(x) = f(x), \quad \xi(x)g'(x) + \eta(x) = g(x). \qquad (12.20)$$

Integrating the linear first-order ordinary differential equations (12.20) for unknown functions f and g and substituting into (12.17), we transform the operators (12.19) to the canonical form III:

$$X_1 = \frac{\partial}{\partial u}, \quad X_2 = t\frac{\partial}{\partial t} + u\frac{\partial}{\partial u}.$$

12.2. INTEGRATION BY MEANS OF TWO SYMMETRIES

Type IV. Proceeding as in the previous case, we obtain from the conditions $[X_1, X_2] = X_1,\ \xi_1\eta_2 - \eta_1\xi_2 = 0$:

$$X_1 = \frac{\partial}{\partial y}, \quad X_2 = \left(y + \eta(x)\right)\frac{\partial}{\partial y}. \tag{12.21}$$

It is transparent that the change of variables

$$t = x,\ u = y + \eta(x) \tag{12.22}$$

transforms (12.21) to the canonical form IV:

$$X_1 = \frac{\partial}{\partial u}, \quad X_2 = u\frac{\partial}{\partial u}.$$

12.2.3 Lie's integration algorithm

Now we can find the general second-order equations that are invariant under Lie algebras L_2 of types I–IV and integrate them. Any equation (12.1) admitting L_r, $r \geq 2$, can be integrated by reducing it to one of these integrable forms.

Type I. To construct all second-order equations admitting an algebra L_2, one can find a basis of invariants and differential invariants of the first and second orders. The extended operators $X_1 = \partial/\partial x$ and $X_2 = \partial/\partial y$ coincide with the operators themselves, so that the desired differential invariants are y' and y''. Hence, the general second-order equation admitting L_2 of type I has the form

$$y'' = f(y'). \tag{12.23}$$

It can be integrated in quadrature:

$$\int \frac{dy'}{f(y')} = x + C_1, \quad \text{or explicitly} \quad y' = \varphi(x + C_1),$$

whence $y = \int \varphi(x + C_1) d(x + C_1) + C_2$.

Type II. A basis of differential invariants of the algebra L_2 of type II is given by x and y'. Hence, the general form of the invariant differential equation is

$$y'' = f(x). \tag{12.24}$$

Its solution is given by two quadratures:

$$y = \int \left(\int f(x)dx\right)dx + C_1 x + C_2.$$

Type III. Here the invariant equation has the form

$$y'' = \frac{1}{x}f(y'). \tag{12.25}$$

It is also solved by two quadratures:

$$\int \frac{dy'}{f(y')} = \ln x + C_1, \quad \text{or explicitly} \quad y' = \varphi(\ln x + C_1),$$

whence $y = \int \varphi(\ln x + C_1) dx + C_2$.

Type IV. Here we have the invariant equation

$$y'' = f(x) y' \tag{12.26}$$

with the general solution

$$y = C_1 \int e^{\int f(x) dx} dx + C_2.$$

Table 12.1 Four types of second-order equations admitting L_2

	Structure of L_2	Canonical form of L_2	Equation
I	$[X_1, X_2] = 0$, $\xi_1\eta_2 - \eta_1\xi_2 \neq 0$	$X_1 = \frac{\partial}{\partial x}$, $X_2 = \frac{\partial}{\partial y}$	$y'' = f(y')$
II	$[X_1, X_2] = 0$, $\xi_1\eta_2 - \eta_1\xi_2 = 0$	$X_1 = \frac{\partial}{\partial y}$, $X_2 = x\frac{\partial}{\partial y}$	$y'' = f(x)$
III	$[X_1, X_2] = X_1$, $\xi_1\eta_2 - \eta_1\xi_2 \neq 0$	$X_1 = \frac{\partial}{\partial y}$, $X_2 = x\frac{\partial}{\partial x} + y\frac{\partial}{\partial y}$	$y'' = \frac{1}{x}f(y')$
IV	$[X_1, X_2] = X_1$, $\xi_1\eta_2 - \eta_1\xi_2 = 0$	$X_1 = \frac{\partial}{\partial y}$, $X_2 = y\frac{\partial}{\partial y}$	$y'' = f(x) y'$

The preceding results can be summarized in the following five-step algorithm for the integration of second-order ordinary differential equations (12.1).

Step 1. Calculate the admitted Lie algebra L_r.

Step 2. If $r = 2$, advance to the next step; if $r > 2$, distinguish a two-dimensional subalgebra $L_2 \subset L_r$ (see Theorem 7.3.3); if $r < 2$, the equation cannot be completely integrated by the Lie group method.

Step 3. Determine the type of your L_2 by Table 12.1; if necessary, change the basis of L_2 in accordance with Lemma 12.2.2.

Step 4. Find a change of variables transforming L_2 to its canonical form in accordance with Theorem 12.2.2. Upon rewriting your differential equation in the new variables, you will obtain a particular case of one of the integrable equations (12.23)-(12.26) presented in Table 12.1.

Step 5. Integrate the equation written in the new variables and rewrite the solution in the original variables, thus completing the integration procedure.

12.2. INTEGRATION BY MEANS OF TWO SYMMETRIES

12.2.4 A sample for implementation of the algorithm

Let us integrate by Lie's algorithm the equation (its frame is given in Section 8.5)

$$y'' = \frac{y'}{y^2} - \frac{1}{xy}. \qquad (12.27)$$

Step 1. Determination of the admissible algebra L_2. Proceeding as in Example 2 of Section 9.3.1, we split the determining equation (9.16) into the following four equations:

$$(y')^3 : \ \xi_{yy} = 0,$$
$$(y')^2 : \ y^2(\eta_{yy} - 2\xi_{xy}) - 2\xi_y = 0,$$
$$(y')^1 : \ y^3(2\eta_{xy} - \xi_{xx}) - y\xi_x + 2\eta + 3(y^2/x)\xi_y = 0,$$
$$(y')^0 : \ x^2 y^2 \eta_{xx} - x^2 \eta_x + xy(2\xi_x - \eta_y) - x\eta - y\xi = 0.$$

The first two equations yield, upon integration with respect to y:

$$\xi = p(x)y + a(x), \quad \eta = -p(x)\ln(y^2) + p'(x)y^2 + q(x)y + b(x).$$

We substitute these expressions for ξ and η into the third and fourth equations. The left-hand sides of these equations will contain, along with powers of y, also the terms with $\ln(y^2)$. Equating the latter to zero, we get $p(x) = 0$. Hence, $\xi = a(x)$, $\eta = q(x)y + b(x)$. Now the third and fourth equations readily yield

$$\xi = C_1 x^2 + C_2 x, \quad \eta = \left(C_1 x + \frac{1}{2}C_2\right)y.$$

Thus, equation (12.27) admits two linearly independent operators:

$$X_1 = x^2 \frac{\partial}{\partial x} + xy \frac{\partial}{\partial y}, \quad X_2 = x \frac{\partial}{\partial x} + \frac{y}{2}\frac{\partial}{\partial y}.$$

According to the algorithm, we advance directly to the third step.

Step 3. Determination of the type of the algebra L_2. Here

$$[X_1, X_2] = -X_1, \quad \xi_1 \eta_2 - \eta_1 \xi_2 = -\frac{x^2 y}{2} \neq 0.$$

Hence, X_1 and X_2 span an algebra of type III. For the correspondence to be complete, one has merely to change the sign of X_2 so that the basis

$$X_1 = x^2 \frac{\partial}{\partial x} + xy \frac{\partial}{\partial y}, \quad X_2 = -x \frac{\partial}{\partial x} - \frac{y}{2}\frac{\partial}{\partial y} \qquad (12.28)$$

has exactly the structure of type III in Lemma 12.2.2.

Step 4. Determination of the integrating change of variable. Upon introducing the canonical variables for X_1 given by

$$t = \frac{y}{x}, \quad u = -\frac{1}{x}, \qquad (12.29)$$

we transform the operators (12.28) to the form

$$X_1 = \frac{\partial}{\partial u}, \quad X_2 = \frac{t}{2}\frac{\partial}{\partial t} + u\frac{\partial}{\partial u}.$$

Their difference from the corresponding operators of type III of Theorem 12.2.2 by the factor $1/2$ in X_2 does not hinder the integration. Excluding the solution

$$y = Kx, \qquad (12.30)$$

we rewrite equation (12.27) in the new variables (cf. equation (12.25)):

$$u'' + \frac{1}{t^2}u'^2 = 0.$$

Whence, integrating once, we obtain $u' = t/(C_1 t - 1)$. If $C_1 = 0$, then

$$u = -\frac{t^2}{2} + C, \qquad (12.31)$$

while for $C_1 \neq 0$ we have

$$u = \frac{t}{C_1} + \frac{1}{C_1^2}\ln|C_1 t - 1| + C_2. \qquad (12.32)$$

Step 5. Solution in the original variables. Substituting (12.29) into the solutions (12.31) and (12.32), and invoking (12.30), we obtain all solutions of equation (12.27) given by two explicit formulae depending on the arbitrary constants K and C, respectively, and by an algebraic equation defining the solution as an implicit function depending upon two arbitrary constants C_1 and C_2:

$$y = Kx, \quad y = \pm\sqrt{2x + Cx^2}, \qquad (12.33)$$

$$C_1 y + C_2 x + x\ln\left|C_1\frac{y}{x} - 1\right| + C_1^2 = 0. \qquad (12.34)$$

12.2.5 Application to the non-homogeneous linear equation

Consider the non-homogeneous linear equation, i.e. equation (2.50) for $n = 2$:

$$y'' + a_1(x)y' + a_2(x)y = f(x). \qquad (12.35)$$

Recall that Lagrange's method [2.9] of variation of parameters suggests seeking the solution in the form $y = u_1 y_1(x) + u_2 y_2(x)$, where $y_1(x)$ and $y_2(x)$ are two linearly independent solutions of the homogeneous equation (12.35),

$$y_1'' + a_1(x)y_1' + a_2(x)y_1 = 0, \quad y_2'' + a_1(x)y_2' + a_2(x)y_2 = 0, \qquad (12.36)$$

12.2. INTEGRATION BY MEANS OF TWO SYMMETRIES

and u_1 and u_2 are unknown functions of x to be found from the system (2.54):

$$y_1(x)\frac{du_1}{dx} + y_2(x)\frac{du_2}{dx} = 0, \quad y_1'(x)\frac{du_1}{dx} + y_2'(x)\frac{du_2}{dx} = f(x). \quad (12.37)$$

Equations (12.37) yield, upon solving for du_1/dx, du_2/dx and integrating:

$$u_1 = -\int \frac{y_2(x)f(x)}{W(x)}dx + C_1, \quad u_2 = \int \frac{y_1(x)f(x)}{W(x)}dx + C_2, \quad (12.38)$$

where C_1 and C_2 are constants of integration, and

$$W(x) \equiv W[y_1(x), y_2(x)] = y_1(x)y_2'(x) - y_2(x)y_1'(x)$$

is the determinant of the system (12.37) and is called the *Wronskian* [12.7] of the functions y_1 and y_2. Hence, the method of variation of parameters provides the following representation of the general solution to equation (12.35):

$$y = y_2(x)\int \frac{y_1(x)f(x)}{W(x)}dx - y_1(x)\int \frac{y_2(x)f(x)}{W(x)}dx + C_1 y_1(x) + C_2 y_2(x). \quad (12.39)$$

I will employ here the group method and give an alternative representation of the solution. The approach is similar to that used in Section 11.1.3 and can be easily extended to higher-order equations by means of consecutive reduction of order. However, let us use, in the case of second-order equations, the method of canonical representations.

Given linearly independent solutions $y_1(x)$, $y_2(x)$ of the homogeneous equation, the linear superposition principle asserts that equation (12.35) admits a two-parameter group generated by

$$X_1 = y_1(x)\frac{\partial}{\partial y}, \quad X_2 = y_2(x)\frac{\partial}{\partial y}. \quad (12.40)$$

Operators (12.40) span a Lie algebra of type II of Table 7.1, Section 7.3.8. Accordingly, they can be transformed to the canonical form

$$X_1 = \frac{\partial}{\partial u}, \quad X_2 = t\frac{\partial}{\partial u}. \quad (12.41)$$

The corresponding canonical variables t, u are determined by first solving the equations $X_1(t) = 0$, $X_1(u) = 1$, whence $t = x$, $u = y/y_1(x)$ (cf. Section 11.1.3). After these variables have been introduced, X_1 takes the required form (12.41). Next, consider X_2. Now it has the form

$$X_2 = \frac{y_2(x)}{y_1(x)}\frac{\partial}{\partial u}.$$

Hence, both operators (12.40) assume the canonical form (12.41) in the variables

$$t = \frac{y_2(x)}{y_1(x)}, \quad u = \frac{y}{y_1(x)}. \quad (12.42)$$

The first equation (8.49) yields

$$D_x = \frac{W(x)}{y_1^2(x)} D_t, \quad \text{where} \quad W(x) = y_1(x)y_2'(x) - y_2(x)y_1'(x). \tag{12.43}$$

By setting $y' = D_x(y)$, $u' = D_t(u)$ and applying the calculus of Sections 8.3.1 and 8.3.3 to the second equation (12.42), written $y = u\,y_1(x)$, one obtains:

$$y' = \frac{W(x)}{y_1(x)} u' + y_1'(x)\,u, \quad y'' = \frac{W^2(x)}{y_1^3(x)} u'' + \frac{W'(x)}{y_1(x)} u' + y_1''(x)\,u.$$

Substitution of these expressions into (12.35), invoking (12.36), yields:

$$\frac{W^2(x)}{y_1^3(x)} u'' = f(x). \tag{12.44}$$

After expressing x in terms of t from the first equation (12.42), equation (12.44) takes the form integrable by quadratures, $u'' = g(t)$.

However, it is simpler to carry out the practical integration in the original variables. To this end, let us denote the total differentiations D_x and D_t by d/dx and d/dt, respectively. Then, using (12.43) in the form

$$\frac{d}{dt} = \frac{y_1^2(x)}{W(x)} \frac{d}{dx}$$

and invoking the definition of u, equation (12.44) is written:

$$\frac{d}{dx}\left[\frac{y_1^2(x)}{W(x)} \frac{d}{dx}\left(\frac{y}{y_1(x)}\right)\right] = \frac{y_1(x)f(x)}{W(x)}. \tag{12.45}$$

Next, integrate the latter equation twice,

$$\frac{y_1^2(x)}{W(x)} \frac{d}{dx}\left(\frac{y}{y_1(x)}\right) = \int \frac{y_1(x)f(x)}{W(x)} dx + C_1, \tag{12.46}$$

$$\frac{y}{y_1(x)} = \int \frac{W(x)}{y_1^2(x)}\left(\int \frac{y_1(x)f(x)}{W(x)} dx + C_1\right) dx + C_2, \tag{12.47}$$

to obtain the following representation of the solution to equation (12.35):

$$y = y_1(x) \int \frac{W(x)}{y_1^2(x)}\left(\int \frac{y_1(x)f(x)}{W(x)} dx\right) dx + C_1 y_1(x) \int \frac{W(x)}{y_1^2(x)} dx + C_2 y_1(x). \tag{12.48}$$

Example. Let us apply both formula (12.39) and an alternative representation (12.48) to the equation

$$y'' + y = \sin x.$$

12.3. LIE'S LINEARIZATION TEST

We have (see Problem 2.16) $y_1(x) = \cos x$, $y_2(x) = \sin x$, and hence $W(x) = W[y_1(x), y_2(x)] = 1$. Therefore formula (12.39) is written

$$y = \sin x \int \cos x \sin x \, dx - \cos x \int \sin^2 x \, dx + C_1 \cos x + C_2 \sin x.$$

Working out the integrals,

$$\int \cos x \sin x \, dx = -\frac{1}{2}\cos^2 x, \quad \int \sin^2 x \, dx = \frac{x}{2} - \frac{1}{4}\sin 2x,$$

we obtain, after elementary simplification:

$$y = -\frac{1}{2}\cos x + C_1 \cos x + C_2 \sin x.$$

Let us use now the representation (12.48) for the solution. Since here

$$\int \frac{y_1(x) f(x)}{W(x)} dx \equiv \int \cos x \sin x \, dx = -\frac{1}{2}\cos^2 x,$$

we readily evaluate the first integral in (12.48):

$$\int \frac{W(x)}{y_1^2(x)} \left(\int \frac{y_1(x) f(x)}{W(x)} dx \right) dx = \int \frac{1}{\cos^2 x} \left(-\frac{1}{2}\cos^2 x \right) dx = -\frac{x}{2}.$$

Likewise, we find the second integral:

$$\int \frac{W(x)}{y_1^2(x)} dx = \int \frac{1}{\cos^2 x} dx = \tan x.$$

Upon substituting these expressions, (12.48) yields:

$$y = -\frac{1}{2}\cos x + C_1 \cos x + C_2 \sin x.$$

12.3 Lie's linearization test

In applications of differential equations, it is useful to have simple criteria of their linearization. The main result on linearization (equations (12.56)) of second-order equations is due to Lie [5.17], Part III, §1. See also [5.16] and [5.36](i).

12.3.1 Main theorem

Lemma 1. The general linear homogeneous equation of the second order

$$y'' + a(x)y' + b(x)y = 0 \tag{12.49}$$

can be reduced, by a change of variables, to the simplest form

$$y'' = 0. \tag{12.50}$$

Proof. The change of variables $(x, y) \mapsto (t, u)$ (cf. Exercise 10.2.1):

$$t = \int \frac{e^{-a(x)dx}}{h^2(x)} dx, \quad u = \frac{y}{h(x)}, \tag{12.51}$$

where $h(x) \neq 0$ is any solution of equation (12.49), reduces (12.49) to $d^2u/dt^2 = 0$.

Lemma 2. A necessary condition that a second-order equation

$$y'' = f(x, y, y') \tag{12.52}$$

be linearizable by a change of variables

$$t = \varphi(x, y), \quad u = \psi(x, y), \tag{12.53}$$

is that equation (12.52) is a cubic in y', i.e. has the form (see (10.46)):

$$y'' + F_3(x, y)y'^3 + F_2(x, y)y'^2 + F_1(x, y)y' + F(x, y) = 0. \tag{12.54}$$

Proof. According to Lemma 1, we can assume that equation (12.52) linearizes, by a suitable change of variables (12.53), to the simplest equation (12.50), viz. $u'' = 0$. Recall now that a change of variables (12.53) carries with it a transformation of derivatives y' and y'' (Section 8.3.1):

$$u' = \frac{D_x(\psi)}{D_x(\varphi)} = P(x, y, y'), \quad u'' = \frac{D_x(P)}{D_x(\varphi)}.$$

Hence, the equation $u'' = 0$ is written $[D_x(\varphi)]^{-3}[D_x(\varphi) \cdot D_x^2(\psi) - D_x(\psi) \cdot D_x^2(\varphi)] = 0$, or $D_x(\varphi) \cdot D_x^2(\psi) - D_x(\psi) \cdot D_x^2(\varphi) = 0$. The latter equation, after substituting

$$D_x(\varphi) = \varphi_x + y'\varphi_y, \quad D_x^2(\varphi) = \varphi_{xx} + 2y'\varphi_{xy} + y'^2\varphi_{yy} + y''\varphi_y$$

and the similar expressions for $D_x(\psi)$ and $D_x^2(\psi)$, becomes:

$$(\varphi_x\psi_y - \varphi_y\psi_x)y'' + (\varphi_x\psi_{yy} - \psi_y\varphi_{yy})y'^3 + (\varphi_x\psi_{yy} + 2\varphi_y\psi_{xy} - \psi_x\varphi_{yy} - 2\psi_y\varphi_{xy})y'^2$$

$$+ (\varphi_y\psi_{xx} + 2\varphi_x\psi_{xy} - \psi_y\varphi_{xx} - 2\psi_x\varphi_{xy})y' + \varphi_x\psi_{xx} - \psi_x\varphi_{xx} = 0. \tag{12.55}$$

Upon dividing by the Jacobian $\varphi_x\psi_y - \varphi_y\psi_x \neq 0$ of (12.53), it has the form (12.54).

Equation (12.54) contains four arbitrary functions $F_i(x, y)$, whereas (12.55) involves only two functions, $\varphi(x, y)$ and $\psi(x, y)$. Hence, the coefficients of those equations (12.54) linearizable by a change of variables should be restricted by two relations. These relations (namely, (12.57)) are given in the following theorem.

Theorem. The following statements are equivalent:
   (1) A second-order ordinary differential equation (12.52) is linearizable by a change of variables (12.53).
   (2) Equation (12.52) admits an eight-dimensional Lie algebra.

12.3. LIE'S LINEARIZATION TEST

(3) Equation (12.52) has the form (12.54) with coefficients F_3, F_2, F_1, F satisfying the integrability conditions of the auxiliary system:

$$\begin{aligned}
\frac{\partial w}{\partial x} &= zw - FF_3 - \frac{1}{3}\frac{\partial F_1}{\partial y} + \frac{2}{3}\frac{\partial F_2}{\partial x}, \\
\frac{\partial w}{\partial y} &= -w^2 - F_2 w + F_3 z + \frac{\partial F_3}{\partial x} - F_1 F_3, \\
\frac{\partial z}{\partial x} &= z^2 - Fw - F_1 z + \frac{\partial F}{\partial y} + FF_2, \\
\frac{\partial z}{\partial y} &= -zw + FF_3 - \frac{1}{3}\frac{\partial F_2}{\partial x} + \frac{2}{3}\frac{\partial F_1}{\partial y};
\end{aligned} \qquad (12.56)$$

in other words, $F_i(x,y)$ satisfy the following equations (cf. (10.48)–(10.49)):

$$3(F_3)_{xx} - 2(F_2)_{xy} + (F_1)_{yy} = 3(F_1 F_3)_x - 3(FF_3)_y - (F_2^2)_x - 3F_3 F_y + F_2(F_1)_y,$$

$$3F_{yy} - 2(F_1)_{xy} + (F_2)_{xx} = 3(FF_3)_x - 3(FF_2)_y + (F_1^2)_y + 3F(F_3)_x - F_1(F_2)_x. \quad (12.57)$$

(4) Equation (12.52) admits an algebra L_2 spanned by *connected* operators $X_1 = \xi_1 \partial/\partial x + \eta_1 \partial/\partial y$, $X_2 = \xi_2 \partial/\partial x + \eta_2 \partial/\partial y$, i.e. such that (Remark 12.2.2)

$$\xi_1 \eta_2 - \eta_1 \xi_2 = 0. \qquad (12.58)$$

Remark. Statement (3) and equations (12.57) are used to recognize linerizable equations, while statement (4) furnishes a practical method of linearization by transforming the equation to type II or IV of Table 12.1.

12.3.2 Examples

Example 1. The equation

$$y'' = (yy'^2 - xy'^3)f(y) \qquad (12.59)$$

with an arbitrary function $f(y)$ has the form (12.54). Its coefficients are $F_3 = xf(y)$, $F_2 = -yf(y)$, $F_1 = 0$, $F = 0$. They obviously satisfy the system (12.57). Hence, equation (12.59) is linearizable.

Example 2. Consider the equation

$$y'' + F(x,y) = 0. \qquad (12.60)$$

Here $F_3 = F_2 = F_1 = 0$. Therefore, the linearization test given by the system (12.57) reduces to the single equation $F_{yy} = 0$, i.e. $F(x,y) = A(x)y + B(x)$. Hence, equation (12.60) cannot be linearized unless it is already linear.

Example 3. The equation

$$y'' = \frac{(y-xy')^3}{x^3} f\left(\frac{y}{x}\right) \qquad (12.61)$$

admits L_2 of type II spanned by

$$X_1 = x^2 \frac{\partial}{\partial x} + xy \frac{\partial}{\partial y}, \quad X_2 = xy \frac{\partial}{\partial x} + y^2 \frac{\partial}{\partial y}.$$

Transformation (12.53) to the canonical form II defined by $X_1(t) = 0$, $X_1(u) = 1$; $X_2(t) = 0$, $X_2(u) = t$, yields $t = y/x$, $u = -1/x$. This change of variables linearizes (12.61) to $u'' + f(t) = 0$.

Example 4. Consider equations (12.23), $y'' = f(y')$. Lemma 2 of Section 12.3.1 singles out the candidates for linearization:

$$y'' + A_3 y'^3 + A_2 y'^2 + A_1 y' + A_0 = 0, \quad A_i = \text{const.} \tag{12.62}$$

Equations (12.57) are obviously satisfied. Thus, $y'' = f(y')$ is linearizable if and only if it has the form (12.62).

Example 5. Let us single out all linearizable equations (12.25):

$$y'' = \frac{1}{x} f(y'). \tag{12.63}$$

Lemma 2 of Section 12.3.1 provides the candidates for linearization:

$$y'' + \frac{1}{x}(A_3 y'^3 + A_2 y'^2 + A_1 y' + A_0) = 0, \quad A_i = \text{const.} \tag{12.64}$$

The linearization test (12.57) yields:

$$3A_3(1 + A_1) - A_2^2 = 0, \quad A_2(2 - A_1) + 9A_0 A_3 = 0.$$

Whence, setting $A_3 = -a$, $A_2 = -b$ and excepting the case $A_3 = A_2 = 0$ when (12.64) is already linear, one obtains (see also [12.8]):

$$A_1 = -\left(1 + \frac{b^2}{3a}\right), \quad A_0 = -\left(\frac{b}{3a} + \frac{b^3}{27a^2}\right).$$

Thus, equation (12.63) is linearizable if and only if it has the following form:

$$y'' = \frac{1}{x}\left[ay'^3 + by'^2 + \left(1 + \frac{b^2}{3a}\right)y' + \frac{b}{3a} + \frac{b^3}{27a^2}\right], \quad a, b = \text{const.} \tag{12.65}$$

Exercise. Linearize the following particular equation (12.65) (cf. equation (1.21)):

$$y'' = \frac{1}{x}\left(y' + y'^3\right).$$

Solution. Consider the symmetry algebra L_8 (see Problem 9.3(iii)) and take, e.g. its subalgebra L_2 of type II spanned by

$$X_7 = \frac{1}{x} \frac{\partial}{\partial x}, \quad X_8 = \frac{y}{x} \frac{\partial}{\partial x}.$$

The canonical form is obtained by going over to the new variables $t = y$, $u = x^2/2$. Aside from the particular solution $y = \text{const.}$, this transforms our equation into $u'' + 1 = 0$. The general solution $u = -(t^2 + C_1 t + C_2)/2$ of the latter provides the implicit representation of the solution of the original equation, $x^2 + y^2 + C_1 y + C_2 = 0$.

12.4 Integration using approximate symmetries

Lie group methods of reduction of order and integration of differential equations can be extended to equations with a small parameter admitting approximate groups. The following example illustrates the method of consecutive integration using infinitesimal approximate symmetries. The method of Lie's canonical forms is not appropriate, in general, for two-dimensional approximate Lie algebras.

Example [12.9]. The second-order equation

$$y'' - x - \varepsilon y^2 = 0 \tag{12.66}$$

has no exact point symmetries if $\varepsilon \neq 0$ is regarded as a constant coefficient (Problem 9.20), and hence cannot be integrated by the Lie method. However, it possesses approximate symmetries if ε is treated as a small parameter, e.g.

$$\begin{aligned} X_1 &= \frac{2}{3}\varepsilon x^3 \frac{\partial}{\partial x} + \left[1 + \varepsilon\left(yx^2 + \frac{11}{60}x^5\right)\right]\frac{\partial}{\partial y}, \\ X_2 &= \frac{1}{6}\varepsilon x^4 \frac{\partial}{\partial x} + \left[x + \varepsilon\left(\frac{1}{3}yx^3 + \frac{7}{180}x^6\right)\right]\frac{\partial}{\partial y}. \end{aligned} \tag{12.67}$$

The operators (12.67) span a two-dimensional Abelian approximate Lie algebra, i.e. $[X_1, X_2] \approx 0$, and can be used for consecutive integration of equation (12.66). Equations $X_1(t) \approx 1$, $X_1(u) \approx 0$ yield canonical variables for X_1:

$$t = y - \varepsilon\left(\frac{1}{2}x^2 y^2 + \frac{11}{60}yx^5\right), \quad u = x - \frac{2}{3}\varepsilon y x^3. \tag{12.68}$$

Now the first operator (12.67) becomes $X_1 \approx \partial/\partial t$, while equation (12.66) reads

$$u'' + uu'^3 + \varepsilon\left[3u^2 u' + \frac{1}{6}(u^2 u')^2 - \frac{11}{60}(u^2 u')^3\right] = 0.$$

Integration by the standard substitution $u' = p(u)$ yields:

$$p' + up^2 + \varepsilon\left(3u^2 + \frac{1}{6}u^4 p - \frac{11}{60}u^6 p^2\right) = 0. \tag{12.69}$$

The second operator (12.67) is written

$$X_2 = \frac{\varepsilon}{2}u^4 \frac{\partial}{\partial u} + \left[p^2 + \varepsilon\left(2u^3 p - \frac{13}{15}u^5 p^2\right)\right]\frac{\partial}{\partial p}$$

and transforms to the infinitesimal translation, $X_2 \approx \partial/\partial q$, by $(t, u) \mapsto (z, v)$:

$$z = u + \varepsilon\frac{u^4}{2p}, \quad v = -\frac{1}{p} + \varepsilon\left(\frac{u^3}{p^2} - \frac{13u^5}{15p}\right). \tag{12.70}$$

Then (12.69) reads $v' + z + (11/60)\varepsilon z^6 = 0$, whence $v = -z^2/2 - (11/420)\varepsilon z^7 + C$.

Problems

12.1. Lower the order of the equations of Table 9.2, Section 9.3.4, using their symmetries.

12.2. Determine the type (by Lemma 12.2.2) of the Lie algebra spanned by

$$X_1 = x^2 \frac{\partial}{\partial x} + xy \frac{\partial}{\partial y}, \quad X_2 = xy \frac{\partial}{\partial x} + y^2 \frac{\partial}{\partial y}$$

and transform it to the canonical form given in Theorem 12.2.2.

12.3. Integrate the nonlinear equation $y'' + y'/x = e^y$, using its symmetries (9.25).

12.4. Integrate the following equation (S. Lie, Example 1 in [5.14], Chap. 20, §5):

$$y'' = \frac{(y - xy')^3}{x^3} f\left(\frac{y}{x}\right)$$

using its two infinitesimal symmetries:

$$X_1 = x^2 \frac{\partial}{\partial x} + xy \frac{\partial}{\partial y}, \quad X_2 = xy \frac{\partial}{\partial x} + y^2 \frac{\partial}{\partial y}.$$

12.5. Apply Lie's integration algorithm to the equation

$$y'' + e^{3y} y'^4 + y'^2 = 0.$$

12.6*. Solve the second-order equation (12.35) by the consecutive reduction of order using the generators (12.40). In this way, extend the group approach to non-homogeneous linear equations of the third and higher orders.

12.7. Show that the equation $y'' + 3yy' + y^3 = 0$ is lenearizable and find a linearizing transformation.

12.8. Equation (12.59) admits two operators, $X_1 = y\partial/\partial x$ and $X_2 = x\partial/\partial x$. Linearize the equation using L_2 spanned by X_1, X_2.

12.9. Linearize equation (12.62).

12.10. Show that the general solution of the linear equation $y'' + a(x)y' + b(x)y = 0$ is provided by its invariant solutions. In other words, every solution can be obtained as an invariant solution via some infinitesimal symmetries [12.10].

12.11. Test for linearization the equation (1.21), $y'' = -(y' + y'^3)/x$.

12.12**. Obtain all non-similar two-dimensional approximate Lie algebras in the plane. For a definition of approximate Lie algebras, see [5.35], vol. 3, Chap. 2. Compare the result with Table 7.1, Section 7.3.8.

Chapter 13

Higher-order equations

Lie's general integration theory of ordinary differential equations states that an equation of order n with n known infinitesimal symmetries $X_1, \ldots, X_n$ can be solved by n quadratures provided that $X_1, \ldots, X_n$ span a *solvable Lie algebra* L_n. The group method is illustrated here by third-order equations admitting L_3. Lie also noticed, in the case of third-order equations, that if L_3 is not solvable, then his theory of integration requires, along with quadratures, the solution of a Riccati equation [13.1]. The concluding Section 13.2 of this chapter provides a group theoretical background of Euler's method for solution of linear equations with constant coefficients discussed in Section 2.2.8.

13.1 Third-order equations

In the case of third-order equations

$$y''' = f(x, y, y', y''), \tag{13.1}$$

one can use, following Lie [13.2], the transformation of three-dimensional Lie algebras L_3 to canonical forms presented in Theorem 2, Section 7.3.8. After reducing L_3 to its canonical form, the equation admitting this algebra assumes a rather simple form convenient for the purposes of integration. This approach is similar to that used for integration of second-order equations and allows one to avoid the lengthy calculations encountered in the procedure of successive integrations.

13.1.1 Equations admitting a solvable Lie algebra

Let an equation (13.1) admit a three-dimensional Lie algebra L_3, e.g. of type 10 by Lie's classification (Theorem 2 of Section 7.3.8). The generators of L_3 can be taken, in appropriate variables, in the form:

$$X_1 = \frac{\partial}{\partial x}, \quad X_2 = \frac{\partial}{\partial y}, \quad X_3 = x\frac{\partial}{\partial x}. \tag{13.2}$$

To obtain the invariant third-order equations, we take the thrice-extended operators (13.2) and find the differential invariants y''/y'^2 and y'''/y'^3 (cf. Section 12.2.3). Hence, the general form of third-order equations admitting (13.2) is $\Omega(y''/y'^2, y'''/y'^3) = 0$, or upon solving for y''' :

$$y''' = y'^3 \omega(y''/y'^2). \tag{13.3}$$

Let us proceed to the integration of this equation using the Lie algebra L_3. The commutators of (13.2) are $[X_1, X_2] = 0$, $[X_1, X_3] = X_1$, $[X_2, X_3] = 0$. It follows that, e.g., the two-dimensional subalgebra $L_2 = \langle X_1, X_3 \rangle$ of L_3 spanned by X_1 and X_3 is an ideal. Furthermore, the one-dimensional subalgebra $L_1 = \langle X_1 \rangle$ spanned by X_1 is an ideal in L_2. These subalgebras provide the sequence (7.80) of the solvable Lie algebra L_3 :

$$L_3 \supset L_2 \supset L_1. \tag{13.4}$$

As a first step of the integration, let us reduce (13.3) to a differential equation of the first order by using the ideal L_2. We will employ now the method of invariant differentiation. Two lower-order differential invariants of L_2 are

$$u = y, \quad v = \frac{y''}{y'^2}. \tag{13.5}$$

We use Theorem 1 of Section 8.3.5 and proceed as in Theorem 12.1.2, to obtain:

$$\frac{dv}{du} = \frac{D_x(y''/y'^2)}{D_x(y)} = \frac{y'''}{y'^3} - 2\frac{y''^2}{y'^4},$$

whence

$$\frac{y'''}{y'^3} = \frac{dv}{du} + 2v^2.$$

Hence, (13.3) reduces [13.3] to the following first-order equation:

$$\frac{dv}{du} = \omega(v) - 2v^2. \tag{13.6}$$

According to Lie's integration theory, the first-order equation (13.6) admits the quotient algebra L_3/L_2, i.e. the operator X_2 (see Section 7.3.4). Here it is obvious since X_2 (13.2) is written in the variables u and v (13.5) as the u-translation, $X_2 = \partial/\partial u$. Therefore, equation (13.6) is directly integrable. The quadrature

$$\int \frac{dv}{\omega(v) - 2v^2} = u + a \tag{13.7}$$

defines v as a function $v = \varphi(u + a)$, where a is the constant of integration. Invoking (13.5), we obtain the second-order differential equation

$$\frac{y''}{y'^2} = \varphi(y + a). \tag{13.8}$$

13.1. THIRD-ORDER EQUATIONS

Equation (13.8) admits the Lie algebra $L_2 = \langle X_1, X_3 \rangle$ because it involves only the invariants u and v of L_2. Hence, it can be integrated by the group methods. Moreover, the basis of L_2 is well suited to the application of the method of consecutive integration (Section 12.2.1) since the basis X_1 of the ideal L_1 in the sequence (13.4) is written in canonical variables. Consequently, equation (13.8) does not involve the independent variable x and can be integrated once by the standard substitution $y' = p(y)$. Then $y'' = pp'$, and (13.8) becomes $p'/p = \varphi(y+a)$, or $\ln|p| = \int \varphi(y+a) d(y+a) + B$. Hence, we arrive at the equation of the first order,

$$y' = b e^{\int \varphi(y+a) d(y+a)} \equiv \psi(y, a, b), \qquad (13.9)$$

containing two arbitrary constants of integration a and b. The final quadrature,

$$\int \frac{dy}{\psi(y, a, b)} = x + c, \qquad (13.10)$$

completes the integration procedure.

Example. The equation (cf. (1.19)) $y''' = y''^2/y'$ has the form (13.3) with $\omega(v) = v^2$. The formulae (13.7)–(13.10) yield the general solution $y = c e^{cx} - a$.

13.1.2 Equations admitting a non-solvable Lie algebra

Let an equation (13.1) admit a non-solvable three-dimensional Lie algebra L_3 of type 3 (Theorem 2 of Section 7.3.8):

$$X_1 = \frac{\partial}{\partial y}, \quad X_2 = y \frac{\partial}{\partial y}, \quad X_3 = y^2 \frac{\partial}{\partial y}. \qquad (13.11)$$

These operators have two independent differential invariants of order ≤ 3:

$$x \quad \text{and} \quad \frac{y'''}{y'} - \frac{3}{2} \frac{y''^2}{y'^2}.$$

Consequently, the general third-order equation admitting (13.11) is (cf. (1.19)):

$$y''' = \frac{3}{2} \frac{y''^2}{y'} + \omega(x) y'. \qquad (13.12)$$

Let us reduce (13.12) to a first-order equation by the method of invariant differentiation. The first two operators (13.11) span a two-dimensional subalgebra $L_2 = \langle X_1, X_2 \rangle$ of L_3. Taking its differential invariants $u = x$, $v = y''/y'$, we have

$$\frac{dv}{du} = \frac{y'''}{y'} - \frac{y''^2}{y'^2}.$$

Hence, equation (13.12) with an arbitrary function $\omega(x)$ reduces to the general Riccati equation written in the form (2.37):

$$\frac{dv}{du} = \frac{1}{2} v^2 + \omega(u). \qquad (13.13)$$

13.2 Group theoretical background of Euler's method for linear equations with constant coefficients

13.2.1 Symmetries and invariant solutions

The general linear homogeneous equation with constant coefficients (2.55),

$$y^{(n)} + a_1 y^{(n-1)} + \cdots + a_{n-1} y' + a_n y = 0, \quad a_1, \ldots, a_n = \text{const.}, \quad (13.14)$$

admits the $(n+2)$-parameter group consisting of the translation of the independent variable x, multiplication of the dependent variable y by any constant, and the linear superposition. The respective generators of these transformations are:

$$X_1 = \frac{\partial}{\partial x}, \quad X_2 = y\frac{\partial}{\partial y}, \quad Y_i = \eta_i(x)\frac{\partial}{\partial y}, \quad i = 1, \ldots, n, \quad (13.15)$$

where $\eta_i(x)$ are n linearly independent solutions of the equation (13.14).

Theorem. The general solution of the equation (13.14) is composed of its invariant solutions (cf. Problem 12.10).

Proof. Given any n linearly independent solutions $\eta_i(x)$, $i = 1, \ldots, n$, of (13.14), the operators X_2 and Y_i (13.14) generate the linear superposition formula (2.53) and hence the general solution of the equation (13.14),

$$y = C_1 \eta_1(x) + \cdots + C_n \eta_n(x). \quad (13.16)$$

If all $\eta_i(x)$ are invariant solutions, then every particular solution obtained from (13.14) by assigning to arbitrary constants $C_1, \ldots, C_n$ definite values, will be an invariant solution since (13.14) is obtained by a group transformation. Hence, to prove the Theorem, it suffices to find n linearly independent invariant solutions. The required solutions should be invariant with respect to the operators X transversal to X_2 and Y_i. But the operators Y_i involve unknown solutions $\eta_i(x)$ and provide the superposition formula (13.16). Therefore, the transversal operators are taken in the form $X = X_1 + \lambda X_2$:

$$X = \frac{\partial}{\partial x} + \lambda y \frac{\partial}{\partial y}, \quad \lambda = \text{const.} \quad (13.17)$$

The equation $dy/y = \lambda dx$ yields the only independent invariant $u = y e^{-\lambda x}$. Invariant solutions are obtained by setting $u = C$. Hence, *Euler's ansatz* (2.62):

$$y = C e^{\lambda x}. \quad (13.18)$$

Substituting (13.18) into (13.14), one obtains the characteristic equation (2.56),

$$P_n[\lambda] \equiv \lambda^n + a_1 \lambda^{n-1} + \cdots + a_{n-1} \lambda + a_n = 0. \quad (13.19)$$

Provided that the roots $\lambda_1, \ldots, \lambda_n$ of the characteristic equation are distinct, one obtains n linearly independent invariant solutions $\eta_1 = e^{\lambda_1 x}, \ldots, \eta_n = e^{\lambda_n x}$. This completes the proof in the case of distinct roots. Multiple roots are discussed in the next section.

13.2.2 The case of multiple roots

The preceding group method of derivation of Euler's solution required a tacit assumption that all roots of the algebraic equation (13.19) are different. Consider now equations with repeated roots whose discussion discloses advantages gained from the use of algebraic and differential invariants in the integration theory of differential equations. Note that the seminvariants of the differential equation (13.14) with constant coefficients and the algebraic invariants of its characteristic equation (13.19) coincide (cf., e.g., (10.34) and (10.7)).

The problem is that, in the case when multiple roots occur, the set $\eta_1 = e^{\lambda_1 x}, \ldots, \eta_n = e^{\lambda_n x}$ contains repeated solutions and hence their superposition (13.16) does not provide the general solution. Let us solve this problem for second-order equations (13.14) written in the standard form (10.33):

$$L_2[y] \equiv y'' + 2a_1 y' + a_2 y = 0. \tag{13.20}$$

Since the seminvariant (10.34) of the differential equation (13.20) and the discriminant (10.7) of its characteristic equation $P_2[\lambda] \equiv \lambda^2 + 2a_1\lambda + a_2 = 0$ coincide and have the form $H_1 = a_2 - a_1^2$, a differential operator of the second order L_2 (13.20) is the square of a differential operator of the first order, viz.

$$L_2 \equiv \frac{d^2}{dx^2} + 2a_1 \frac{d}{dx} + a_2 = \left(\frac{d}{dx} - \lambda\right)^2, \quad \lambda = \text{const.}, \tag{13.21}$$

if and only if its seminvariant vanishes, $H_1 = 0$.

On the other hand, equation (13.20) can be reduced to $z'' = 0$ by a linear transformation of the dependent variable (10.31),

$$y = \sigma(x) z, \quad \sigma(x) \neq 0, \tag{13.22}$$

if and only if the seminvariant of (13.20) vanishes:

$$H_1 \equiv a_2 - a_1^2 = 0. \tag{13.23}$$

Indeed, provided that equation (13.20) satisfies the condition (13.23), its transformation (13.22) to $z'' = 0$ is given by

$$y = z e^{-a_1 x}. \tag{13.24}$$

The converse statement is evident since H_1 (13.23) is valid for the equation $z'' = 0$ and hence for all equations obtained by transformation (13.22).

Thus, if the characteristic equation $P_2[\lambda] = 0$ has equal roots, then equation (13.20) reduces to $z'' = 0$. Substituting the solution $z = C_1 + C_2 x$ of $z'' = 0$ in the transformation (13.24), one obtains the general solution of (13.20) (cf. (2.59)):

$$y = (C_1 + C_2 x) e^{\lambda x}, \quad \lambda = -a_1. \tag{13.25}$$

Likewise, one can treat higher-order equations (13.14) via their seminvariants. Consider, e.g., a third-order equation

$$L_3[y] \equiv y''' + 3a_1 y'' + 3a_2 y' + a_3 y = 0. \tag{13.26}$$

Let its seminvariant (10.34) vanish, $H_1 \equiv a_2 - a_1^2 = 0$. Then the characteristic equation $P_3[\lambda] \equiv \lambda^3 + 3a_1 \lambda^2 + 3a_2 \lambda + a_3 = 0$ has one repeated root, $\lambda = \lambda_1$, and

$$L_3 = \left(\frac{d}{dx} - \lambda_1\right)^2 \left(\frac{d}{dx} - \lambda_2\right).$$

The function $u = (d/dx - \lambda_2)y \equiv y' - \lambda_2 y$ solves the second-order equation

$$\left(\frac{d}{dx} - \lambda_1\right)^2 u = 0$$

and therefore has the form (13.25), $u = (K_1 + K_2 x)e^{\lambda_1 x}$. Hence,

$$y' - \lambda_2 y = (K_1 + K_2 x)e^{\lambda_1 x}.$$

Integration of this non-homogeneous linear equation of the first order yields:

$$y = (C_1 + C_2 x)e^{\lambda_1 x} + C_3 e^{\lambda_2 x}.$$

Problems

13.1–13.13*. For each of 13 canonical forms of three-dimensional Lie algebras L_3 given in Theorem 2 of Section 7.3.8, find the general third-order equation $y''' = f(x, y, y', y'')$ admitting the algebra L_3. Integrate those equations admitting solvable Lie algebras.

Hint: See [5.14], Chap. 25, and Ibragimov and Nucci's paper cited in [4.5].

13.14*. Let both invariants (10.34) and (10.35) of a third-order equation (13.26) vanish, $H_1 \equiv a_2 - a_1^2 = 0$, $H_2 \equiv a_3 - 3a_1 a_2 + 2a_1^3 = 0$. Integrate this equation.

Answers

Answers and hints for the solution of selected problems are presented here.

Chapter 1

1.3. Rewrite equation (1.21) in the form

$$\frac{dy'}{y'(1+y'^2)} + \frac{dx}{x} = 0,$$

use the expansion

$$\frac{1}{y'(1+y'^2)} = \frac{1}{y'} - \frac{y'}{1+y'^2}$$

and integrate once to obtain

$$\frac{xy'}{\sqrt{1+y'^2}} = C_1,$$

where obviously the constant of integration C_1 satisfies the condition $|C_1| < |x|$. Solve the above equation with respect to y' and integrate, to obtain: $y = C_2 \pm C_1 \operatorname{arccosh}(x/C_1)$ with two constants of integration C_1 and C_2, where cosh denotes the hyperbolic cosine.

1.4. The meteoroid falls to earth with the *escape velocity*, $v_* \approx 11$ km/s.

1.5. Write equation (1.27) in the form $kd(r^3v) = -gr^3dx$ and integrate to obtain its general solution:

$$v = -\frac{gr}{4k} + \frac{C}{r^3}.$$

1.The 6. Solution to Problem 1.5 and the initial condition (1.27) yield

$$v_0 = -\frac{gr_0}{4k} + \frac{C}{r_0^3}, \quad \text{whence} \quad C = \left(v_0 + \frac{gr_0}{4k}\right).$$

Invoking the solution formula of Problem 1.5 one obtains (1.28).

1.7. The general solution to the equation $dv/dt = -g - (\alpha/m)v$ is given by

$$v = -\frac{mg}{\alpha} + Ce^{-\frac{\alpha}{m}t}.$$

The initial condition (1.31) determines the constant of integration

$$C = \left(v_* + \frac{mg}{\alpha}\right)e^{\frac{\alpha}{m}t_*},$$

whence:
$$v = -\frac{mg}{\alpha} + \left(v_* + \frac{mg}{\alpha}\right)e^{\frac{\alpha}{m}(t_* - t)}.$$

It follows that the velocity approaches the constant value $-mg/\alpha$ when $t \gg t_*$ regardless of the initial velocity v_*.

1.8. *Hint:* Rewrite the differential equation
$$\frac{dv}{dt} = -g + \frac{\beta}{m}v^2$$
in the form
$$\left(\frac{dv}{1-bv} + \frac{dv}{1+bv}\right) = -2g\,dt,$$
where $b = \sqrt{\beta/mg}$, and integrate it.

1.10. The equation is similar to that in Problem 1.7. Hence, its solution is written $\tau = T - Be^{-kt}$, where the constant of integration B is positive for heating ($\tau < T$) and negative for cooling ($\tau > T$). It is in the nature of the problem that $\tau \to T$ when $t \to \infty$.

1.11. *Hint:* The coefficient k in Newton's law of cooling (1.35) is independent of T.

1.12. The outside temperature has the form $T(t) = -10 + 3(t - t_1)$, $t_1 \le t \le t_2$, where $t_1 = 6$, $t_2 = 12$. Therefore equation (1.38) is written
$$\frac{d\tau}{dt} = \frac{1}{6}[-10 + 3(t - t_1) - \tau]$$
and is solved by $\tau = -10 - (3/k) + 3(t - t_1) - Be^{-kt}$, or taking into account that $k = 1/6$ and using the initial condition $\tau\big|_{t-t_1} = -5, 25$,
$$\tau = -28 + 3(t - t_1) + 22,75e^{-(t-t_1)/6}, \quad t_1 \le t \le t_2.$$

1.13. Here $\tau = -28 + 3(t - t_1) + 28e^{-(t-t_1)/6}$, $t_1 \le t \le t_2$.

1.14. The reckoning yields: $\tau(t) = -10 + 35e^{-(t-t_0)/18}$ when $t_0 \le t \le t_1$, and $\tau(t) = -64 + 3(t - t_1) + 72e^{-(t-t_1)/18}$ when $t_0 \le t \le t_1$. See Fig. 1.5.

Chapter 2

2.1. *Hint:* Use the method of variation of parameters.
(v) $y = Ax(\ln x)^2 + Bx^3 + Cx\ln x + Kx$, $K = \text{const}$.

2.2. Cf. equation (1.41).

2.3. *Hint:* Recall that the constant C in (2.7) is arbitrary.

2.4. *Hint:* See Example 2 in Section 2.1.2.

2.5. Take $C = 1$ in the solution formula to Example 2 of Section 2.1.2 and denote by $y_0(x)$ the corresponding particular solution. Then $(x^2 - k^2)(y_0 - l^2) = 1$. Upon substitution $(x^2 - k^2) = (y_0 - l^2)^{-1}$ in the general solution, one arrives at the formula
$$y^2 - l^2 = C(y_0(x) - l^2)$$
equivalent to (2.14). If $y_1(x)$ is another particular solution corresponding to $C = C_1 \ne 0$, then $x^2 - k^2 = C_1(y_1 - l^2)^{-1}$. Hence, $(y^2 - l^2) = CC_1(y_1 - l^2)$, or
$$y^2 = Ky_1^2(x) + (1 - K)l^2,$$

ANSWERS

which is (2.14) with a new arbitrary constant $K = CC_1$.

2.6. Formula (2.17) and the law of exterior multiplication (2.16) yield:

$$d(M\,dx+N\,dy) = \frac{\partial M}{\partial x}dx\wedge dx+\frac{\partial M}{\partial y}dy\wedge dx+\frac{\partial N}{\partial x}dx\wedge dy+\frac{\partial N}{\partial x}dy\wedge dy = \left(\frac{\partial N}{\partial x}-\frac{\partial M}{\partial y}\right)dx\wedge dy.$$

2.7. *Hint:* Use the condition (2.25).

2.8. Here

$$M = \frac{1}{x} - \frac{y^2}{x^2}, \quad N = \frac{2y}{x}$$

and

$$\frac{\partial N}{\partial x} = \frac{\partial M}{\partial y} = -\frac{2y}{x^2}.$$

Hence, the equation is exact. The formula (2.27) yields:

$$\Phi(x,y) = \int_{x_0}^{x}\left(\frac{1}{z}-\frac{y^2}{z^2}\right)dz + \int_{y_0}^{y}\frac{2z}{x_0}dz = \ln x + \frac{y^2}{x} + K(x_0, y_0).$$

Thus, the solution is given implicitly by $\ln x + y^2/x = C$.

2.9. Here $P(x) = e^{-x}$, $Q(x) = -e^x$, $\bar{x} = \phi(x) \equiv e^x$, $x = \phi^{-1}(\bar{x}) \equiv \ln\bar{x}$, $\phi'(x) = e^x$. Invoking (2.36), one obtains: $\bar{P}(\bar{x}) = 1/\bar{x}^2$, $\bar{Q}(\bar{x}) = -1$. Whence the transformed equation

$$\frac{dy}{d\bar{x}} + y^2 = \frac{1}{\bar{x}^2}.$$

2.10. *Hint:* The inverse to (2.35) is also a linear-rational transformation. Therefore, find the inverse and substitute it into the Riccati equation (2.31).

2.12. *Hint:* Carry out the calculations sketched in solution of Exercise 2, Section 2.1.8.

2.14. Separation of variables yields $\int(P+Qy+Ry^2)^{-1}dy = x + C$.

2.15*. The expression $y = u_1(x)y_1(x) + \cdots + u_n(x)y_n(x)$ in Theorem 2.2.7 contains n undetermined functions $u_i(x)$, restricted by the equation $L_n[y] = f(x)$ only. Consequently, one can choose any $n - 1$ "complementary" equations provided that all the relations are consistent with one another. In Lagrange's method, the complementary relations are chosen so that to get "simple" expressions for derivatives $y', y'', \ldots$.

(i) Let $n = 2$, and let $y_1(x)$, $y_2(x)$ be a fundamental system for the homogeneous equation (2.52), $L_2[y] = 0$. The solution of $L_2[y] = f(x)$ is taken in the form

$$y = u_1(x)y_1(x) + u_2(x)y_2(x).$$

Its first derivative,

$$y' = [u_1(x)y_1'(x) + u_2(x)y_2'(x)] + [y_1(x)u_1'(x) + y_2(x)u_2'(x)],$$

is simplified by equating to zero the second term, i.e. by chosing the complementary relation $y_1(x)u_1'(x) + y_2(x)u_2'(x) = 0$. Then $y' = u_1(x)y_1'(x) + u_2(x)y_2'(x)$, whilst

$$y'' = [u_1(x)y_1''(x) + u_2(x)y_2''(x)] + [y_1'(x)u_1'(x) + y_2'(x)u_2'(x)].$$

It follows

$$L_2[y] = u_1 L_2[y_1] + u_2 L_2[y_2] + y_1'(x)u_1'(x) + y_2'(x)u_2'(x).$$

Hence, the equation $L_2[y] = f(x)$ implies $y_1'(x)u_1'(x) + y_2'(x)u_2'(x) = f(x)$ since $L_2[y_1] = L_2[y_2] = 0$. Thus, one arrives at the relations (2.54) with $n = 2$:

$$y_1(x)\frac{du_1}{dx} + y_2(x)\frac{du_2}{dx} = 0, \quad y_1'(x)\frac{du_1}{dx} + y_2'(x)\frac{du_2}{dx} = f(x).$$

(ii) In the case $n > 2$, the relations (2.54) are derived by merely applying the first step of the preceding procedure to the successive derivatives $y'', \ldots, y^{(n-1)}$.

2.16. (i) The characteristic equation $\lambda^2 + \omega^2 = 0$ has two distinct roots, namely $\lambda_1 = i\omega$ and $\lambda_2 = -i\omega$, and formula (2.53) yields $y = C_1 \cos(\omega t) + C_2 \sin(\omega t)$;

(ii) By virtue of the preceding result, the solution to the equation in question is taken in the form $y = u_1(t)\cos(\omega t) + u_2(t)\sin(\omega t)$, where the functions $u_i(t)$ are defined by the equations $\cos(\omega t)u_1' + \sin(\omega t)u_2' = 0$, $-\omega\sin(\omega t)u_1' + \omega\cos(\omega t)u_2' = f(t)$ (cf. Exercise 2.2.8). Resolving the latter with respect to u_1', u_2',

$$u_1' = -\frac{1}{\omega}f(t)\sin(\omega t), \quad u_2' = \frac{1}{\omega}f(t)\cos(\omega t),$$

and integrating, one ultimately reaches the solution by two quadratures:

$$y = -\frac{\cos(\omega t)}{\omega}\int f(t)\sin(\omega t)dt + \frac{\sin(\omega t)}{\omega}\int f(t)\cos(\omega t)dt + C_1\cos(\omega t) + C_2\sin(\omega t).$$

2.17. $y = C_1 e^{-x} + \left(C_2 \cos(\sqrt{3}x/2) + C_3 \sin(\sqrt{3}x/2)\right)e^{x/2}$.

Chapter 3

3.1. Let $|f_y(x,y)| \leq N$, where f_y is the partial derivative of $f(x,y)$ with respect to y. The theorem of mean value, $f(x, y_1) - f(x, y_2) = (y_1 - y_2)f_y(x, \tilde{y})$, where $\tilde{y}$ is between y_1 and y_2, yields the Lipschitz condition: $|f(x, y_1) - f(x, y_2)| \leq N|y_1 - y_2|$.

3.2. Since $||y_1| - |y_2|| \leq |y_1 - y_2|$, the Lipschitz condition is satisfied with $K = 1$.

3.7. (i) $y = e^t x$; (ii) $y^1 = e^t x^1$, $y^2 = e^{kt} x^2$.

3.8. $y^1 = x^1 \cosh t + x^2 \sinh t$, $y^2 = x^1 \sinh t + x^2 \cosh t$.

3.9. It is evident that both equations $y' = 1$ and $y' = 2xy - 1$ satisfy the Lipschitz condition on the (x, y) plane. Letting $x = x_0, y = y_0$ in (3.40), one obtains $C_1 = y_0 - x_0$ and $C_2 = y_0 e^{-x_0^2}$. Hence

$$y = y_0 + (x - x_0) \quad \text{and} \quad y = e^{x^2}\left(y_0 e^{-x_0^2} - \int_{x_0}^{x} e^{-t^2} dt\right).$$

3.10. The y-derivative of the right-hand sides of the equations $y' = \pm\sqrt{1 - y^2}$ is bounded when $|y| \leq \delta < 1$ and is infinite at $|y| = 1$. Hence the Lipschitz condition is satisfied at (x_0, y_0) with $|y_0| < 1$. Letting $x = x_0, y = y_0$ in (3.42), one obtains $C_1 = -x_0 + \arcsin y_0$ and $C_2 = x_0 + \arcsin y_0$. Hence

$$y = \sin(x - x_0 + \arcsin y_0), \quad -\frac{\pi}{2} - \arcsin y_0 < x - x_0 < \frac{\pi}{2} - \arcsin y_0,$$

and

$$y = \sin(x_0 - x + \arcsin y_0), \quad -\frac{\pi}{2} - \arcsin y_0 < x_0 - x < \frac{\pi}{2} - \arcsin y_0.$$

ANSWERS

3.12. The parametric representation (3.46) yields $x = p + e^p$, $y = (p-1)e^p + \frac{1}{2}p^2 + C$.

3.13. Here the equation (3.52) is written $2p(dx/dp) + x = 2\sqrt{p}$, whence upon integration $x = \sqrt{p} + C/\sqrt{p}$. Hence, the parametric representation (3.53) of the general solution is

$$x = \sqrt{p} + \frac{C}{\sqrt{p}}, \quad y = -xp + \frac{4}{3}p\sqrt{p}.$$

After elimination of the parameter p, by taking into account the condition $p > 0$, the solution is written in the explicit form:

$$y = \frac{1}{6}\left(x^3 + (x^2 - 4C)^{3/2}\right) - Cx.$$

3.14. The general solutions for both equations are readily obtained by the formula (3.57). The singular solutions are: (i) the parabola $4x = y^2$, (ii) the parabola $4y = (x+1)^2$.

3.16. The solution of the Riccati equation in question comprises the general integral

$$w = \frac{2z^3 + C}{(z^3 - C)z}, \quad C = \text{const.},$$

and its envelope [3.7] $w = -1/z$. Consequently, the integral curve passing through (z_0, w_0) such that $z_0 \neq 0$, $1 + z_0 w_0 \neq 0$, is

$$w = \frac{2(1 + z_0 w_0)z^3 + (z_0 w_0 - 2)z_0^3}{[(1 + z_0 w_0)z^3 - (z_0 w_0 - 2)z_0^3]z}.$$

Hence, the singular points are the fixed pole $z = 0$ and the movable pole $z = z_*$ defined by $(1 + z_0 w_0)z_*^3 = (z_0 w_0 - 2)z_0^3$.

Chapter 4

4.1. The system is written $dt = dx/x^2 = dy/(xy)$. The first equation yields $t + 1/x = C_1$. From the second equation, $dx/x^2 = dy/(xy)$, multiplying through by x and integrating, we find $y/x = C_2$. These two first integrals are independent and provide the general integral. Consequently, the general solution is given by $x = (C_1 - t)^{-1}$, $y = C_2(C_1 - t)^{-1}$.

4.2. Rewrite the equation in question in the form $3x^2 dx - 2y dy = 0$ and integrate to obtain the first integral $x^3 - y^2 = C$.

4.3. (i) The system (4.20) is written $dx^1/x^1 = dx^2/x^2 = \cdots = dx^n/x^n$. Assuming $x^n \neq 0$, one obtains the first integrals $x^1/x^n = C_1$, $x^2/x^n = C_2$, ..., $x^{n-1}/x^n = C_{n-1}$. Hence, according to Theorem 4.2.2, the general solution is given by

$$u = F\left(x^1/x^n, x^2/x^n, \ldots, x^{n-1}/x^n\right);$$

(ii) $u = F(x^2 + y^2)$; (iii) $u = F(x^2 - y^2)$; (iv) (see Problem 4.2) $u = F(x^3 - y^2)$.

4.4. The equation has the form (4.24). Its general solution is $u = \ln x + F\left(y/x^2, x^2 z\right)$.

4.5. *Hint:* Use, e.g., the method given in Exercise 4.2.3. (i) $u = -y + F\left(x^2 + y^2\right)$;
(ii) $u = \int g(x)dx + F\left(x^2 + y^2\right)$; (iii) $u = -\int h(y)dy + F\left(x^2 + y^2\right)$.

4.6. According to Lemma 1 of Section 4.2.2, the function $\psi(x, y, u) = u + \arctan(y/x)$

provides a first integral of the system $dx/y = -dy/x = du$ if it solves the equation $X(\psi) = 0$, where

$$X = y\frac{\partial}{\partial x} - x\frac{\partial}{\partial y} + \frac{\partial}{\partial u}.$$

The verification of the desired equation $X(\psi) = 0$ is straightforward.

4.7. Use Laplace's method, e.g., take Example 4.3.1 as a sample. By virtue of the change of variables $x' = x^2 + y^2$, $y' = y$, the equation in question is written in the form

$$\frac{\partial u}{\partial y} = -\sqrt{x' - y^2}.$$

Upon integration, $u = -[y\sqrt{x' - y^2} - x'\arcsin(y/\sqrt{x'})]/2 + F(x')$, or returning to the original variables and invoking the trigonometrical identity $\arcsin t = \arctan(t/\sqrt{1 - t^2})$,

$$u = -\frac{1}{2}\left[xy + (x^2 + y^2)\arctan(y/x)\right] + F\left(x^2 + y^2\right).$$

4.8. Proceeding as in Example 4.3.2, one readily obtains

$$u = (x^n)^\sigma F\left(x^1/x^n, x^2/x^n, \ldots, x^{n-1}/x^n\right).$$

4.9. It suffices to prove that $\phi(x) = x^n$ cannot be a function of $\psi_1(x), \ldots, \psi_{n-1}(x)$. This follows immediately from Theorem 4.2.2, since $X(\phi) = \xi^n \neq 0$ according to the assumption of Theorem 4.3.2, whereas any function $u(x) = F(\psi_1(x), \ldots, \psi_{n-1}(x))$ solves the homogeneous equation (4.18), $X(u) = 0$.

4.10. Proceed as in Exercise 4.4.2 to obtain the complete integral $u = ax + by + \psi(a)$.

4.11. Here, one can start as in Exercise 4.4.2 to obtain a simple first integral, $p = a$, of the characteristic system (4.70). Hence, the complete integral is to be found from the system $p = a$, $u = ax + yq + \psi(a, q)$. The first equation yields $u = ax + v(y)$, and the second one reduces to Clairaut's equation (3.55), $v = yq + \psi(a, q)$, with an independent variable y and with an unknown function v. Now the formula (3.57), where the arbitrary constant C is denoted by b, yields $v = by + \psi(a, b)$. Thus, one arrives at the complete integral $u = ax + by + \psi(a, b)$.

4.12. We have $P = x + q$, $Q = y + p$, $pP + qQ = xp + yq + 2pq$, $X + pU = Y + qU = 0$ (see Exercise 4.4.2). Hence the equations (4.80) are written

$$\frac{dx}{d\tau} = x + q, \quad \frac{dy}{d\tau} = y + p, \quad \frac{du}{d\tau} = xp + yq + 2pq, \quad \frac{dp}{d\tau} = \frac{dq}{d\tau} = 0.$$

Starting the integration with the two last, simplest, equations and using the initial conditions, $x = x_0, y = y_0, u = u_0, p = p_0, q = q_0$ at $\tau = 0$, one readily obtains the solution (4.84) in the following form:

$$x = x_0 e^\tau + q_0(e^\tau - 1), \quad y = y_0 e^\tau + p_0(e^\tau - 1),$$

$$u = u_0 + (x_0 p_0 + y_0 q_0 + 2p_0 q_0)(e^\tau - 1), \quad p = p_0, \quad q = q_0.$$

The initial curve γ in question can be represented in the parametric form (4.85) by the equations $x_0 = 0, y_0 = s, u_0 = s^2$. Then equations (4.86) become $p_0 q_0 + s q_0 - s^2 = 0$ and $q_0 = 2s$, whence $p_0 = -s/2, q_0 = 2s$. Upon substitution of these values of p_0 and q_0, the

ANSWERS

above expressions for x, y, u become $x = 2s(e^\tau - 1)$, $y = s(e^\tau + 1)/2$, $u = s^2$. Eliminating the parameters s and τ, we arrive at the solution of the Cauchy problem, $u = (x - 4y)^2$.

4.13. For the operators

$$X_1 = z\frac{\partial}{\partial y} - y\frac{\partial}{\partial z}, \qquad X_2 = y\frac{\partial}{\partial x} + z\frac{\partial}{\partial y},$$

formula (4.95) yields $[X_1, X_2] = X_3$, where

$$X_3 = z\frac{\partial}{\partial x} - y\frac{\partial}{\partial y} + z\frac{\partial}{\partial z}.$$

The operators X_1, X_2, X_3, are unconnected since the determinant $|\xi_\alpha^i|$ of their coefficients is not zero:

$$\begin{vmatrix} 0 & z & -y \\ y & z & 0 \\ z & -y & z \end{vmatrix} = y^3.$$

Therefore, the system of equations $X_1(u) = 0, X_2(u) = 0$ is not complete. However, the system obtained by adding to them the equation $X_3(u) = 0$ is complete because more than three operators cannot be unconnected in the space of $n = 3$ variables x, y, z.

4.14. The reckoning shows that the commutator of the operators

$$X_1 = z\frac{\partial}{\partial y} - y\frac{\partial}{\partial z}, \qquad X_2 = \frac{\partial}{\partial x} + t\frac{\partial}{\partial y} + y\frac{\partial}{\partial t}$$

has the form $[X_1, X_2] = X_3$, where

$$X_3 = t\frac{\partial}{\partial z} + z\frac{\partial}{\partial t}.$$

The commutators of X_1 and X_2 with X_3 are $[X_1, X_3] = -X_4$, $[X_2, X_3] = -X_1$, and hence provide another new operator, viz.

$$X_4 = t\frac{\partial}{\partial y} + z\frac{\partial}{\partial t}.$$

The commutators of X_1, X_2, X_3, X_4 do not generate new operators. Indeed, the whole set of commutators is as follows:

$$[X_1, X_2] = X_3, \quad [X_1, X_3] = -X_4, \quad [X_2, X_3] = -X_1,$$
$$[X_1, X_4] = X_3, \quad [X_2, X_4] = 0, \quad [X_3, X_4] = X_1.$$

Furthermore, X_1, X_2, X_3, X_4 are connected, viz. $tX_1 + 0 \cdot X_2 + yX_3 + zX_4 = 0$.

Since the operator $X_3 = [X_1, X_2]$ is independent of X_1 and X_2, the system of equations $X_1(u) = 0, X_2(u) = 0$ is not complete.

Chapter 5

5.2. The reckoning shows that $d\overline{u} - \overline{p}_i d\overline{x}^i = -(du - p_i dx^i)$.

5.3. Here

$$d\overline{x} = dx + \frac{tdp}{\sqrt{1+p^2}} - \frac{tp^2 dp}{(1+p^2)^{3/2}}, \quad d\overline{y} = dy + \frac{tpdp}{(1+p^2)^{3/2}}, \quad \overline{p} = p.$$

Whence $d\bar{y} - \bar{d}d\bar{x} = dy - pdx$.

5.4. The transformation T_a (5.22) is written $\bar{u} \approx u + au'' + a^2 u^{iv}/2$. Substitute now $\bar{u}$ in (5.22) with the parameter b to obtain $\bar{\bar{u}} \approx \bar{u} + b\bar{u}'' + b^2 \bar{u}^{iv}/2$. It follows

$$\bar{\bar{u}} \approx u + au'' + \frac{a^2}{2}u^{iv} + bu'' + abu^{iv} + \frac{b^2}{2}u^{iv} = u + (a+b)u'' + \frac{(a+b)^2}{2}u^{iv}.$$

Whence the group property $T_b T_a = T_{a+b}$ in the second order of precision.

5.5. It follows from $T_b T_a = T_{a+b}$ that $T_{-a} T_a = T_0 = I$. Hence $T_a^{-1} = T_{-a}$. Thus, the inverse to the Lie-Bäcklund transformation (5.22) has the form

$$x = \bar{x}, \quad u = \bar{u} + \sum_{s=1}^{\infty}(-1)^s \frac{\bar{u}^{(2s)}}{s!} a^s.$$

5.6. The conserved quantity (5.24) can be written, ignoring the incidental sign, as:

(i) the energy $E = \frac{m}{2}|v|^2 + U(r)$, (ii) the anglular momentum $M = mx \times v$.

5.7. The equation (5.32), $v'' - \lambda v = 0$, yields $v_1 = e^{\sqrt{\lambda}x}$, $v_2 = e^{-\sqrt{\lambda}x}$. Hence, Lie's solution (5.33) for the heat equation has the form

$$u = \int_{\alpha_1}^{\beta_1} f_1(\lambda) e^{\lambda t + \sqrt{\lambda}x} d\lambda + \int_{\alpha_2}^{\beta_2} f_2(\lambda) e^{\lambda t - \sqrt{\lambda}x} d\lambda.$$

5.8*. Write down the prolonged operator (5.42) using the equations $R_x = \rho$, $R_t = -\rho v$:

$$Y = \frac{t^2}{2}\frac{\partial}{\partial x} + t\frac{\partial}{\partial v} - R\frac{\partial}{\partial p} + (1 - tv_x)\frac{\partial}{\partial v_t} - t\rho_x\frac{\partial}{\partial \rho_t} + (\rho v - t p_x)\frac{\partial}{\partial p_t} - \rho\frac{\partial}{\partial p_x},$$

and verify the invariance test for equations (5.40) with $A(p,\rho) = f(\rho)$, viz.

$$Y(v_t + vv_x + \rho^{-1}p_x) = 0, \quad Y(\rho_t + v\rho_x + \rho v_x) = 0, \quad Y(p_t + vp_x + f(\rho)v_x) = 0.$$

Chapter 6

6.1. Calculate $\bar{b}^2 - 4\bar{a}\bar{c}$ invoking (6.4) to obtain $\bar{b}^2 - 4\bar{a}\bar{c} = b^2 - 4ac$.

6.3*. See Chapter 10.

6.4*. See Chapter 10.

6.5. Use the geometrical representation of complex numbers

$$x + iy = r(\cos\theta + i\sin\theta)$$

with $r = \sqrt{x^2 + y^2}$ and $\theta = \arctan(y/x)$, where $-\pi \leq \theta \leq \pi$. Then an nth root of $x + iy$ is given by

$$r^{1/n}[\cos(\theta/n) + i\sin(\theta/n)].$$

In our case, $2(-1+i) = 2\sqrt{2}[\cos(3\pi/4) + i\sin(3\pi/4)]$. Hence

$$\sqrt[3]{2(-1+i)} = \sqrt{2}[\cos(\pi/4) + i\sin(\pi/4)] = 1 + i.$$

Likewise, $2(-1-i) = 2\sqrt{2}[\cos(-3\pi/4) + i\sin(-3\pi/4)]$, and hence

$$\sqrt[3]{2(-1-i)} = \sqrt{2}[\cos(-\pi/4) + i\sin(-\pi/4)] = 1 - i.$$

ANSWERS

6.6. Expand $\cos(\theta - a)$ and $\sin(\theta - a)$ in the transformation formulas
$$\overline{x} = r\cos(\theta - a), \quad \overline{y} = r\sin(\theta - a)$$
and substitute $r\cos\theta = x, r\sin\theta = y$.

6.7. The definition of the composition (6.18) yields:
$$T_1(T_2 T_3)(z) = T_1((T_2 T_3)(z)) = T_1(T_2(T_3(z)))$$
and
$$(T_1 T_2)T_3(z) = (T_1 T_2)(T_3(z)) = T_1(T_2(T_3(z)))$$
for any $z \in \mathbb{R}^n$. This precisely means that $T_1(T_2 T_3) = (T_1 T_2)T_3$.

6.8. According to definition (6.17), the composition $T_b T_a$ of translations $T_a : \overline{x} = x + a$ and $T_b : \overline{x} = x + b$ has the form
$$\overline{\overline{x}} = \overline{x} + b = (x + a) + b = x + (a + b).$$
Hence, $T_b T_a = T_{a+b}$.

6.9. To verify that the set $\{I, T_1, \ldots, T_5\}$ is a group, check that
$$T_1^{-1} = T_1, \ T_2^{-1} = T_2, \ T_3^{-1} = T_5, \ T_4^{-1} = T_4, \ T_5^{-1} = T_3,$$
and
$$T_1^2 = I, \ T_2^2 = I, \ T_3^2 = T_5, \ T_4^2 = I, T_5^2 = T_3, \ T_2 T_1 = T_3, \ T_1 T_2 = T_4, \ T_3 T_1 = T_2,$$
$$T_1 T_3 = T_4, \ T_4 T_1 = T_5, \ T_1 T_4 = T_3, \ T_5 T_1 = T_4, \ T_1 T_5 = T_2, \ T_3 T_2 = T_4, \ T_2 T_3 = T_1,$$
and so on. The statement that $H = \{I, T_1\}$ is a subgroup of the group $G = \{I, T_1, \ldots, T_5\}$, i.e. that $T_1^{-1}, T_1^2 \in H$, results from $T_1^{-1} = T_1$ because then $T_1^2 \equiv T_1 T_1 = T_1 T_1^{-1} = I$.

6.10. *Hint:* Calculate the composition of two transformations of the form (6.25).

6.11. Let G be the set of all rotations (6.30). The consecutive application of two rotations, $T_a, T_b \in G$, yields:
$$\overline{x} = x \cos b + y \sin a, \quad \overline{y} = y \cos b - \overline{x} \sin b.$$

Invoking (6.30), one obtains
$$\overline{\overline{x}} = (x \cos a + y \sin a) \cos b + (y \cos a - x \sin a) \sin b,$$
$$\overline{\overline{y}} = (y \cos a - x \sin a) \cos b - (x \cos a + y \sin a) \sin b,$$
or
$$\overline{\overline{x}} = x(\cos a \cos b - \sin a \sin b) + y(\sin a \cos b + \cos a \sin b),$$
$$\overline{\overline{y}} = y(\cos a \cos b - \sin a \sin b) - x(\sin a \cos b + \cos a \sin b).$$

Ultimately:
$$\overline{\overline{x}} = x \cos(a+b) + y \sin(a+b), \quad \overline{\overline{y}} = y \cos(a+b) - x \sin(a+b).$$

Hence $T_b T_a = T_{a+b} \in G$. It follows that $T_a^{-1} = T_{-a} \in G$. It is evident that $I = T_0 \in G$.

6.12. The consecutive application of two transformations, T_a and T_b, yields:

$$\bar{\bar{x}} = \frac{x}{1-(a+b)x}, \quad \bar{\bar{y}} = \frac{y}{1-(a+b)y},$$

i.e. $T_b T_a = T_{a+b}$. Other properties of a group are transparent.

6.15. Let G be the set of transformations (6.36), $C_a = S_1 T_a S_1$. We have $I = C_0 \in G$. Furthermore, the composition of two transformations is written $C_b C_a = S_1 T_b S_1 S_1 T_a S_1$. Invoking the equation $S_1 S_1 = I$ and the group property $T_b T_a = T_{a+b}$ of the translation group, we have:

$$C_b C_a = S_1 T_b T_a S_1 = S_1 T_{a+b} S_1 = C_{a+b} \in G.$$

6.16. Let G be the set of all transformations T_a:

$$\bar{x} = x + a, \quad \bar{y} = y + a + a^3,$$

with an arbitrary real parameter a. It is evident that G contains the identity $I = T_0$ and the inverse transformation $T_a^{-1} = T_{-a}$: $x = \bar{x} - a, y = \bar{y} - a - a^3$. It remains to check whether or not G contains the composition of its elements. The consecutive application of two transformations, T_a and T_b, yields:

$$\bar{\bar{x}} = x + a + b, \quad \bar{\bar{y}} = y + a + b + a^3 + b^3.$$

The first equation has the form $\bar{\bar{x}} = x + c$ with $c = a + b$, i.e. the form required by the group property $T_b T_a \in G$. However, the second equation is not compatible with this property, for $\bar{\bar{y}} = y + c + c^3$ would mean that $c^3 = a^3 + b^3$, i.e. that $(a+b)^3 = a^3 + b^3$. Hence, G is not a group.

6.17. Let G be the set of all transformations of the form $\bar{x} = f(x)$ with arbitrary $f(x)$. The consecutive application of two transformations, $T_f : \bar{x} = f(x)$ and $T_g : \bar{x} = g(x)$, where $f(x)$ and $g(x)$ any given functions, yields:

$$\bar{\bar{x}} = g(f(x)) \stackrel{\text{def}}{=} (g \circ f)(x).$$

Hence, the product of transformations is defined by the composition of functions, viz. $T_g T_f = T_{g \circ f} \in G$. It follows that $T_f^{-1} = T_{f^{-1}} \in G$. It is evident that G contains the identical transformation when $f(x) = x$.

6.18. *Hint:* First check that the composition of any transformations (6.51) is again a transformation of the form (6.51) and that the conditions (6.52) hold for the composition. Then show that the composition of transformations of types T and R obeys the scheme $TT = T, TR = R, RT = R, RR = T$, where T and R do not necessarily denote the same transformation at each occurrence.

6.19. *Hint:* Investigate the invertibilty of transformations (6.30) and (6.37).

Chapter 7

7.1. One can set $a = a' + 1$. An alternative choice is $a = e^{\tilde{a}}$. The dilation $\bar{x} = ax$ is rewritten in the new parameters, after denoting them again by a, in the following forms: $\bar{x} = x + ax$, or $\bar{x} = e^a x$. In both cases the identical transformation corresponds to $a = 0$.

7.2. (i) $\phi(a,b) = ab$, $a^{-1} = 1/a$; (ii) $\phi(a,b) = a + b + ab$, $a^{-1} = -a/(1+a)$; (iii) $\phi(a,b) = a + b$, $a^{-1} = -a$.

ANSWERS

7.3. Equation $T_{a_0} = I$ yields $T_{a_0}T_a = T_a$, $T_b T_{a_0} = T_b$. Invoking (7.4), one obtains (7.8).

7.4. (i) $X = (y^2 - x^2)\dfrac{\partial}{\partial x} - 2xy\dfrac{\partial}{\partial y}$; (ii) $X = (y^2 + z^2 - x^2)\dfrac{\partial}{\partial x} - 2xy\dfrac{\partial}{\partial y} - 2xz\dfrac{\partial}{\partial z}$.

7.5. The transformation $\bar{x} = x\sqrt{1-a^2} + ya$, $\bar{y} = y\sqrt{1-a^2} - xa$ forms a local group with $a_0 = 0$ and with the following composition law (7.7):
$$\phi(a,b) = a\sqrt{1-b^2} + b\sqrt{1-a^2}.$$
Formula (7.21) yields $w(s) = \sqrt{1-s^2}$, whence the canonical parameter $\tilde{a} = \arcsin a$. The group in question is represented in the parameter $\tilde{a}$ as the rotation group (6.30).

7.6*. It is convenient to use the notation:
$$F = F(x), \quad \overline{F} = F(\bar{x}), \quad X = \xi^i(x)\dfrac{\partial}{\partial x^i}, \quad \overline{X} = \xi^i(\bar{x})\dfrac{\partial}{\partial \bar{x}^i}.$$

In this notation, the equation (see Proof of Theorem 7.1.7)
$$\dfrac{dF(f(x,a))}{da} = \xi^i(\bar{x})\dfrac{\partial F(\bar{x})}{\partial \bar{x}^i}$$
is written $d\overline{F}/da = \overline{X}(\overline{F})$. Furthermore, $[\overline{F}]_0 = F$, $[d\overline{F}/da]_0 = XF$, where the zero means evaluated at $a = 0$. The iteration yields:
$$\dfrac{d^2\overline{F}}{da^2} = \dfrac{d}{da}(\overline{X}\overline{F}) = \overline{X}(\overline{X}\overline{F}) = \overline{X}^2(\overline{F}), \quad \ldots, \quad \dfrac{d^s\overline{F}}{da^s} = \overline{X}^s(\overline{F}); \quad \left[\dfrac{d^s\overline{F}}{da^s}\right]_0 = X^s F.$$

To obtain (7.25), substitute these expressions into the Taylor formula:
$$F(\bar{x}) = [\overline{F}]_0 + a\left[\dfrac{d\overline{F}}{da}\right]_0 + \dfrac{a^2}{2!}\left[\dfrac{d^2\overline{F}}{da^2}\right]_0 + \cdots + \dfrac{a^s}{s!}\left[\dfrac{d^s\overline{F}}{da^s}\right]_0 + \cdots.$$

7.7. We have:
$$\bar{u} = \bar{y}\bar{x}^{-2} = (ye^{2a})(xe^a)^{-2} = yx^{-2} = u,$$
$$\bar{v} = \bar{z}\bar{x}^2 = (ze^{-2a})(xe^a)^2 = zx^2 = v,$$
$$\bar{w} = \ln|\bar{x}| = \ln|xe^a| = \ln|x| + a = w + a.$$

7.8. The group transformations for (i) to (v) are to be found in Section 7.1.10.
(iii) The Lie equations $d\bar{x}/da = \bar{y}$, $d\bar{y}/da = -\bar{x}$ yield (see Exercise 2.3)
$$\bar{x} = C_1 \cos a + C_2 \sin a, \quad \bar{y} = C_2 \cos a - C_1 \sin a.$$
Letting here $a = 0$ and invoking the initial conditions, one obtains $C_1 = x, C_2 = y$. Hence, one arrives at the rotation group (6.30). Use of the exponential map is similar to that in Example 7.1.9. For canonical variables, see Example 4.3.1 and Problem 6.6.

(iv) Consider, e.g. introduction of canonical variables. The equation $X(u) = 0$ yields $u = y^2 - x^2$. To solve the equation $X(t) = 1$, one can employ, e.g. Laplace's method. Namely, in semi-canonical variables u, x, the equation $X(t) = 1$ is written (Remark 4.3.2)
$$\sqrt{u + x^2}\,\dfrac{dt}{dx} = 1,$$
whence $t = \ln|x + y| + C(u)$, provided that $x + y \neq 0$. Hence,
$$t = \ln|x + y|, \quad u = y^2 - x^2$$

are canonical variables, so that the group transformation is written $\bar{t}=t+a, \bar{u}=u$, i.e.

$$\ln|\bar{x}+\bar{y}|=\ln|x+y|+a, \quad \bar{y}^{2}-\bar{x}^{2}=y^{2}-x^{2}.$$

It follows from continuity that $\bar{x}+\bar{y}\neq 0$ in a neighborhood of $a=0$, whenever $x+y\neq 0$. Furthermore, in this neighborhood, $x+y$ and $\bar{x}+\bar{y}$ have the same sign. Therefore, the above equations can be rewritten in the form

$$\bar{x}+\bar{y}=(x+y)e^{a}, \quad (\bar{x}+\bar{y})(\bar{y}-\bar{x})=(x+y)(y-x),$$

or $\bar{x}+\bar{y}=(x+y)e^{a}, \; \bar{y}+\bar{x}=(y-x)e^{-a}$. Hence, upon solving with respect to $\bar{x}$ and $\bar{y}$, one arrives at the Lorentz transformation (6.37):

$$\bar{x}=x\cosh a+y\sinh a, \quad \bar{y}=y\cosh a+x\sinh a.$$

(vi) The Lie equations have the form

$$\frac{d\bar{x}}{da}=\bar{x}^{2}, \quad \frac{d\bar{y}}{da}=3\bar{x}\bar{y}+\bar{x}^{2}; \quad \bar{x}\big|_{a=0}=x, \; \bar{y}\big|_{a=0}=y.$$

The first equation yields (see the solution of equations (7.44)) $\bar{x}=x/(1-ax)$. Now the second equation becomes a non-homogeneous linear equation of the first order:

$$\frac{d\bar{y}}{da}=\frac{3x}{1-ax}\bar{y}+\frac{x^{2}}{(1-ax)^{2}}.$$

It follows upon integration (e.g. by the methods of Section 2.1.1):

$$\bar{x}=\frac{x}{1-ax}, \quad \bar{y}=\frac{y}{(1-ax)^{3}}+\frac{x}{2(1-ax)^{3}}-\frac{x}{2(1-ax)}.$$

Introduction of canonical variables leads to this result immediately. Indeed, proceeding as in Example 1 (iii) of Section 7.1.9 and invoking (7.41), we take the canonical variables $u=(2y+x)/(2x^{3})$ and $t=-1/x$. Then $\bar{t}=t+a, \bar{u}=u$. Hence, the above transformation is obtained from the equations:

$$-\frac{1}{\bar{x}}=-\frac{1}{x}+a, \quad \frac{2\bar{y}+\bar{x}}{2\bar{x}^{3}}=\frac{2y+x}{2x^{3}}.$$

7.9. Let us consider here the shortest way – via canonical variables. Integrating the characteristic system (7.27),

$$\frac{dt}{0}=\frac{dx}{2t}=-\frac{du}{xu},$$

and the equation $X(s)=1$, we get the following canonical variables:

$$t, \quad z=ue^{x^{2}/(4t)}, \quad s=\frac{x}{2t}.$$

Then $\bar{t}=t, \bar{s}=s+a, \bar{z}=z$, or

$$\bar{t}=t, \quad \frac{\bar{x}}{2\bar{t}}=\frac{x}{2t}+a, \quad \bar{u}e^{\bar{x}^{2}/(4t)}=ue^{x^{2}/(4t)}.$$

The last two equations are written by taking into account the first one, $\bar{t}=t$. Upon solving with respect to $\bar{x}$ and $\bar{u}$, we arrive at the following transformation (cf. (5.43)):

$$\bar{t}=t, \quad \bar{x}=x+2at, \quad \bar{u}=ue^{-(ax+a^{2}t)}.$$

ANSWERS

Here the parameter a ranges over all real numbers. The group is *global*.

7.10*. *Hint:* One can proceed from the following. If $f(x)$ is an analytic function such that $f(0) = 0$, then $f(x) = xg(x)$, where $g(x)$ is bounded near $x = 0$. One can use, e.g. the Taylor expansion: $f(x) = c_0 x + c_1 x^2 + \cdots = x(c_0 + c_1 x + \cdots) = \lambda(x) x$. One needs to justify the convergence of the series in brackets. Likewise, one can consider the case of many variables x. The problem of equation (7.56) can be reduced to the above situation, e.g. by a change of variables (7.51).

7.11. Let $F = x^2 + y^2 - e^{2\arctan(y/x)}$. Then $XF = X(x^2+y^2) - \frac{2x^2}{x^2+y^2} e^{2\arctan(y/x)} X(y/x)$. Here $X(x^2+y^2) = 2x(x-y) + 2y(x+y) = 2(x^2+y^2)$ and $X(y/x) = (x^2+y^2)/x^2$. Therefore $XF = 2F$, and hence $XF\big|_{F=0} = 0$. An invariant representation: take the generator X (7.67) to get the invariant in polar coordinates, $\psi = re^{-\theta}$, so that the induced invariant on the spiral (7.64) is $\widetilde{\psi} = e^{\theta} e^{-\theta} = 1$, i.e. (7.60) is of the form $\widetilde{\psi} = 1$ (cf. footnote to relations (7.60)); hence, an invariant representation (7.57) of the spiral is $\psi = 1$, i.e. $re^{-\theta} = 1$; in Cartesian coordinates, it becomes $(x^2+y^2)e^{-2\arctan(y/x)} = 1$.

7.12. *Hint:* Calculations can be simplified by employing the polar coordinates and then returning to rectangular ones.

7.13. *Hint:* Use formula (7.30).

7.14*. *Hint:* Proceed as in Example 4 in Section 7.2.3.

7.16. If X_1, X_2 is a basis of a two-dimensional Lie algebra L_2, the derived algebra $L_2^{(1)} = \langle [X_1, X_2] \rangle$. We have either $[X_1, X_2] = 0$, then $L_2^{(1)} = 0$, or $[X_1, X_2] \neq 0$, then $\dim L_2^{(1)} = 1$. In the latter case, $L_2^{(2)} = 0$ since the derived algebra of any one-dimensional Lie algebra is manifestly zero. Thus, the first or second derivative of L_2 vanishes. Hence, any two-dimensional Lie algebra is solvable. The sequence (7.79) can be easily constructed following the proof of Theorem 7.3.6.

7.17. In each case, X_1, X_2, X_3 span a Lie algebra L_3 such that $L_3^{(1)} = L_3$. Consequently, $L_3^{(s)} = L_3$, hence $L_3^{(s)} \neq 0$ for any s. By Theorem 7.3.6, L_3 is a non-solvable Lie algebra.

7.18. Let $[X_1, X_2] = \alpha_1 X_1 + \alpha_2 X_2 \neq 0$. The relation $[X_1', X_2'] = X_1'$ is obtained by taking a new basis $X_1' = \alpha_1 X_1 + \alpha_2 X_2$, $X_2' = \beta_1 X_1 + \beta_2 X_2$ with any β_1, β_2 satisfying the condition $\alpha_1 \beta_2 - \alpha_2 \beta_1 = 1$. For example, if $\alpha_2 \neq 0$, one can take $\beta_1 = -1/\alpha_2$, $\beta_2 = 0$.

7.20*. *Hint:* For Problems 7.20* and 7.21*, see [5.14], Chap. 22, §1. See also Exercise 1 in Section 7.3.8. Constant factors of operators are immaterial and can be neglected.

The changes of variables: $\bar{x} = x$, $\bar{y} = y - x$; $\widetilde{x} = x+y$, $\widetilde{y} = 2xy$.

7.21*. The changes of variables: $\bar{x} = x$, $\bar{y} = y^2$; $\widetilde{x} = 1/y$, $\widetilde{y} = -x/y$.

7.22. $T_a = T_{a^3} T_{a^2}$ with $a = (a^2, a^3)$ is written $\bar{x} = x + ya^3 + a^2 a^3$, $\bar{y} = y + a^2$. Let $b = (b^2, b^3)$. The composition $T_b T_a$ has the form

$$\bar{\bar{x}} = x + y(a^3 + b^3) + a^2(a^3 + b^3) + b^2 b_3, \quad \bar{\bar{y}} = y + a^2 + b^2.$$

In order that this be a transformation T_c of the same family, i.e., that

$$\bar{\bar{x}} = x + yc^3 + c^2 c^3, \quad \bar{\bar{y}} = y + c^2,$$

the vector-parameter $c = (c^2, c^3)$ has to satisfy the following three equations:

$$c^2 = a^2 + b^2, \quad c^3 = a^3 + b^3, \quad c^2 c^3 = a^2(a^3 + b^3) + b^2 b^3.$$

However, this overdetermined system of equations is not compatible. Indeed, the substitution of c^2 and c^3 given by the first two equations into the third equation yields $b^2 a^3 = 0$. Thus, the two-parameter family of transformations T_a is not a group.

Chapter 8

8.3. $D^4(f) = f^{(4)} y'^4 + 6 f''' y'^2 y'' + 4 f'' y' y''' + 3 f'' y''^2 + f' y^{(4)}$,

$D_x^5(f) = f^{(5)} y'^5 + 10 f^{(4)} y'^3 y'' + 10 f''' y'^2 y''' + 15 f''' y' y''^2 + 5 f'' y' y^{(4)} + 10 f'' y'' y''' + f' y^{(5)}$.

8.4. The equations (8.8) and (8.8) yield:

1) $l_1 = 6, p = 6$; 2) $l_1 = 4, l_2 = 1, p = 5$; 3) $l_1 = 3, l_3 = 1, p = 4$; 4) $l_1 = 2, l_2 = 2, p = 4$;
5) $l_1 = 2, l_4 = 1, \ p = 3$; 6) $l_1 = 1, \ l_2 = 1, \ l_3 = 1, \ p = 3$; 7) $l_1 = 1, \ l_5 = 1, \ p = 2$;
8) $l_2 = 3, \ p = 3$; 9) $l_2 = 1, \ l_4 = 1, \ p = 2$; 10) $l_3 = 2, \ p = 2$; 11) $l_6 = 1, \ p = 1$.

Hence, $D^6(f) = f^{(6)} y'^6 + 15 f^{(5)} y'^4 y'' + 20 f^{(4)} y'^3 y''' + 45 f^{(4)} y'^2 y''^2 + 15 f''' y'^2 y^{(4)}$

$+ 60 f''' y' y'' y''' + 15 f''' y''^3 + 6 f'' y' y^{(5)} + 15 f'' y'' y^{(4)} + 10 f'' y'''^2 + f' y^{(6)}$.

8.5. $\bar{x} = x \cos a + y \sin a, \ \bar{y} = y \cos a - x \sin a, \ \bar{y}' = (y' - \tan a)/(1 + y' \tan a)$.

8.8*. *Hint:* Cf. proof of Theorem 8.3.1. See also [7.4](i), Section 26.

8.9. Formula (8.21) yields (see also Example 8.3.1):

$$X_{(2)} = y \frac{\partial}{\partial x} - x \frac{\partial}{\partial y} - (1 + y'^2) \frac{\partial}{\partial y'} - 3 y' y'' \frac{\partial}{\partial y''}.$$

8.11*. *Hint:* Equation (8.39) follows from the linearity of the prolongation formula (8.30) with respect to ξ and η. For the proof of the more complicated property (8.40), see [5.14], Chap. 17, §1; see also [5.33](i), §9.

8.13. Taking $u = y$ and $v = (y/y') - x$ from Example 8.3.5, one obtains the invariant differentiation and the differential invariants w and ψ of orders 2 and 3:

$$\mathcal{D} = \frac{1}{y'} D_x, \quad w = \mathcal{D}(v) = -\frac{y y''}{y'^3}, \quad \psi = \mathcal{D}(w) = \frac{y'''}{y'^4} - 3 \frac{y''^2}{y'^5}.$$

8.15. If $f(x, y, y') = D_x(g(x,y)) \equiv g_x + y' g_y$, then Lemma 1 of Section 8.4.1 yields:

$$\frac{\delta f}{\delta y} = \frac{\partial}{\partial y} D_x(g) - D_x(g_y) = \left(\frac{\partial}{\partial y} D_x - D_x \frac{\partial}{\partial y} \right)(g) = 0.$$

Conversely, let $f(x, y, y')$ be a differential function such that $\delta f/\delta y \equiv f_y - D_x(f_{y'}) = 0$. In other words, let $f \in \mathcal{A}$, $\mathrm{ord}(f) = 1$, satisfy the equation

$$f_y - f_{xy'} - y' f_{yy'} - y'' f_{y'y'} = 0$$

identically in x, y, y', y''. Since f is independent of y'', it follows from this equation that $f_{y'y'} = 0$. Hence, $f = a(x,y) y' + b(x,y)$. Upon substituting into $f_y - f_{xy'} - y' f_{yy'} = 0$, one obtains $b_y - a_x = 0$. This is precisely the integrability condition for the system $\partial g/\partial x = b(x,y), \ \partial g/\partial y = a(x,y)$ (see Section 2.1.4). Letting $g(x,y)$ be its solution, we get $f = g_y y' + g_x \equiv D_x(g)$.

8.16. Formula (8.77) yields: (i) $Y = (lu - x u_x - k y u_y) \dfrac{\partial}{\partial u}$; (ii) $Y = (1 - t u_x) \dfrac{\partial}{\partial u}$.

8.17*. *Hint:* See the proof of Lemma 2, Section 8.4.1. See also [8.18].

8.18. The frame of $xy' = 0$ is composed of the (x, y) and (y, y') planes.

Chapter 9

9.1. To deduce the differential equation of conics, proceed as in Section 1.2.2. Further, application of the dilation (9.8) yields:

$$\bar{y}'' + \frac{\bar{y}'^2}{\bar{y}} - \frac{\bar{y}'}{\bar{x}} = \frac{l}{k^2}\left(y'' + \frac{y'^2}{y} - \frac{y'}{x}\right).$$

It follows that the differential equation of the conics is invariant under the dilation with arbitrary k, l. Hence, it admits two operators, $X_1 = x\partial/\partial x$ and $X_2 = y\partial/\partial y$.

9.2. The generator of the translation group, $X = \partial/\partial x$, coincides with its extension to $y', \ldots$. Consequently, the infinitesimal test (9.5) is written

$$X\left[y^{(n)} - f(x, y, \ldots, y^{(n-1)})\right]_{y^{(n)}=f} \equiv -\frac{\partial f}{\partial x} = 0.$$

Hence, the variable x does not actually occur in f. The case of the y-translations is similar.

9.3. (iii) Equation $y'' = (y' + y'^3)/x$ admits an L_8 spanned by the following operators:

$$X_1 = \frac{\partial}{\partial y}, \quad X_2 = x\frac{\partial}{\partial x} + y\frac{\partial}{\partial y}, \quad X_3 = \frac{x^4 - y^4}{x}\frac{\partial}{\partial x} + 2(x^2 + y^2)y\frac{\partial}{\partial y}, \quad X_4 = \frac{x^2 + y^2}{x}\frac{\partial}{\partial x},$$

$$X_5 = \left(3xy + \frac{y^3}{x}\right)\frac{\partial}{\partial x} - 2x^2\frac{\partial}{\partial y}, \quad X_6 = 2xy\frac{\partial}{\partial x} + (y^2 - x^2)\frac{\partial}{\partial y}, \quad X_7 = \frac{1}{x}\frac{\partial}{\partial x}, \quad X_8 = \frac{y}{x}\frac{\partial}{\partial x}.$$

9.4. (i) Equation $y''' = 0$ admits L_7 of point symmetries (cf. Example 3, Section 9.2.1):

$$X_1 = \frac{\partial}{\partial x}, \quad X_2 = \frac{\partial}{\partial y}, \quad X_3 = x\frac{\partial}{\partial x}, \quad X_4 = x\frac{\partial}{\partial y},$$

$$X_5 = x^2\frac{\partial}{\partial y}, \quad X_6 = y\frac{\partial}{\partial y}, \quad X_7 = x^2\frac{\partial}{\partial x} + 2xy\frac{\partial}{\partial y}.$$

9.6. Consider the general heat equation with n spatial variables, $u_t = \Delta u$, where Δ denotes the n-dimensional Laplacian in the variables $x = (x^1, \ldots, x^n) \in \mathbf{R}^n$. The Lie algebra admitted by this equation is composed of the finite-dimensional subalgebra spanned by

$$X_0 = \frac{\partial}{\partial t}, \quad X_i = \frac{\partial}{\partial x^i}, \quad X_{ij} = x^j\frac{\partial}{\partial x^i} - x^i\frac{\partial}{\partial x^j}, \quad X_{0i} = 2t\frac{\partial}{\partial x^i} - x^i u\frac{\partial}{\partial u},$$

$$Z_1 = 2t\frac{\partial}{\partial t} + x^i\frac{\partial}{\partial x^i}, \quad Z_2 = u\frac{\partial}{\partial u}, \quad Y = t^2\frac{\partial}{\partial t} + tx^i\frac{\partial}{\partial x^i} - \frac{1}{4}(2nt + |x|^2)u\frac{\partial}{\partial u}$$

and the infinite-dimensional ideal consisting of $X_\tau = \tau(t,x)\partial/\partial u$, where $\tau(t,x)$ is any solution of the heat equation, $\tau_t = \Delta \tau$. The answers to questions (i), (ii) and (iii) are obtained by letting $n = 1, 2$ and 3, respectively. Note that $X_{ij} = 0$ when $n = 1$.

9.7. The Lie algebra admitted by the Black-Scholes equation (1.5) is spanned by:

$$X_1 = \frac{\partial}{\partial t}, \quad X_2 = x\frac{\partial}{\partial x}, \quad X_3 = 2t\frac{\partial}{\partial t} + (\ln x + Kt)x\frac{\partial}{\partial x} + 2Ctu\frac{\partial}{\partial u},$$

$$X_4 = A^2 tx\frac{\partial}{\partial x} + (\ln x - Kt)u\frac{\partial}{\partial u}, \quad X_5 = u\frac{\partial}{\partial u}, \quad X_\tau = \tau(t,x)\frac{\partial}{\partial u},$$

$$X_6 = 2A^2t^2\frac{\partial}{\partial t} + 2A^2tx\ln x\frac{\partial}{\partial x} + \left((\ln x - Kt)^2 + 2A^2Ct^2 - A^2t\right)u\frac{\partial}{\partial u},$$

where $K = B - A^2/2$ and $\tau(t,x)$ is any solution of equation (1.5). See R.K. Gazizov and N.H. Ibragimov, 'Lie symmetry analysis of differential equations in finance', to appear in *Nonlinear Dynamics*.

9.9. The shallow-water equations admit the Lie algebra L_9 spanned by (here $i = 1, 2$):

$$X_0 = \frac{\partial}{\partial t},\ X_i = \frac{\partial}{\partial x^i},\ Y_i = t\frac{\partial}{\partial x^i} + \frac{\partial}{\partial v^i},\ X_3 = t^2\frac{\partial}{\partial t} + tx^i\frac{\partial}{\partial x^i} + (x^i - tv^i)\frac{\partial}{\partial v^i} - 2th\frac{\partial}{\partial h},$$

$$X_4 = x^2\frac{\partial}{\partial x^1} - x^1\frac{\partial}{\partial x^2} + v^2\frac{\partial}{\partial v^1} - v^1\frac{\partial}{\partial v^2},\ X_5 = t\frac{\partial}{\partial t} + x^i\frac{\partial}{\partial x^i},\ X_6 = x^i\frac{\partial}{\partial x^i} + v^i\frac{\partial}{\partial v^i} + 2h\frac{\partial}{\partial h}.$$

9.11. The system of equations (9.64),

$$X_4(J) \equiv t\frac{\partial J}{\partial t} - \frac{1}{4}(Me^{4\psi} - 1)\frac{\partial J}{\partial \psi} = 0,$$

$$X_5(J) \equiv \sin x\, e^{-z}\frac{\partial J}{\partial x} - \cos x\, e^{-z}\frac{\partial J}{\partial z} + \frac{1}{2}\cos x\, e^{-z}(Me^{4\psi} - 1)\frac{\partial J}{\partial \psi} = 0$$

is a Jacobian system (see Definition 2 in Section 4.5.2). Hence, one can carry out the successive integration in any order (cf. Remark 4.5.3). The first equation, $X_4(J) = 0$, yields three functionally independent solutions, $J_1 = x$, $J_2 = z$, $J_3 = t(e^{-4\psi} - M)$. Hence, the common solution $J(t, x, z, \psi)$ of the system should be a function of J_1, J_2, J_3 only. Therefore we rewrite X_5 in new variables $J_1, J_2, J_3, \tau(t, x, z, \psi)$ by formula (7.30),

$$X_5 = X_5(J_1)\frac{\partial}{\partial J_1} + X_5(J_2)\frac{\partial}{\partial J_2} + X_5(J_3)\frac{\partial}{\partial J_3} + X_5(\tau)\frac{\partial}{\partial \tau},$$

and keep the terms acting on J_1, J_2, J_3 to obtain:

$$X_5 = \sin J_1\, e^{-J_2}\frac{\partial}{\partial J_1} - \cos J_1\, e^{-J_2}\frac{\partial}{\partial J_2} + 2\cos J_1\, e^{-J_2}J_3\frac{\partial}{\partial J_3}.$$

It is easy to solve the equation $X_5(J) = 0$ and obtain two independent invariants:

$$\lambda = e^{J_2}\sin J_1 \equiv e^z\sin x,\quad v = J_3\, e^{2J_2} \equiv te^{2z}\left(e^{-4\psi} - M\right).$$

9.12. Cf. [5.28], Section 5.4.

9.14**. *Hint:* Let L_r be the Lie algebra of an irreducible contact transformation group admitted by an ordinary differential equation of order $n \geq 3$. First show that $r < \infty$. Then use Theorem 8.3.8, according to which r can assume only the values 6, 7 or 10. It remains to investigate regular and singular invariant ordinary differential equations for each of the algebras (8.56), (8.57), and (8.58).

9.15*. The most general transformation mapping any equation of the form $y'' = F(x, y)$ into a similar equation, $\bar{y}'' = \Phi(\bar{x}, \bar{y})$, is given by $x = f(\bar{x})$, $y = C\sqrt{f'(\bar{x})}\,\bar{y} + g(\bar{x})$, where f and g are arbitrary functions, $C =$const. See [5.17], Part IV, §2 (reprinted in [5.12], vol. 5, paper XVI, p. 441).

9.17. The group classification of nonlinear wave equations due to W.F. Ames, R.J. Lohner, and E. Adams, 'Group properties of $u_{tt} = [f(u)u_x]_x$', *Int. J. Non-Linear Mechanics*, **16**(5-6), 1981, p.439-447 (see also [5.35], vol. 1, Section 12.4.1) singles out three types of symmetries for equation (9.88) corresponding to the following σ:

(i) for arbitrary $\sigma \neq 0$ equation (9.88) admits L_4 of operators

$$X^0 = \left(C_1 + C_3 t\right)\frac{\partial}{\partial t} + \left(C_2 + C_3 x + C_4 x\right)\frac{\partial}{\partial x} + 2C_4 \frac{u}{\sigma}\frac{\partial}{\partial u};$$

(ii) for $\sigma = -4/3$ equation (9.88) admits L_5 of operators

$$X^0 = \left(C_1 + C_3 t\right)\frac{\partial}{\partial t} + \left(C_2 + C_3 x + C_4 x + C_5 x^2\right)\frac{\partial}{\partial x} - \left(\frac{3}{2}C_4 + 3C_5 x\right)u\frac{\partial}{\partial u};$$

(iii) for $\sigma = -4$ equation (9.88) admits L_5 of operators

$$X^0 = \left(C_1 + C_3 t + C_5 t^2\right)\frac{\partial}{\partial t} + \left(C_2 + C_3 x + C_4 x\right)\frac{\partial}{\partial x} + \left(-\frac{1}{2}C_4 + C_5 t\right)u\frac{\partial}{\partial u}.$$

9.18. The proof follows from (9.84) and (9.79).

9.19. The maximal symmetry algebra L_5 of equation (9.88) with $\sigma = -4/3$ is stable, i.e. (9.87) inherits the symmetries of (9.88).

9.22. The canonical form of an ellipse is obtained from (9.114) by rotating the coordinate system, namely by setting $\theta = \phi + \arctan(A_2/A_1)$. Then equation (9.114) takes the form $r = p/(1 + e\cos\phi)$ with $p = -M^2/(\mu m)$ and $e = -A/\mu$, where $A = |\mathbf{A}| = \sqrt{A_1^2 + A_2^2}$.

9.24*. Given an increment $\delta \boldsymbol{x}$, the corresponding increments of the velocity $\boldsymbol{v}$ and of a Lagrangian $L = m|\boldsymbol{v}|^2/2 - U(r)$ are determined as follows:

$$\delta \boldsymbol{v} = \frac{\mathrm{d}}{\mathrm{d}t}\delta \boldsymbol{x}, \quad \delta L = m(\boldsymbol{v}\cdot\delta\boldsymbol{v}) - \frac{U'(r)}{r}(\boldsymbol{x}\cdot\delta\boldsymbol{x}).$$

An infinitesimal canonical Lie–Bäcklund transformation $\bar{\boldsymbol{x}} \approx \boldsymbol{x} + \delta\boldsymbol{x}$ is a Noether symmetry (see Definition 2 in Section 9.7.2) if $\delta L = \mathrm{d}F/\mathrm{d}t$, $F \in \mathcal{A}$. In this notation, the constant of motion (9.107) associated with a Noether symmetry is written $T = m(\boldsymbol{v}\cdot\delta\boldsymbol{v}) - F$. For the increment $\delta\boldsymbol{x} = \boldsymbol{x}\times(\boldsymbol{v}\times\boldsymbol{a}) + (\boldsymbol{x}\times\boldsymbol{v})\times\boldsymbol{a}$ representing (9.113), we have

$$\delta L = 2\left[\frac{\mathrm{d}}{\mathrm{d}t}\bigl((\boldsymbol{x}\cdot\boldsymbol{a})U\bigr) - (\boldsymbol{v}\cdot\boldsymbol{a})(rU' + U)\right].$$

Hence, the test for a Noether symmetry, $\delta L = \mathrm{d}F/\mathrm{d}t$, is satisfied if and only if the expression $\Phi = (\boldsymbol{v}\cdot\boldsymbol{a})(rU' + U)$ is a total derivative. According to Theorem 8.4.1, it means that the variational derivative of Φ vanishes,

$$\frac{\delta\Phi}{\delta\boldsymbol{x}} \equiv \frac{\partial\Phi}{\partial\boldsymbol{x}} - \frac{\mathrm{d}}{\mathrm{d}t}\frac{\partial\Phi}{\partial\boldsymbol{v}} = 0,$$

or

$$\frac{\delta\Phi}{\delta\boldsymbol{x}} = (rU' + U)'(\boldsymbol{v}\cdot\boldsymbol{a})\frac{\boldsymbol{x}}{r} - \frac{\mathrm{d}}{\mathrm{d}t}\bigl(rU' + U\bigr)\boldsymbol{a} = \frac{1}{r}\bigl(rU' + U\bigr)'\boldsymbol{v}\times(\boldsymbol{x}\times\boldsymbol{a}) = 0.$$

It follows that $(rU' + U)' = 0$. The solution of the latter equation yields Newton's potential $U = \mu/r$ to within an immaterial additive constant.

Chapter 10

10.1. Equation (10.37) has the form $h'' + h = 0$, whence $h = \sin x$. Hence, the change of variables (10.38) is given by

$$\bar{x} = \int\frac{\mathrm{d}x}{\sin^2 x} = -\frac{\cos x}{\sin x}, \quad y = z\sin^2 x.$$

In these variables, the equation $y''' + 4y' = 0$ becomes $z''' = 0$, where $z''' = \mathrm{d}^3 z/\mathrm{d}\bar{x}^3$.

10.3. The transformations (10.18)–(10.19) form a group with the multiplication rules:

$$T_{\alpha_2}T_{\alpha_1} = T_{\alpha_1}T_{\alpha_2} = T_{\phi(\alpha_1,\alpha_2)}, \quad S_\beta T_\alpha = S_{\phi(\alpha,\beta)}, \quad T_\alpha S_\beta = S_{\phi(\beta,-\alpha)},$$

$$S_{\beta_2}S_{\beta_1} = T_{\phi(\beta_1,-\beta_2)}, \quad S_{\beta_1}S_{\beta_2} = T_{\phi(\beta_2,-\beta_1)},$$

where $\phi(a,b) = (a+b)/(1-ab)$.

10.9. It is convenient to deal with third-order equations (10.38) written in Laguerre's canonical form (10.36), i.e. $y''' + c(x)y = 0$. Then $\lambda = -2c(x)$, and the statement of Example 1 is an immediate consequence of the invariance of equation (10.44) under the equivalence transformations (10.31)–(10.32). To prove the statement of Example 2, note that $\theta = 0$ for the equation $y''' + y = 0$, and then use the invariance of the function (10.45). The equation $y''' + x^{-6}y = 0$ is transformed to the form $z''' + z = 0$ by $\bar{x} = -1/x, y = x^2 z$.

Chapter 12

12.2. Type II. A transformation of $X_1 = x^2 \partial/\partial x + xy \partial/\partial y, X_2 = xy \partial/\partial x + y^2 \partial/\partial y$ to the canonical form $X_1 = \partial/\partial u, X_2 = t\partial/\partial u$ is given by $t = y/x, u = -1/x$.

12.4. See the linearization given in Example 3 of Section 12.3.2. The solution is also given in [5.14], Chap. 20, §5, Example 1.

12.5. The equation $y'' + e^{3y}y'^4 + y'^2 = 0$ admits L_2 spanned by $X_1 = e^{-y}\partial/\partial y$, $X_2 = \partial/\partial x$. This is an algebra of type I (Table 7.1, Section 7.3.8), and the canonical variables are $t = e^y, u = x$. In these variables, the equation in question becomes $u'u'' = 1$, where $u' = du/dt$. Upon integration and substitution of the expressions for t and u, the solution of the initial equation is written $y = \ln|(3x - C_2)^{2/3} - C_1| - \ln 2$.

12.7. The equation $y'' + 3yy' + y^3 = 0$ linearizes to $u'' = 0$ via $t = x - 1/y$, $u = x^2/2 - x/y$.

12.8. The operators $X_1 = y\partial/\partial x$, $X_2 = x\partial/\partial x$ satisfy $[X_1, X_2] = X_1$, $\xi_1\eta_2 - \eta_1\xi_2 = 0$, and hence span an L_2 of type IV. A transformation to the canonical form of Table 12.2.3 $X_1 = \partial/\partial u, X_2 = u\partial/\partial u$ is given by $t = y, u = x/y$. This change of variables linearizes the equation $y'' = (yy'^2 - xy'^3)f(y)$ to $tu'' + [t^2 f(t) + 2]u' = 0$.

12.10. According to Lemma 1 of Section 12.3.1, it suffices to check this property for the simplest linear equation $y'' = 0$. For the latter, it is easy to do using the known infinitesimal symmetries (9.21). For example, show that $y = x$ is an invariant solution for generators (9.21). Then the general solution $y = C_1 x + C_2$ is obtained by group transformations (e.g. dilation and translation), and hence is a group invariant solution.

Notes

These notes provide a brief account of relevant historical circumstances as well as biographical and bibliographical notes for each chapter. Titles of all books and papers are given in English translation. References to *Notes* are given in square brackets, e.g. the symbol [1.6] refers to item 6 in the notes to Chapter 1.

Chapter 1

1. Newton, Isaac (1642-1727), a natural philosopher and mathematician; developed the universal gravitation law in his great work [1.3]; invented the differential and integral calculus (1666, 1684) which he called *fluxions* and the *inverse method of fluxions*.

2. Leibnitz, Gottfried Wilhelm (1646-1716), a philosopher and mathematician; discovered, independently of Newton, the differential and integral calculus (1673, 1675) and introduced the basic notation commonly used in *Calculus*.

3. I. Newton, *Mathematical principles of natural philosophy*, 1st ed., 1687; 2nd ed., 1713; 3rd ed., 1726. Translated into English by Andrew Motte, to which are added, *The laws of the Moon's motion, according to gravity* by John Machin, in two volumes, Printed for Benjamin Motte, Middle-Temple-Gate, in Fleet Street, 1929.

4. d'Alembert, Jean Le Rond (1717-1783), a philosopher and mathematician; he was a founder, together with D. Diderot, of *Dictionnaire encyclopédique*; the general mechanical principle, known as *d'Alembert's principle*, first formulated in 1742 and published in his *Treatise on dynamics* (1743) is a mainstay of classical mechanics; in the theory of differential equations, his name is associated with equation (1.3) formulated and solved by d'Alembert in 1747.

5. The differential expression (1.19) was introduced by H.A. Schwarz in 1869; it has remarkable properties, e.g. it is invariant under the three-parameter group of projective transformations of the dependent variable y.

6. In the Middle Ages, a popular belief was that the Universe was made up of a number of spheres contained within one another, with the Earth at the center. The spheres were said to make musical sounds as they moved. These sounds were in harmony and were known as the *music of the spheres*.

7. Kepler, Johann (1571-1630), astronomer; published in 1609 the result of about twenty-two years of calculation in the form of two cardinal principles of astronomy – the laws of elliptical orbits and of equal areas, known since then as Kepler's first and second laws; his third law relating the planetary periods and distances was published in 1619 and based on his observations on three comets of 1618.

8. Galileo Galilei (1564-1642), astronomer and experimental philosopher, universally recognized as the founder of modern science. Galileo was the first man to study the stars with a telescope. Through it he demonstrated that the so-called perfect Sun had spots.

9. See, e.g. **R.K. Nagle and E.B. Saff,** *Fundamentals of differential equations,* Menlo Park, California: Benjamin/Cummings, 1986.

10. Formulated by G.S. Ohm (1787-1854) in 1827 and generalized by G.R. Kirchhoff (1824-1887).

11. In fact, van der Pol's equation was the first nonlinear differential equation of real physical significance having periodic solutions, the latter property being originally recognized by van der Pol from his experiences with electrical circuits.

12. Krylov, Aleksei Nikolaievich (1863-1945), mathematician and naval architect; the material for Section 1.4.4 is taken from his book *Differential equations of mathematical physics applicable to technical problems,* 1912; 2nd ed., 1950, §23 (in Russian).

13. The phenomenon of radioactivity was discovered in 1896 by H. Becquerel in uranium, the element known since 1789. A simple theory of radioactivity commonly used in the literature is due to E. Rutherford and F. Soddy (1903).

14. Malthus, Thomas Robert (1766–1834), the pioneer in the mathematical treatment of demographic problems, the author of *An essay on the principle of population as it affects the future improvement of society, with remarks on the speculations of Mr. Godwin, M. Condorcet, and other writers,* 1st ed., 1798; 2nd revised ed., 1803.

15. See, e.g. **M.W. Hirsch and S. Smale,** *Differential equations, dynamical systems, and linear algebra,* New York: Academic Press, 1974, Ch. 12.

Chapter 2

1. See any university text on ordinary differential equations. I consulted the following classical texts (ordered according to the frequency of use):

(i) **E.L. Ince,** *Ordinary differential equations,* New York: Dover, 1944 (1st ed., Longmans, Green and Co., 1927),

(ii) **E. Goursat,** *Differential equations* (English translation by E.R. Hedrick and O. Dunkel of Goursat's famous *Course of mathematical analysis,* vol. 2, part II), Boston: Ginn and Co., 1917,

(iii) **V.V. Stepanov,** *Course of differential equations,* Moscow: State publisher of phys.-math. literature, 7th ed., 1958 (in Russian, contains a concise historical survey written by A.P. Yushkevich),

(iv) **A.R. Forsyth,** *A treatise on differential equations,* London: Macmillan and Co., 5th ed., 1921 (1st ed., 1885).

2. The method was used by Jean Bernoulli (1667-1748) in 1697. The solution of the linear equations by quadratures was, however, known to Leibnitz earlier.

3. Due to Gottfried Leibnitz (1691) and Jean Bernoulli (1694).

4. See the classical book **de Rahm,** *Differentiable manifolds,* Paris: Hermann, 1955, or appropriate university texts, e.g., **Y. Choquet-Bruhat, C. DeWitt-Morette, with M. Dillard-Bleick,** *Analysis, manifolds and physics,* Amsterdam: North-Holland, 1882, **M. Spivak,** *Calculus on manifolds,* New York: Benjamin, Inc., 1965.

5. The integrability test (2.25) was obtained in 1740 by A. Clairaut and L. Euler.

6. Integrating factors were invented by A. Clairaut (1713-1765) in 1739. Integrating factors for systems of first-order equations are known as *Jacobi's multipliers* (see [2.1](ii)).

7. Riccati, J. Francesco (1676-1754) investigated in 1724 the equation $y' = ay^2 + bx^m$.

Riccati and Daniel **Bernoulli** (1700-1782) noted independently that this equation is integrable in finite form if $m = -4k/(2k \pm 1)$ where k is any integer. The general equation (2.31) was studied by J. d'Alembert in 1763.

8. Bernoulli, Jacques (1654-1705) discovered equation (2.39) in 1695. The method of solution is due to Leibnitz (1696).

9. The method of variation of parameters for higher-order linear equations was developed by J.L. Lagrange in 1774/75. However, it was used by L. Euler and D. Bernoulli in 1740 in the case of second-order equations. For derivation of equations (2.54), see, e.g., books cited in [2.1]; see also Problem 2.15.

10. Euler, Leonard (1707-1783), the greatest analyst in history and a man of wide culture, he contributed to many branches of pure and applied mathematics: algebra, geometry and trigonometry treated as a branch of analysis, differential and integral calculus, differential equations, calculus of variations, hydrodynamics, optics and astronomy; the method of solution of linear equations discussed in Section 2.2.8 was published in 1743.

11. It will be shown in Section 13.2 that Euler's method has a group theoretical nature.

12. A detailed discussion of Euler's method for systems of first-order equations with constant coefficients, in particular the case of multiple roots of the characteristic equation, is to be found, e.g., in Chap. 7, Section 2.4 of Stepanov's book [2.1] (iii).

Chapter 3

1. Cauchy, Augustin Louis (1789-1857), one of the initiators of present standards of rigor in mathematical analysis; Cauchy developed original methods for investigating the general properties of differential equations in his lectures delivered during 1820–1830 at the École Polytechnique and presented his renowned existence theorems in subsequent publications.

2. Cauchy's proof of existence first presented in his lectures (see the preceding note) and summarized in his 1835 lithographed memoir *On integration of differential equations* (in French, Prague, 1835), was supplemented by R. Lipschitz in 1876. The method is known in the literature as the *Cauchy-Lipschitz method*.

3. The method of majorants was developed by A.L. Cauchy in his papers of 1839-46 published in *C.R. Acad. Sci. Paris* (reprinted in his *OEuvres*) and was apparently independently discovered by K. Weierstrass in 1842 (*Math. Werke*). A thorough presentation of the method of majorants is to be found, e.g., in Chap. II of Goursat's book [2.1] (ii).

4. The method of *successive approximations* was used by J. Liouville (1809-1882) in 1838 in the case of linear equations of the second order; a general theory is due to E. Picard (1856-1941) who published his approach in 1893, e.g. in his *Treatise on analysis*, vol. 2.

5. The Lipschitz condition is sufficient but not necessary for the uniqueness of the solution of the Cauchy problem. The necessary and sufficient conditions were investigated by W.F. Osgood (*Monatsh. Math. Phys.*, **9**, 1898, p. 331) who showed that the existence and uniqueness theorem remains true if the Lipschitz condition (3.3) is replaced by less restrictive conditions, e.g. by $|f(x, y_1) - f(x, y_2)| \leq K|y_1 - y_2| \ln |y_1 - y_2|^{-1}$. More details are to be found in Sec. 3·21 of Ince's book [2.1] (i).

6. This statement is true for classical solutions. However, the solution can involve more constants if one allows generalized solutions. For example, the generalized solution to the first-order equation $xy' = 0$ contains two constants and is given by $y = C_1 + C_2\theta(x)$, where $\theta(x)$ is the Heaviside function: $\theta(x) = 1$ for $x > 0$ and $\theta(x) = 0$ for $x < 0$.

7. The *envelope* of a one-parameter family of curves $F(x, y, p) = 0$ is a line tangent to any curve of the family. The envelope is obtained by eliminating the parameter p from the system of equations $F = 0$, $\partial F/\partial p = 0$.

8. For discussions of singular solutions the reader is referred to the texts [2.1]. Sections 71–74 of Goursat's book [2.1](ii) contain a profound general presentation of the topic.

9. This classification, given by Poincaré (**Poincaré**, Jules Henri, 1854-1912) in 1881 and continued by I. Bendixson in 1898-1900, is a part of the qualitative theory of differential equations created by Poincaré and Liapunov (**Liapunov**, Alexander Mikhailovich, 1857-1918). A good presentation of singularity analysis of equations $y' = g(x, y)/h(x, y)$, where $g(x, y) = ax + by + cx^2 + dxy + ey^2 + \cdots$, $h(x, y) = kx + ly + mx^2 + nxy + ry^2 + \cdots$, is given by Frommer (*Math. Ann.*, **99**, 1928).

10. See, e.g., **E. Picard**, *Treatise on analysis*, vol. 2, Paris, 1893; **P. Painlevé**, *Lectures on the analytic theory of differential equations*, delivered in Stockholm in 1895 (in French, lithographed, Paris, 1897); see also Part II of Ince's book [2.1](i).

11. See, e.g., Section 12.5 of Ince's book [2.1](i) or Section 67 of Goursat's book [2.1](ii).

12. For the proof, see Section 67 of Goursat's book [2.1](ii).

Chapter 4

1. For detailed discussions, see classical university texts, e.g. Chapter V in Goursat [2.1](ii), Chapters VIII-IX in Stepanov [2.1](iii) or Chapter IX in Forsyth [2.1](iv). A concise presentation of the topic is given in Chapter III: Fundamental theory of partial differential equations, of **V.I. Smirnov**, *A Course of higher mathematics*, vol. IV (English translation by D.E. Brown, edited by I.N. Sneddon), Pergamon Press, 1964.

2. The solution of the general linear equation (4.4) with two independent variables (written in the notation of equation (4.2)), $p + \alpha(x, y)q + c(x, y)z + T(x, y) = 0$, was given by d'Alembert [1.4] in vol. 4 of his *Opuscules mathématiques* (1761-80).

3. The first solution of a particular quasi-linear equation (4.6) is due to P.S. Laplace. Namely, in Section III of his fundamental paper 'Studies on integral calculus of partial differences' (Recherches sur le calcul intégral aux différences partielles, *Mémoires de l'Académie royale des Sciences de Paris*, 1773/77, p. 341-402; reprinted in Laplace's Œuvres complètes, vol. IX, Gauthier-Villars, Paris, 1893, pp. 5-68; English translation, New York, 1966) Laplace developed new methods of integration of partial differential equations of the first and second orders, and gave, *inter alia*, a generalization of d'Alembert's solution [4.2] to the equations of the form $p + \alpha(x, y)q + V(x, y, z) = 0$.

4. See, e.g., Section 99 of Smirnov's book cited in [4.1].

5. Semi-canonical variables were also employed in the problem of integration of ordinary differential equations, see N.H. Ibragimov and M.C. Nucci 'Integration of third order ordinary differential equations by Lie's method', *Lie Groups and Their Applications*, **1**(2), 1994, pp. 49-64. See also [7.3].

6. The fundamental concept of characteristics is due to J.L. Lagrange (1736-1813) and G. Monge (1746-1818). Namely, Lagrange reduced (in 1776) the integration of the partial differential equation (4.44) to the solution of the system of ordinary differential equations (4.49) sometimes called Lagrange's equations. Lagrange also generalized (in 1781-87) his method to the case of many variables and emphasized that reduction to ordinary differential equations is a general method for solving first-order partial differential equations.

NOTES

Monge, supplementing Lagrange's method in his works of 1795-1807, disclosed the geometric significance of Lagrange's results and introduced the term *characteristics*.

7. An alternative, geometric, proof of Theorem 4.3.4 is to be found in Goursat [2.1](ii), Section 76, or Stepanov [2.1](iii), Chap. VIII, Section 3.4.

8. Lagrange, Joseph Louis (1736-1813), by 1761 was recognized as the greatest living mathematician; his name is associated, e.g., with *Lagrangian mechanics* due to his great work *Analytical mechanics* (1788) and with *Euler-Lagrange equations* in the calculus of variations, as well as with *Lagrange's method* for the solution of nonlinear partial differential equations of the first order; his *Theory of analytic functions*, 1797 (in French) contains many other important results.

9. Supplementing Lagrange's approach to partial differential equations developed in the 1770s, P. Charpit gave a general method applicable to an arbitrary partial differential equation of the first order with two independent variables. He presented (30 June 1784) a memoir containing the method to *Académie des Sciences, Paris*. But Charpit died soon afterwards and his memoir was not published. However, the method was mentioned by **Lacroix**, *Treatise on differential and integral calculus*, 2nd ed., vol, 2, 1814 (in French), and became known in the literature as the Lagrange-Charpit method.

10. Detailed calculations are to be found in Goursat's book [2.1](ii), Sections 81, 83.

11. Monge, Gaspar (1746-1818), the founder of the new geometry and the geometrical theory of differential equations; at the age of 16 being required to work out the *défilement* of a fortress he invented a new geometrical method and obtained the result so quickly that the commandant refused to accept it, but upon examination the method was adopted and Monge, continuing his research, developed a general method of the application of geometry to the arts of construction which became known as *descriptive geometry*; in 1795-1807 he developed a geometric approach to partial differential equations in the framework of the theory of surfaces.

12. Jacobi, Karl Gustav Jacob (1804-1851), advanced the theory of partial differential equations and developed mathematical methods of classical mechanics (*Lectures on dynamics, 1866*); his name is associated, e.g., with the *Hamilton-Jacobi theory* in mechanics.

Chapter 5

1. See, e.g., **S. Barnard and J.M. Child**, *Higher algebra*, London: Macmillan and Co., 1936, Chap. XII.

2. Galois, Évariste (1811-1832), invented group theory in a hurried synopsis of his epochal work on algebraic equations written the night before his fatal duel; Galois emphasized that his theory is not intended to be a practical method for solving equations.

3. Sylow, Ludvig (1832-1918), proved one of the basic theorems on finite groups (1872) named after him and mentioned in most texts on group theory.

4. Chasles, Michel (1793-1880), one of the creators of projective geometry.

5. Poncelet, Jean Victor (1788-1867), a follower of G. Monge in the development of the new geometry; Poncelet began his work on projective geometry when he was taken prisoner during the Russian campaign and confined at Saratov on the Volga until 1814, and published the results in *Treatise on projective properties of figures*, 1822 (in French).

6. Plücker, Julius (1801-1868), a mathematician and physicist; invented the principle of duality in geometry; a standard notation of analytical geometry was developed in his

Analytic geometrical developments, 1831 (in German); he made important discoveries in the spectroscopy of gases and chemical analysis, e.g. observed the three lines of the hydrogen spectrum which were recognized later in the spectrum of the solar protuberances.

7. Klein, Felix (1849-1925), a successor of Plücker, an enthusiast and great organizer of mathematics and its applications to practical affairs; he is widely known as the author of the *Erlangen program* in geometry (1872) and of a large number of influential books.

8. Jordan, Camille (1838-1922), was awarded the Poncelet prize for his fundamental *Treatise on substitutions and algebraic equations*, Paris, 1870 (in French), containing a comprehensive account of results of Lagrange, Abel and Galois' theory as well as his own results on the solution of algebraic equations by radicals; the book appeared during the stay of Klein and Lie in Paris and they spent many hours reading this exciting work.

9. Darboux, Jean Gaston (1842-1917), an excellent mathematician, teacher and organizer of science; he published his *Lectures on general theory of surfaces* in four volumes in 1887-96 (in French); in the years since then, this fundamental treatise has proved to be one of the best sources in differential geometry and the theory of differential equations.

10. Kummer, Ernst Eduard (1810-1893), his main field was the theory of numbers.

11. S. Lie, 'On general theory of partial differential equations of an arbitrary order', 1895; reprinted in [5.12], vol. 4, paper IX, pp. 320-384 (in German).

12. Sophus Lie, *Collected Works*, Leipzig-Oslo: B.G. Teubner-H. Aschehoug, 1922-37.

13. S. Lie with **F. Engel**, *Theory of transformation groups*, vol. 1-3, Leipzig: B.G. Teubner, 1888, 1890, 1893 (in German).

14. S. Lie with **G. Scheffers**, *Lectures on differential equations with known infinitesimal transformations*, Leipzig: B.G. Teubner, 1891 (in German). This is the bible of group analysis of ordinary differential equations, and is still unsurpassed for its completeness.

15. S. Lie with **G. Scheffers**, *Lectures on continuous groups with geometrical and other applications*, Leipzig: B.G. Teubner, 1893 (in German).

16. S. Lie with **G. Scheffers**, *Geometry of contact transformations*, Leipzig: B.G. Teubner, 1896 (in German).

17. S. Lie, 'Classification and integration of ordinary differential equations between x, y, admitting a group of transformations', Parts I–IV, 1883-84; reprinted in [5.12], vol. 5 (in German); the citation (Section 5.2.3) of Lie's historical remarks is taken from p. 240.

18. S. Lie, 'On integration of a class of linear partial differential equations by means of definite integrals', 1881; reprinted in [5.12], vol. 3, paper XXXV, pp. 492-524 (in German). English translation is published in [5.35] vol. 2.

19. The problem of nonlinear superposition was solved in 1893 due to works of E. Vessiot (*Ann. Sci. École Norm. Sup.*, **10**, 1893, p. 53), A. Guldberg (*C. R. Acad. Sc., Paris*, **116**, 1893, p. 964) and S. Lie. A historical survey, the proof of the main theorem and numerous applications are to be found in [5.15], Chap. 24.

20. The following papers of S. Lie contain interesting notes:

(i) 'Infinitesimal contact transformations of mechanics', 1889; reprinted in [5.12], vol. 6, paper VI, pp. 237-247 (in German),

(ii) 'Infinitesimal contact transformations of optics', 1896; reprinted in [5.12], vol. 6, paper XXIV, pp. 615-617 (in German); see also [5.16], p. 97.

21. S. Lie, 'Foundations of a theory of invariants of contact transformations', 1774;

reprinted in [5.12], vol. 4, paper I, pp. 1-96 (in German).

22. Bäcklund, Albert Victor (1845-1922), nowadays his name is associated with *Bäcklund transformations* and *Lie-Bäcklund transformation groups*; however, he contributed (1878) one more result of fundamental mathematical importance, viz. the theory of characteristics for second-order partial differential equations with an arbitrary number of independent variables (for two variables, characteristics were known since Monge and Ampère). See also an excellent biographical paper by C.W. Oseen 'Albert Victor Bäcklund', Jahresber. Dtsch. Math. Vereinigung, **38**, 1929, pp. 113-153 (in German).

23. A.V. Bäcklund, 'On surface transformations', *Math. Annalen*, **9**, 1876, pp. 297-320 (in German).

24. Laplace used in 1773 (in his paper cited in [4.3]) transformations of linear hyperbolic second-order equations $s + a(x,y)p + b(x,y)q + c(x,y)z = 0$ given by the relations $\bar{x} = x$, $\bar{y} = y$, $\bar{z} = q + az$, $\bar{p} = (a_x - c)z - bq$. This is precisely what is called a Bäcklund transformation. Here the partial derivatives are written in the classical notation, $p = z_x$, $q = z_y$, $s = z_{xy}$, $\bar{p} = \bar{z}_{\bar{x}}$.

25. See, e.g., F. Klein's *Lectures on higher geometry*, §76.

26. N.H. Ibragimov and R.L. Anderson, 'Groups of Lie-Bäcklund tangent transformations', Soviet Math. Dokl., **17**, 1976, p. 437. See also [5.35], vol. 3, Chap. 1 and 6.

27. N.H. Ibragimov, 'On the theory of Lie-Bäcklund transformation groups', *Mat. Sbornik*, **109**(2), 1979, pp. 229-253. English translation: *Math. USSR Sbornik,*, **37**(2), 1980, pp. 205-226.

28. N.H. Ibragimov, *Transformation groups applied to mathematical physics*, Moscow: Nauka, 1983. English translation by D. Reidel, Dordrecht, 1985.

29. By the time the fundamental principles of mechanics were established, L. Euler [2.10] and D. Bernoulli [2.7] supplementing Newton's principles, independently derived the conservation of areas (equivalent to the angular momentum) in Kepler's problem. Daniel Bernoulli, in a letter addressed to Euler in 1744, discussed the idea of conservation of the angular momentum and its possible dependence on the linear momentum. It seems that Euler regarded these two conservation laws to be independent. These investigations originated the concept of conservation laws in connection with symmetries.

30. Noether, A.E. (Emmy) (1882-1935), a central figure in the development of the theory of ideals in modern algebra. Her paper on conservation laws, 'Invariant variational problems', *König. Gesell. Wissen., Göttingen, Math.-Phys. Kl.*, 1918, pp. 235-257, is available in English in *Transport Theory and Statistical Physics*, **1**, 1971, pp. 186-207.

31. N.H. Ibragimov, 'The Noether identity', *Continuum Dynamics*, **38**, Institute of Hydrodynamics, Novosibirsk, 1979, pp. 26-32 (in Russian). The operators N^i (8.78) and the identity (8.82) (see Chapter 8) were introduced in this paper and were given the names *Noether operators* and *Noether identity* because of their relation to Noether's theorem. The use of this new identity in conservation theorems is to be found in [5.28], Chap. 5.

32. Ovsyannikov, Lev Vasil'evich (born 1919), was the spearhead in the restoration of group analysis of differential equations in the 1960s.

33. L.V. Ovsyannikov, (i) *Group properties of differential equations*, Novosibirsk: Siberian Branch of the USSR Ac. Sci., 1962 (in Russian). English translation by G.W. Bluman, 1967, unpublished); (ii) *Group analysis of differential equations*, Moscow: Nauka, 1978, published in English by Academic Press, 1982.

34. L.V. Ovsyannikov, 'Group properties of the Chaplygin equation', *Journal of Applied Mechanics and Technical Physics* (in Russian), No. 3, 1960, pp. 126-145.

35. N.H. Ibragimov (ed.), *CRC Handbook of Lie group analysis of differential equations*, Boca Raton, Florida: CRC Press. Consists of three volumes: vol. 1. 'Symmetries, exact solutions and conservation laws', 1994; vol. 2. 'Applications in engineering and physical sciences', 1995; vol. 3. 'New trends in theoretical developments and computational methods', 1996. Contains an extensive compilation and systematization of the results on group analysis of ordinary and partial differential equations obtained by Lie and his followers during the period of over one hundred years.

36. On the invariance principle in initial-value problems, see N.H. Ibragimov, (i) 'Group analysis of ordinary differential equations and the invariance principle in mathematical physics (for the 150th anniversary of Sophus Lie)', *Uspekhi Matem. Nauk*, **47**(4), 1992, pp. 83-144. English translation: *Russian Math. Surveys*, **47**(4), 1992, pp. 89-156;
(ii) *Primer of the group analysis*, Moscow: Znanie, No. 8, 1989 (in Russian); elementary introduction to Lie group analysis, contains first steps towards the group analysis of equations with distributions and a new approach to the Galois group.
(iii) 'Sophus Lie and harmony in mathematical physics, on the 150th anniversary of his birth', *The Mathematical Intelligencer*, **16**(1), 1994, pp. 20-28;

37. Laplace, Pierre-Simon (1749-1827), a mathematician and astronomer. In the beginning of his career, Laplace had strong support from d'Alembert [1.4]. His famous *Celestial mechanics* in five volumes (1799-1825, English transl., New York, 1966) systematized the work of three generations of illustrious mathematicians. The vector *A* is derived in *Celestial mechanics*, vol. 1, Book II, Chap. III, Section 18. Laplace taught mathematics at the École Militaire in Paris and was an examiner of young Napoleon. Later, Napoleon nominated him a Minister of Home Affairs. Laplace used this opportunity to teach mathematics in his Ministry. Napoleon said that Laplace introduced infinitesimals into politics and dismissed him.

38. (i) I.Sh. Akhatov, R.K. Gazizov and N.H. Ibragimov, 'Nonlocal symmetries. Heuristic approach', Moscow: VINITI, 1989 (English translation in *Journal of Soviet Mathematics*, **55**(1), 1991, pp. 1401-1450); (ii) V.A. Baikov, R.K. Gazizov and N.H. Ibragimov, 'Perturbation methods in group analysis', *ibid.*, pp. 1450-1490.

39. On group theoretic modeling, see N.H. Ibragimov, (i) 'Seven miniatures on group analysis', *Differential Equations*, **29**(10), 1993, p. 1511-1520; (ii) 'Small effects in physics hinted by the Lie group philosophy: Are they observable?', *Lie groups and their applications*, **1**(1), 1994, pp. 113-123, and 'Galilean principle in thermal diffusion', *Symmetri*, **2**(3), 1995, pp. 56-62; (iii) 'Group theoretic modeling', in: [5.35], vol. 3, Chap. 6; (iv) 'Massive galactic halos and dark matter: Unusual properties of neutrinos in the de Sitter universe', in: *New Extragalactic Perspectives in the New South Africa*, Proc. Int. Conference on "Cold Dust and Galaxy Morphology", Johannesburg, 22-26 January 1996, D.L. Block and J.M. Greenberg (eds.), Dordrecht: Kluwer, 1996, pp. 544-546.

40. P.A.M. Dirac, 'The electron wave equation in the de Sitter space', *Annals of Mathematics*, **36**(3), 1935, pp. 657-669.

41. A.V. Bobylev and N.H. Ibragimov, 'Relationships between the symmetry properties of the equations of gas kinetics and hydrodynamics', *J. Mathematical Modeling*, **1**(3), 1989, pp. 100-110. English translation: *Mathematical Modeling and Computational Experiment*, **1**(3), 1993, pp. 291-300.

Chapter 6

1. For a discussion of the physical significance of this duality, see **V.A. Fock**, *The theory of space, time, and gravitation*, New York: Pergamon Press, 1959.

2. The transformation (6.3) has long been successfully used in algebra, e.g., in Cardan's book mentioned in Section 5.1.

3. This transformation is defined by a linear equation (similar to $b - 2a\varepsilon = 0$ for (6.3)), the latter being obtained from the requirement to annul the term next to the highest.

4. J.E. Campbell, *Introductory treatise in Lie's theory of finite continuous transformation groups*, Oxford: Clarendon Press, 1903.

5. This important result was first published by Lie in 1874. For the proof of Theorem 6.3.1, see e.g. the book [5.13], vol. 3 (§ 1, pp. 2-6); it also contains Lie's enumeration of all possible continuous groups on the plane which exhibits that the plane, unlike the straight line, contains a large variety of different types of groups.

6. The geometric properties of inversion, namely the property (ii) of Theorem 6.3.2, is used in a mechanical construction for transforming circular motions into rectilinear ones and vice versa. One takes seven rigid rods $AP, AQ, BM, PM, PN, QM, QN$ such that $AP = AQ, PM = PN = QM = QN$. The rods are joined together by hinges as shown in the figure. The points A and B are fixed such that $AB = BM$. Then elementary geometry shows that AMN is a straight line and that the product $AM \cdot AN$ is constant, $AM \cdot AN = AP^2 - PM^2$. Hence the points M and N are connected by an inversion (6.33) with the center A and $R^2 = AP^2 - PM^2$. Thus, when M moves along the circle with the center B, the point N moves along a straight line.

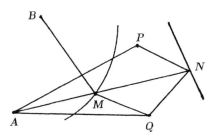

7. Recall that any analytic function furnishes a conformal mapping (see Definition 2 in Section 6.3.3) in $\mathbb{R}^2$. Accordingly, the conformal group on the plane is an infinite group determined by the Cauchy-Riemann equations. Inversion (6.34), and hence the transformation (6.36), are elements of the conformal group. Indeed, one can readily verify the Cauchy-Riemann equations:

$$\frac{\partial \overline{x}}{\partial x} + \frac{\partial \overline{y}}{\partial y} = 0, \quad \frac{\partial \overline{x}}{\partial y} - \frac{\partial \overline{y}}{\partial x} = 0 \quad \text{for} \quad (6.34)$$

and

$$\frac{\partial \overline{x}}{\partial x} - \frac{\partial \overline{y}}{\partial y} = 0, \quad \frac{\partial \overline{x}}{\partial y} + \frac{\partial \overline{y}}{\partial x} = 0 \quad \text{for} \quad (6.36).$$

8. Continuous and discontinuous subgroups of the six-parameter group of motions in $\mathbb{R}^3$ serve in crystallography to characterize symmetries of crystalline substances. Accordingly, group theory furnishes a proper tool for a rigorous mathematical approach to crystallography.

9. Liouville, Josef, 1809-1882, the founder of the *Journal des Mathematiques Pures et Appliquées* and its editor during the period of 40 years since its foundation in 1836. The result on conformal mappings, often referred to as Liouville's theorem, was published in *J. Math. Pures et Appl.*, **12**, 1847, p. 265.

10. Simultaneously, W. Thomson (Lord Kelvin) emphasized the physical significance of the inversion. He showed (see his paper in the same volume of *J. Math. Pures et Appl.* as Liouville's paper, p. 256) that inversion (6.45) leaves invariant the Laplace equation $\Delta u \equiv u_{xx} + u_{yy} + u_{zz} = 0$ provided that the dependent variable u undergoes the transformation $\overline{u} = ru$. This combined transformation is known in the literature as Kelvin's transformation. It is widely used in mathematical physics, e.g. in the theory of potentials for constructing Green's function.

11. Generalization of Liouville's theorem to higher dimensions and a group theoretic treatment of conformal mappings are to be found in [5.16].

Chapter 7

1. See [5.14], Chap. 2, §5, Formula (20).

2. At first, I noticed that it is advantageous to introduce semi-canonical variables for solving quasi-linear partial differential equations by Laplace's method. Indeed, in these variables, any quasi-linear partial differential equation (4.32) reduces to a nonlinear ordinary differential equation (4.35) of the first order (see Section 4.3.2). Subsequently, we employed the concept of semi-canonical variables in the investigation of third-order ordinary differential equations admitting non-solvable Lie algebras [4.5].

3. The proof given here is due to Ovsyannikov [5.33](i), §8.7.

4. For the theory of multi-parameter transformation groups, see, e.g.

(i) **L.P. Eisenhart**, *Continuous groups of transformations*, Princeton: Princeton University Press, 1933,

(ii) **L.S. Pontryagin**, *Continuous groups*, Moscow: Gostekhizdat, 1954 (in Russian). English translation: *Topological groups*, New York: Gordon and Breach, 1966.

5. The term *Lie algebra* is due to H. Weyl. Lie himself referred to an *infinitesimal group*. In modern mathematics, the term *algebra* is used to identify any vector space having a multiplication of its elements that is bilinear. Then an abstract Lie algebra is defined as an algebra such that the multiplication is skew-symmetric and satisfies the Jacobi identity. Group analysis of differential equations engages the services of Lie algebras whose elements are represented by differential operators (7.15).

6. The operators (7.77) span the maximal Lie algebra admitted by the second-order differential equation $u_x u_{xx} + u_{yy} = 0$ describing a steady-state transonic gas flow (see [5.28], Section 5.2, and [5.33](i), §14).

7. Theorems on similarity of continuous groups and Lie algebras are due to S. Lie (see [5.13], vol. 1, p. 256) and L.P. Eisenhart. For a detailed presentation, see [7.4](i), §22.

8. See [5.14], Chap. 21, §2 and §3.

9. See [5.14], Chap. 22.

10. **L. Bianchi**, *Lectures on theory of finite continuous transformation groups*, Pisa: Enrici Spoerri, 1918 (in Italian).

11. Lie consideres these examples in [5.14], Chap. 21, §3. His calculations are based on the geometric interpretation of the triplet (X_1, X_2, X_3) as vertices of a triangle.

12. The algebra (7.93) is one of few three-dimensional algebras that are not similar to any of Lie's canonical forms of Theorem 2 of Section 7.3.8 in the real domain, but they are similar to an appropriate type of Lie's classification in the complex domain (see [5.35], vol. 3, Chap. 8). It was found by F.M. Mahomed and P.G.L. Leach and presented in

NOTES 335

their paper 'Lie algebras associated with scalar second-order ordinary differential equations' (*J. Math. Phys.*, **30**, 1989, pp. 2770-2777) as one of those equations missing in Lie's classification. The calculations of this exercise (taken from [5.36](i)) show that the algebra (7.93) is contained in Lie's classification and is similar, in the complex domain, to type 1) of Theorem 2, Section 7.3.8.

13. V.A. Baikov, R.K. Gazizov and N.H. Ibragimov, (i) *Approximate symmetries of equations with a small parameter*, Preprint 150, Moscow: Institute of Applied Mathematics, USSR Academy of Sciences, 1987 (in Russian); (ii) 'Approximate transformation groups and deformations of symmetry Lie algebras', Chap. 2 in [5.35], contains a detailed discussion of approximate symmetries including the theory of approximate Lie algebras and multi-parameter approximate transformation groups.

14. N. Bourbaki, *Lie Groups and Lie Algebras*, Chap. 1-3, Paris: Hermann, 1975.

15. N.H. Ibragimov, 'Perturbation methods in group analysis', in *Differential equations and chaos: Lectures on selected topics*, N.H. Ibragimov, F.M. Mahomed, D.P. Mason, D. Sherwell (eds.), New Delhi: New Age International Publishers, 1996, pp. 41-60.

Chapter 8

1. Differential algebra was developed by **J.F. Ritt,** *Differential algebra*, New York: AMS Coll. Publ., 33, 1950. However, differential algebraic techniques were known some 200 years ago and were used, e.g., in Lagrange's *Theory of analytic functions* [4.8].

2. Ritt [8.1] introduced and developed an algebraic theory of *differential polynomials*, i.e. polynomials of variables $x, u, u_{(1)}, \ldots$ The concept of arbitrary *differential functions* was introduced for the needs of modern group analysis in N.H. Ibragimov, 'On equivalence of evolutionary equations admitting an infinite Lie-Bäcklund algebra', *C. R. Acad. Sci. Paris*, Sér. I, **293**, 1981, p. 657-660 (in French). See also [5.28], Chap. 3 and 4.

3. Faà de Bruno, 'Note on a new formula of the differential calculus', *Quarterly Journal of Pure and Appl. Math.*, vol. 1, 1857, pp. 359-360 (in French). The formula for higher-order derivatives, given by Faà de Bruno in the case of one independent and one dependent (differential) variable, can be extended to any number of independent and differential variables (see [5.28], Section 19.2).

4. A detailed presentation of extentions of point transformations and infinitesimal generators is to be found in [5.14], Chap. 13. Lie's term (in German) "Erweiterte Gruppe" is translated into English as "extended group" in such classical sources as [6.4], [7.4](i), **A. Cohen,** *An introduction to the Lie theory of one-parameter groups, with applications to the solution of differential equations*, Boston-New York-Chicago: D.C. Heath, 1911, and L.E. Dickson, 'Differential equations from the group standpoint', *Ann. of Math.* (2), **25**, 1925, pp. 287-378. The term "prolongation" for the extension of transformations and infinitesimal generators appeared in the literature in the 1960s.

5. The notation $X_{(p)}$ for p-times extended infinitesimal generators is taken from Eisenhart [7.4](i). Lie [6.4] denotes the symbol of an infinitesimal transformation by U, its first, second, etc. extensions by U', U'', etc. Cohen and Dickson adopt Lie's notation. In examples, we will employ Eisenhart's notation. In differential algebra, however, it is advantageous to consider infinitesimal generators X extended to all derivatives. Then X acts in the universal space $\mathcal{A}$, and there is no ambiguity if the *infinite-order extension* of an operator X (the extension to all derviatives) is denoted by the same symbol X. Thus extended operator X truncates when it acts on differential functions $f \in \mathcal{A}$. Namely, X acts as $X_{(1)}$ if $\text{ord}(f) = 1$, as the twice-extended $X_{(2)}$ if $\text{ord}(f) = 2$, etc.

6. See, e.g., N.H. Ibragimov, 'Group theoretical nature of conservation theorems', *Lett. Math. Phys.*, **1**, 1977, pp. 423-428.

7. See, e.g., [5.14], Chap. 16, §5. See also Section 44 of Dickson's paper cited in [8.4] and **P.J. Olver**, *Applications of Lie groups to differential equations*, New York: Springer-Verlag, 1986, §2.5.

8. A. Tresse, 'On differential invariants of continuous groups of transformations', *Acta Math.*, **18**, 1894, pp. 1-88 (in French). See also [5.33](ii), Chap. 7.

9. The operators (8.47) are termed invariant differentiations because they map any differential invariant of G_r again into a differential invariant. Furthermore, all differential invariants of all orders can be obtained from a finite set of differential invariants through invariant differentiation and functional operations. See the works cited in [8.8].

10. For a detailed discussion of differential substitutions and their use in the theory of differential equations, the reader is referred to [5.28], Chap. 4, in particular Section 19.4.

11. Definition and a criterion of reducible contact tranformation groups are discussed in Lie-Engel [5.13], vol. 2, Chap. 21, §94. Enumeration of all irreducible contact transformation groups in the plane, formulated in Theorem 8.3.8, was first published by Lie in 1874. For the proof, see [5.13], vol. 2, Chap. 23, §104, Theorem 69.

12. See Definition 1, Section 6.3.1.

13. The operator (8.59) (truncated to derivatives of a finite order) is sometimes referred to as the *Euler operator*, named after Euler who first introduced it (1744) for the one-dimensional case. It is also called the *Lagrange operator*, bearing the name of Lagrange who considered the multi-dimensional case (1762) and gave the name *variational derivative*. In the modern literature, the terms *Euler-Lagrange operator* and *variational derivative* are used interchangeably as (8.59) usually arises in variational problems.

14. I.M. Gelfand and L.A. Dikii, 'Asymptotic behavior of the resolvent of Sturm-Liouville equations and the algebra of the Korteweg-de Vries equations', *Uspekhi Matem. Nauk*, **30**(5), 1975. English transl.: *Russian Math. Surveys*, **30**(5), 1975, pp. 77-113.

15. The statement (8.66) is true for any $f \in \mathcal{A}$ with any number of variables x^i and u^α.

16. Noether [5.30] used operators (8.73) for purposes of her conservation theorems, by formally introducing a finite number of higher derivatives $u_{(s)}$ in coefficients ξ^i, η^α of infinitesimal transformations $\bar{x}^i \approx x^i + a\xi^i$, $\bar{u}^\alpha \approx u^\alpha + a\eta^\alpha$. Later, Lie–Bäcklund operators were discovered anew by H.H. Johnson ('Bracket and exponential for a new type of vector fields', *Proc. Amer. Math. Soc.*, **15**, 1964, p. 432, and 'A new type of vector field and invariant differential systems', *ibid.*, p. 675), who treated the generalized vector field in the framework of Ehresmann's jet bundles (C. Ehresmann, 'Introduction to the theory of infinitesimal structures and Lie's pseudo-groups' (in French), in: *Géométrie différentielle*, Colloq. Int. Centre Nat. Recherche Sci., Strasbourg, 1953, p. 97), then in physics by Z.V. Khukhunashvili ('The symmetry of the differential equations of field theory', *Soviet Phys. J.*, **14**, 1971) and by R.L. Anderson, S. Kumei and C.E. Wulfman, 'Generalization of the concept of invariance of differential equations: Results of applications to some Schrödinger equations', *Phys. Rev. Lett.*, **28**, 1972, p. 988).

17. The basis for this definition is amplified by the fact that the ideal L_* is admitted by *any* differential equation. Hence, an operator $X \in L_B$ is admitted by a *given* differential equation if and only if this is true for its canonical representation Y (8.77) (see [5.28], §17). Furthermore, canonical operators are convenient (since they have as their

invariants the independent variables x^i), e.g. for investigating symmetries of integro-differential equations (see [5.35], vol. 3, Chap. 5). However, the canonical representation of infinitesimal generators of Lie point and contact transformation groups involves higher derivatives. Therefore, it was not used in classical Lie theory.

18. The idea to consider differential equations as surfaces (or manifolds in multi-dimensional cases) in the space of independent and dependent variables and their derivatives goes back to Monge's geometric approach to differential equations [4.11]. This idea is intensely employed in works of Lie and his followers, though they treat a differential equation as a natural texture of the surface defined above and solutions of the equation. The notion of the frame was introduced in [5.36](i) with the aim to separate the differential algebraic component (the frame) from the functional analytic ingredient (solutions) of the concept of differential equations.

Chapter 9

1. For different approaches to symmetry analysis of differential equations, see [2.1](i), Chap. IV, [2.1](ii), Chap. II, Section IV, [5.14], [5.33], [5.35], [5.36](i), Cohen's book and Dickson's paper cited in [8.4], Olver's book cited in [8.7], and the following books:

(i) **G. Bluman and J.D. Cole**, *Similarity methods for differential equations*, New York: Springer-Verlag, 1974,

(ii) **G. Bluman and S. Kumei**, *Symmetries and differential equations*, New York: Springer-Verlag, 1989,

(iii) **H. Stephani**, *Differential equations: Their solution using symmetries*, Cambridge: Cambridge University Press, 1989,

(iv) **J.M. Hill**, *Differential equations and group methods for scientists and engineers*, Boca Raton, Florida: CRC Press, 1992.

2. Two definitions of symmetry groups are motivated and illustrated by numerous interesting examples in Lie's lectures [5.14] (they are repeated in Dickson's paper cited in [8.4]). Equivalence of two definitions for locally solvable systems of partial differential equations is discussed in [5.33](i), §15.1; Ovsyannikov's geometrical reasoning is clear and reasonably complete. If a system is not locally solvable, two definitions may not be equivalent; examples are to be found in Olver's book cited in [8.7], §2.6; see also [5.35], vol. 2, Section 1.3.8.

3. Symbolic programs for calculating symmetries are described, e.g., in [5.35], vol. 3, Chap. 13 and 14.

4. S. Lie, 'On infinite continuous groups' (1883; reprinted in [5.12], vol. 5, paper XIII, see p. 359) contains a proof of the statement that the group admitted by an ordinary differential equation $y^{(m)} = F(x, y, y', \ldots, y^{(m-1)})$ of order $m \geq 2$ cannot be infinite and hence may contain at most arbitrary constants, and that the group is infinite if $m = 1$. See also [5.17], Part III (footnote on p. 373 in [5.12], vol. 5).

5. Lie's proof is based on interesting geometrical constructions; see [5.14], Chap. 17, §3, Theorem 39. I follow Dickson's analytical proof (Section 48 of his paper cited in [8.4]).

6. Often, beginners think erroneously of constants involved in the general solution of determining equations (e.g. constant coefficients C_i in (9.20)) as group parameters of the corresponding symmetry group.

7. Equation $y'' = e^{y'} + xy$ is discussed in Dickson's paper [8.4]. It is similar to the equation $y'' - e^{y'} - e^{-y'} - xy = 0$ discussed in [5.14] (Chap. 16, §6) as an example of a second-order equation admitting no infinitesimal point symmetries.

8. Differential invariants can be used also for determining partial differential equations admitting a given infinite group. See, e.g., [5.11], Chap. IV ([5.12], vol. 4, p. 371-384).

9. See [5.28], Section 9.4.

10. See [5.11], Chap. III ([5.12], vol. 4, pp. 352-371). Subsequently, Lie's result was rediscovered several times. Lie provides a proof of Theorem 9.4.1, in particular the representation (9.60) of the general form of invariant solutions, in Section 77 of [5.11], where he considers the case of one partial differential equation (9.1), $F(x, u, u_{(1)}, \ldots, u_{(k)}) = 0$, for a single dependent variable u. Invariant solutions are defined, via the characteristic functions $W_\nu(x, u, u_{(1)}) \equiv \eta_\nu(x, u) - u_i \xi_\nu^i(x, u)$ of (9.57), by the overdetermined system

$$W_1 = 0, \ldots, W_r = 0, \ F = 0.$$

Investigation of the integrability of the latter is an essential part of his proof. Lie considers both point and contact symmetries. A discussion of systems with several variables u^α is to be found in Sections 76 and 77 of [5.11]. For a proof of Theorem 9.4.1 different from that given by Lie, and further developments, see [5.33](i).

11. For a discussion of the physical significance of the model equation (9.62), based on analysis of numerical solutions, and its use in applied agricultural sciences, see G. Vellidis and A.G. Smajstrla, 'Modeling soil water redistribution and extraction patterns of drip-irrigated tomatoes above a shallow water table, *Transactions of the American Society of Agricultural Engineers*, **35**(1), 1992, pp. 183-191. Symmetries and invariant solutions are studied by V.A. Baikov, R.K. Gazizov, N.H.Ibragimov, and V.F. Kovalev, 'Water redistribution in irrigated soil profiles: Invariant solutions of the governing equation', *Nonlinear Dynamics*, **13**(4), 1997, pp. 395-409 (see also [5.35], vol. 2, Section 9.8); a group classification shows that there are 30 distinctly different types of symmetry groups in accordance with the choice of the coefficients $C(\psi)$, $K(\psi)$ and $S(\psi)$ of equation (9.62).

12. Equation (9.62) can be regarded, in fact, as an equation for invariant solutions of a three-dimensional model, the invariance being considered with respect to the group of y-translations. The model (9.62) can be obtained by using the invariance principle (Section 5.5.1) meaning in this particular case that we foresee that a soil moisture ψ assumes the same numerical value for all values of y since the boundary conditions, along with the governing equation, are invariant under y-translations. This philosophy does not apply to the variables t, x and z, because boundary conditions are not invariant under t, x or z-translations even though the differential equation (9.62) is translational invariant.

13. Classification of all non-similar one- and two-dimensional subalgebras of the symmetry Lie algebra of shallow-water equations (9.66) shows that there are 47 distinctly different types of invariant solutions described by ordinary differential equations and 14 types of those described by partial differential equations with two independent variables, instead of three variables in the original system. See N.H. Ibragimov, 'Classification of invariant solutions of equations of two-dimensional non-steady state gas flow', *Journal of Applied Mechanics and Technical Physics* (in Russian), No. 4, 1966, pp. 19-22.

14. For the proof, see [5.33](i), §19.6 or [5.33](ii), §20.4. Cf. also Theorem 12.2.1.

15. G.J. Reid, D.T. Weih, and A.D. Wittkopf, 'A point symmetry group of a differential equation which cannot be found using infinitesimal methods', in: *Modern group analysis: Advanced analytical and computational methods in mathematical physics*, Proc. Int. Workshop, Acireale, Catania, Italy, 27-31 October 1992, N.H. Ibragimov, M. Torrisi and A. Valenti (eds.), Dordrecht: Kluwer, 1993, pp. 311-316.

16. Cf. Section 5.4.1.

NOTES

17. According to E.O. Tuck and L.W. Schwartz, 'A numerical and asymptotic study of some third-order ordinary differential equations relevant to draining and coating flows', *SIAM Rev.*, **32**(3), 1990, pp. 453-469, equation $y''' = y^{-2}$ describes fluid draining problems; specifically, it is relevant to draining on a dry wall if y is small, and on a wet wall if y is large. Its solution given by W.F. Ford (*SIAM Rev.*, **34**(1), 1992, pp. 121-122) can be derived from Lie group methods (see Ibragimov and Nucci's paper cited in [4.5]).

Chapter 10

1. Laplace's invariants appeared in his paper cited in [4.3].

2. Cayley, Arthur (1821-1895), can be compared with L. Euler for his analytical power and fertility of initiating new productive theories; published over 800 works that treat almost all subjects of mathematics: theory of groups, matrices, elliptic functions, algebraic invariants, abstract geometry (the concept of multi-dimensional spaces, introduction of the "absolute"), higher singularities of curves and surfaces (his famous twenty-seven lines that lie on a cubic surface), theoretical dynamics and astronomy (secular acceleration of the moon's mean motion), etc. A new approach to linear transformation is presented in his paper 'On the theory of linear transformations', *Cambridge Math. Journal*, **4**, 1845, pp. 193-209 [*Collected Mathematical papers* of A. Cayley, vol. 1, 1889, pp. 80-94].

3. I consulted the following papers: (1) E. Laguerre, (i) 'Sur les équations différentielles linéaires du troisième ordre', *C.R. Acad. Sci. Paris*, **88**, 1879, pp. 116-119, (ii) 'Sur quelques invariants des équations différentielles linéaires', *ibid.*, pp. 224-227; (2) J.C. Malet, 'On a class of invariants', *Phil. Trans. Royal Soc. London*, **173**, 1882, pp. 751-776; (3) G.H. Halphen, 'Sur les invariants des équations différentielles linéaires du quatrième ordre', *Acta Mathematica*, **3**, 1883, pp. 325-380 (reprinted in *Œuvres d'Halphen*, vol. 3, 921, p. 463); (4) A.R. Forsyth, 'Invariants, covariants, and quotient-derivatives associated with linear differential equations', *Phil. Trans. Royal Soc. London*, **179**, 1888, pp. 377-489 (a detailed presentation of the topic containing a good historical survey and proper references); (5) Valuable comments on group theoretical background of differential invariants are to be found in [5.11], Chap. I, §1.

See also: (6) J. Cockle, (i) 'Correlations of analysis', *Philosophical Magazine*, **24**(4), 1862, p. 532, (ii) 'On linear differential equations of the third order], *Quarterly J. Math.*, **15**, 1876, pp. 340-353; (7) R. Harley, 'Professor Malet's Classes of invariants identified with Sir James Cockle's Criticoids', *Proc. Royal Soc. London*, **38**, (1884), pp. 45-57.

4. The material for Chapter 10 is taken from N.H. Ibragimov, 'Infinitesimal method in the theory of invariants of algebraic and differential equations', *Notices of the South African mathematical Society*, **29**, June 1997, pp. 61-70.

5. An astonishing similarity of the *seminvariants* H_ν and h_ν of algebraic and differential equations, respectively, is deeper than a transparent likeness, e.g., of respective formulas (10.7), (10.9) and (10.34), (10.35). This similarity inspired profound investigations of the theory of invariants by brilliant mathematicians of the nineteenth century.

Chapter 11

1. If $\xi M + \eta N \equiv 0$, then the integral curves of equation (11.1) are identical with the path curves of the group G generated by (11.2).

2. E. Vessiot, 'On a class of differential equations', *Ann. Sci. École Norm. Sup.*, **10**, 1893, p. 53 (in French) developed a theory of nonlinear superposition for first-order equations $f(x, y, y') = 0$ and discovered an exceptional role of the Riccati equation, then he discussed (in his notes in *C.R. Acad. Sci. Paris*, **116**, 1893, pp. 427, 959, 1112) this problem for second-order equations $f(x,y,y',y'') = 0$; A. Guldberg, 'On differential equations possessing a fundamental system of integrals', *C.R. Acad. Sci. Paris*, **116**, 1893, p. 964 (in French) discussed nonlinear superposition for systems $dx^i/dt = F_i(t, x^1, \ldots, x^n)$, $i = 1, \ldots, n$; S. Lie, 'Differential equations possessing fundamental integrals', Leipziger Berichte, 1893, p. 341 (reprinted in [5.12], vol. 4, paper VI) gave elucidating remarks to the works of Vessiot and Guldberg and proved the general result; for Lie's proof of Theorem 11.2.2, see [5.15], Chap. 24, pp. 793-804.

3. See, e.g., R.L. Anderson, 'A nonlinear superposition principle admitted by coupled Riccati equations of the projective type', *Lett. Math. Phys.*, **4**, 1980, pp. 1-7; R.L. Anderson and P. Winternitz, 'A nonlinear superposition principle admitted by coupled Riccati equations of conformal type', *Lecture Notes in Physics*, **135**, 1980, pp. 165-169.

On nonlinear superposition for partial differential equations, see J.M. Goard and P. Broadbridge, 'Nonlinear superposition principles obtained by Lie symmetry methods', *J. Mathematical Analysis and Applications*, **214**, 1997, pp. 633-657.

4. The material of Section 11.2.5 is taken from [5.36](ii), §4. See also [5.36](i).

5. Cf. [2.1](ii), §7.

6. This idea was employed in [5.35], vol. 2, Chap. 6.

7. B. Y. Zel'divitch, N.F. Pilipetskii, and V.V. Shkunov, *Wave front reversal*, Moscow: Nauka, 1985.

Chapter 12

1. Both methods are presented, with numerous examples, in [5.14], Chap. 16, §5 and §6.

2. Theorem 36 in [5.14], Chap. 16, §5.

3. This example was considered by Lie (Example 5 in [5.14], Chap. 16, §5) to explain a group theoretic meaning of the well-known connection between linear second-order and Riccati's equations provided by the substitution $v = y'/y$.

4. Theorem 12.2.1 is contained, as a particular case, in Lie's general theory of integration of complete systems of partial differential equations (equivalent to higher-order ordinary differential equations) with known infinitesimal transformations. This theory is developed in his fundamental papers (i) 'General theory of partial differential equations of the first order', 1876/77; reprinted in [5.12], vol. 4, papers II and III, pp. 97-264, and (ii) 'General investigations on differential equations admitting a finite continuous group', 1885; reprinted in [5.12], vol. 6, paper III, pp. 139-223 (both in German). See also [5.14].

5. For the proof, see [5.14], Chap. 18, §1.

6. Theorem 41 in [5.14], Chap. 18, §2. See also [6.1], Section 52.

7. After **H. Wronski** (1775-1853) who introduced this determinant in 1821 for an arbitrary number of functions, namely the determinant $W[y_1, y_2, \ldots, y_n]$ of the system (2.54). It is important for our purposes that $W[y_1, y_2, \ldots, y_n] \neq 0$ in that interval of x where $y_1(x), y_2(x), \ldots, y_n(x)$ are linearly independent. For details, see books [2.1].

8. The paper of F.M. Mahomed and P.G.L. Leach cited in [7.12] contains another proof.

9. Detailed calculations are to be found in [5.39](i), Section 5.

10. Partial differential equations possess similar property. Namely, every solution of a linear partial differential equation is an invariant solution, provided that the equation has at least one non-trivial Lie's infinitesimal symmetry. See P. Broadbridge, J.M. Goard and D.J. Arrigo, 'Ad-hoc PDE solution methods in the context of Lie symmetries', paper presented at *XXII International Colloquium on Group Theoretical Methods in Physics*, University of Tasmania, Hobart, 13-17 July 1998.

Chapter 13

1. In papers (i), (ii) cited in [12.4], Lie proved that the integration of a third-order equation (specifically, of an equivalent linear partial differential of the first order) admitting a three-dimensional algebra L_3 requires three quadratures if L_3 is solvable, and the solution of a Riccati equation otherwise. This result is repeated in [5.14], Chap. 25, §2, Theorem 49. A simple proof, based on the group classification of third-order equations $y''' = f(x, y, y', y'')$, is to be found in Ibragimov and Nucci's paper cited in [4.5].

2. See [5.14], Chap. 25, §3.

3. Recall that the significance of a reduction of order is not so much with the fact that an equation of the nth order can be rewritten as an equation of order $< n$. The key point is the requirement that the transformation to an equation of a lower order be integrable by quadrature (see Definition 2.2.2).

Index

Abel, N.H., 100
Abelian Lie algebra, 158
    approximate, 299
Action (variational integral), 236
    elementary, 238
    invariant under a group, 238
Admitted operator, 208
Angular momentum, 241
    relativistic, 246
Anharmonic oscillations, 20
Anharmonic ratio (cross-ratio), 34, 275
Approximate group generator, 174
    canonical representative of, 174
Approximate Lie equations, 176
Approximate symmetry, 231
    determining equation for, 231
    infinitesimal, 231
Approximate transformation, 173
Approximate transformation group, 173
Approximately equal functions, 172
    canonical representative of, 173

Bäcklund, A.V., 107, 331
Bäcklund transformations, 108
Bäcklund's non-existence theorem, 108
Basis of invariants, 142, 169
"Beating" and collapse of shafts, 20-22
Bernoulli, Daniel, 327
Bernoulli, Jacques, 327
Bernoulli, Jean, 326
Bernoulli equation, 35, 327
Bianchi, L., 164
Bicharacteristics, 86
Black-Scholes equation, 4, 114, 226
Boole, G., 249
Boole's covariants, 250
Branch point, 62

Canonical forms of L_2, 142, 287
Canonical forms of L_3, 164

Canonical parameter, 140
    determination of, 140
Canonical transformations, 106
Canonical variables, 103, 143
Cardan, H., 99
Cardan's solution, 120
Cauchy, A. L., 45, 327
Cauchy's method, 87
Cauchy problem, 46, 84
    for systems, 52
Cauchy-Riemann system, 218
Cayley, A., 249, 339
Center-of-mass theorem, 241
    relativistic, 246
Chain rule, 34, 185
Characteristics (characteristic curves), 78
Characteristic equation, 41, 43
    for the wave equation, 86
Characteristic polynomial, 41, 43
Characteristic strip, 85
Characteristic system, 70
    for quasi-linear equations, 76
    for nonlinear equations, 85
Charpit, P., 81
Chasles, M., 100, 329
Christoffel symbols, 245
Clairaut, A., 326
Clairaut equation, 58
    envelope of its general integral, 59
Classical solutions, definition of, 45
    for partial differential equations, 65
Cockle, J., 257
Commutator of operators, 92, 155
Complete integral, 80
Complete systems, 92
Completely integrable systems, 81
Composite function, 185
Composition rule in a group, 136
Conformal group, 130
    continuous, 131

Conformal transformation, 128
Connected operators, 89, 287
Consecutive integration, 285
Conservation laws, 110
    of classical mechanics, 240
    of relativistic mechanics, 243
Constant of integration, 5
Contact symmetries, 223
Contact transformations, 105
    in geometrical optics, 106
    infinitesimal, 107
    irreducible and reducible, 197
    trivial, 106
Cotes, R., 117
Covariant formulation, 284

d'Alembert, J., 325
d'Alembert's equation, 4
Darboux, J.G., 100, 330
Decoupling of systems, 279
Deformation of symmetries, 232
    determining equation for, 233
Dependent variables, 3, 184
Derived algebra, 158
De Sitter group, 171
De Sitter universe, 116
    neutrino-antineutrino in, 117
Determining equations, 102, 210
    for approximate symmetries, 231
Differential equation, 3, 207
    algebraic, 55
    of families of curves, 6
Differential form, 30
    closed, exact, locally exact, 30
Differential function, 109, 184
Differential invariants, 192
    of equivalence transformations, 249
Differential substitution, 195
Differential variables, 183
Dirac equation, 116
    in de Sitter universe, 117
Directional differential, 69
Discriminant, 119, 121, 256
Dominant function, 51

Ellipse, 59
    area of, 122
Energy of a free particle, 241
    relativistic, 246

Engel, F., 101
Envelope, 59, 328
Equivalence transformations, 61, 249
    of algebraic equations, 250
    of linear differential equations, 257
    of nonlinear equations, 261
    of partial differential equations, 262
    of Riccati equation, 34
Equivalent equations, 34
Equivalent operators, 89
Equivalent systems, 89
Euclidean group, 126, 129
Euler, L., 41, 99, 327
Euler's method for linear equations, 41
    group theoretical background, 304
Euler-Lagrange equation, 8, 110, 236
Euler-Lagrange operator, 198
Event, 243
Exact equations, 31
Exponential map, 54, 140
    approximate, 177
    differential of, 177
Extended point transformation, 188
Exterior differential calculus, 30

Faà de Bruno, 186, 335
Faà de Bruno's formula, 186
Ferrari, L., 99
Ferro, S., 99
First integrals, 37, 67
    independent, 68
First prolongation, 188, 335
First prolongation formula, 188, 191
Fourier, J.B.J., 4
Frame of differential equations, 204
Free motion of a particle, 240
    relativistic, 244
Functionally independent, 68
Fundamental identity, 203

Galileo Galilei, 13, 325
Galileo's relativity principle, 123
Galilean group, 131
Galilean transformation, 123
Galois, E., 100, 329
Galois group, 100, 253
General integral (solution), 37, 67
    of homogeneous linear partial differential equation, 70

INDEX

of non-homogeneous equation, 70
of nonlinear equation, 70
Gravitational constant, 9
Group classification of differential equations, 104, 113
Group composition law, 136
Group of motions, 126, 129
Group of transformations, 124
    admitted by an equation, 102, 208
    continuous, 102, 132
    discontinuous, 132
    finite continuous, 125
    global, 133
    infinite continuous, 132
    local, 133
    mixed, 132
Guldberg, A., 271

Hamilton's (canonical) equations, 88
Hamilton's principle of least action, 236
Hamilton-Jacobi equation, 88
    complete integral of, 88
Hamilton-Jacobi theory, 106
Harmonic oscillations, 20
Heat conduction equation, 4
Heaviside function, 327
Hilbert, D., 101
Homogeneous equations, 33
Hooke's law, 19
Huygens' construction of wave fronts, 106
Huygens' principle, 114, 249

Ideal of a Lie algebra, 157
Independent variables, 3, 183
Induced generator, 152
Induced group, 152, 253
Inertial system of reference, 243
Infinitesimal generator, 102, 138
Infinitesimal symmetry, 208
Infinitesimal transformation, 102, 138
    symbol of, 102, 138
Inherited symmetries of equations for invariant solutions, 229
Initial value problem, 45-46
Inner derivation ad$X$, 177
Integral (antiderivative), 5
Integral curves, 45
Integral surface, 66
Integrating factor, 32

Intermediate integral, 37
Invariance principle, 114
Invariant differential equations, two definitions, 208, 209
Invariant differentiation, 194, 284
Invariant equations, 150
    infinitesimal test for, 150, 169
Invariant families of curves, 208
Invariant manifolds, 150
    group induced on, 151
    invariant representation of, 152
    regular and singular, 152, 170
Invariant solutions, 103, 225
    equations for, 225
    regular and singular, 225
Invariants, 141, 169
    basis of, 142, 169
    of algebraic equations, 250
Inversion, 127, 130
Isoclines, 204
Isometric motion, 129
Isomorphic Lie algebras, 160

Jacobi, K.G.J., 88, 329
Jacobi identity, 156
Jacobian, 296
Jacobian systems, 92
Jordan, C., 100, 330

Kepler, J., 12, 13, 325
Kepler's laws, 12
    group theoretic background of, 243
Killing, W., 101
Killing equations, 245
Klein, F., 100, 330
Kronecker symbols (deltas), 130, 219
Krylov, A.N., 21, 326
Kummer, E.E., 100, 330

Lagrange, J.L., 40, 99, 329
Lagrange's equation, 58
    parametric solution of, 58
Lagrangian, 8, 236
    relativistic, 244, 245
Lagrange-Charpit method, 82
Laguerre, E., 257
Laguerre's canonical form, 257
Laplace, P.S., 73, 332
Laplace invariants, 261

Laplace method, 73
    extended to many variables, 74
Laplace operator, 116, 280
Laplace vector, 114
Legendre transformation, 106
Leibnitz, G.W., 3, 325
Leibnitz formula, 187
Lie, S., 100
Lie algebra, 103, 155
    basis of, 156
    non-solvable, 303
    solvable, 159
    structure constants of, 155
Lie equations, 138
Lie's characteristic function, 107, 223
    generalization to many variables, 191
Lie's integrating factor, 268
Lie's integration algorithm, 289
Lie's linearization test, 295
Lie-Bäcklund operator, 109, 201
    canonical, 202
Lie-Bäcklund transformation group, 109
Linear differential operators, 39
Linear equations of the first order, 25
    fundamental system of solutions, 26
    superposition of solutions, 26
    systems of, 42
    variation of parameters, 26
Linear equations of the nth order, 39
    fundamental system of solutions, 40
    homogeneous, non-homogeneous, 39
    superposition of solutions, 40
    with constant coefficients, 40
Linear partial differential equation, 66
    characteristic system of, 70
    homogeneous, solution of, 70
    non-homogeneous, solution of, 71
Liouville, J., 131, 333
Liouville's theorem, 131
Lipschitz, R., 327
Lipschitz condition, 46, 49, 50
Local groups, 136, 167
Locally analytic function, 50
Locally solvable system, 207
Logistic law of population growth, 23
Lorentz group, 244
Lorentz transformation, 128

Majorant problem, 51

Majorants, method of, 46, 50, 52
Malthus, T.R., 326
Malthusian principle of population, 22
Maxwell equations, 262
Minkowski space-time, 131, 244
Momentum, 241
    relativistic, 246
Monatomic gas, 113
    in the solar neighborhood, 118
Monge, G., 84, 329
Monge's theory of characteristics, 84
Multi-parameter groups, 167

Newton, I., 3, 325
Newton's *Principles*, 3, 325
Newton's gravitation law, 13
    "hidden" symmetry of, 114
Newton's law of cooling, 14
Newton-Cotes potential, 117
Noether, E., 110, 331
Noether symmetry, 239
Noether theorem, 110, 238
Nonlinear superposition, 28, 105, 271
    for Riccati equation, 34
Non-local symmetries, 115, 144

Ohm's law, generalized, 19
One-parameter local group, 136
Optical radiations in laser systems, 279
Ovsyannikov, L.V., 111, 331
Ovsyannikov invariants, 262

Parametric integration, 56
Partial differential equations, 65
    homogeneous linear, 65
    quasi-linear, nonlinear, 65
Partial differential operators, 69, 89
    connected, equivalent, 89
Path curve of a group (orbit), 132
Perturbed equation, 232
Plücker, J., 100, 329
Poincaré, H., 101
Poincaré group, 128, 131
Poincaré lemma, 30
Point transformations, 187
    extended, 188
Polar coordinates, 126
Poncelet, J.V., 100, 329
Principal Lie algebra, 227

INDEX

Principle of least action (relativistic), 243
Principle of relativity, 243
  Galilean, 243
Product of transformations, 123
Projective group, 125, 126
  special, 127
Propagation of light waves, 86

Quadrature, 6
  integrable by, 6
  integration by, 5
Quasi-linear equations, 66
  characteristic system for, 76
  integration by Laplace's method, 73
  reduction to a homogeneous linear equation, 75
Quotient algebra, 158

Radioactivity, 22
Reduction of order, 36, 283
Riccati, J.F., 326
Riccati equations, 34, 275
Rotation group, 126

Scalar equation, 207
Scheffers, G., 102
Second prolongation formula, 191
Semi-canonical variables, 75, 144
Seminvariants, 250, 257, 262
Separation of variables, 28, 272
Shallow-water equations, 113, 228
Similar figures, areas of, 121
Similar groups, 125
Similar Lie algebras, 160
Singular integral, 81
Singular points, 60
  branch points, poles, 63
  center, focus, node, saddle, 61
  fixed and movable, 62
Singular solutions, 60
Solution of differential equation, 205
Solvable Lie algebra, 159
Space $\mathcal{A}$ of differential functions, 184
Special relativity, 243
Spiral transformation group, 154
  in polar coordinates, 154
Stable symmetry, 232
Stock option pricing, 4
Structure constants, 156

Subalgebra of a Lie algebra, 157
Subgroup, 124
Successive approximations, 46
Sylow, L., 100, 329
Symmetry group, 97, 208
Symmetry Lie algebra, 214
System of ordinary differential equations of the first order, 67
  first integrals of, 67
  general integral of, 67
  symmetric form, 67

Table of commutators, 156
Tangency (contact) condition, 106
  second-order (osculation), 107
  infinite-order, 108
Tangent plane, 78
Tartaglia, N., 99
Thermal diffusion, 4
  dependence of temperature on a choice of inertial reference frame, 116
Total derivative, 37
  test for, 199
Total differentiation D_i, 76, 183
Total differentiation D_x, 7
Translation group, 124
Tschirnhausen, E.W., 251
Tschirnhausen's transformation, 251

Unperturbed equation, 232
  inherited symmetries of, 232

Van der Pol equation, 19
Variation of parameters, 26, 40
  group interpretation of, 271, 292
Variational integral, 237
  extremum of, 237
Velocity of light, 86, 243
Vessiot, E., 271
Vessiot-Guldberg-Lie algebra, 272
Volterra-Lotka equations, 23

Wave equation, 4, 86
  characteristic equation for, 86
Wave front reversal, 279
World line, 243
World points, 243
Wronski, H., 340
Wronskian, 293